Forest Biodiversity, Ecology and Management

Forest Biodiversity, Ecology and Management

Editor: Malcolm Fisher

Callisto Reference
www.callistoreference.com

Callisto Reference,
118-35 Queens Blvd., Suite 400,
Forest Hills, NY 11375, USA

Visit us on the World Wide Web at:
www.callistoreference.com

ISBN: 978-1-64116-832-8 (Hardback)

Cataloging-in-Publication Data

Forest biodiversity, ecology and management / edited by Malcolm Fisher.
p. cm.
Includes bibliographical references and index.
ISBN 978-1-64116-832-8
1. Forest biodiversity. 2. Forest ecology. 3. Forest management. I. Fisher, Malcolm.
QH86 .F67 2023
577.34--dc23

Table of Contents

Preface

Forests refer to areas that contain a large number of trees. They are classified into three main types, including boreal, temperate and tropical. Forests exhibit a high level of biodiversity and have the most diverse ecosystems on land, as they are home to a large number of the terrestrial species found in the world. Forest biodiversity is inherently linked to the ecosystem and human well-being, and it is important from a number of socio-economic aspects as well as. It also provides a variety of goods and services including timber and non-timber forest resources, reduces the effects of climate change, and helps conserve genetic resources. Forest management refers to the process of protecting and maintaining a forested area. It encompasses the administrative, economic, legal and social aspects. This book unravels the recent studies related to forest biodiversity, ecology and management. It is a resource guide for experts as well as students.

This book unites the global concepts and researches in an organized manner for a comprehensive understanding of the subject. It is a ripe text for all researchers, students, scientists or anyone else who is interested in acquiring a better knowledge of this dynamic field.

I extend my sincere thanks to the contributors for such eloquent research chapters. Finally, I thank my family for being a source of support and help.

Editor

1

A Disturbance Ecology Perspective on Silvicultural Site Preparation

Jéssica Chaves Cardoso [1,*], Philip J. Burton [2] and Che M. Elkin [3]

1 Natural Resources and Environmental Studies-Forestry, University of Northern British Columbia, Prince George, BC V2N 4Z9, Canada
2 Ecosystem Science & Management Program, University of Northern British Columbia, Terrace, BC V5G 1K7, Canada; phil.burton@unbc.ca
3 Ecosystem Science & Management Program, University of Northern British Columbia, Prince George, BC V2N 4Z9, Canada; che.elkin@unbc.ca
* Correspondence: jessica.chavescardoso@alumni.unbc.ca

Abstract: Silvicultural site preparation methods are used as planned disturbances for counteracting soil and vegetation constraints, as well as facilitating successful tree regeneration and growth. Understanding the possible effects of silvicultural site preparation on the ecosystem and evaluating site preparation as an ecological disturbance can help guide the selection and application of site preparation techniques for forest management goals. This review evaluates silvicultural site preparation techniques that are commonly used in boreal mixedwood ecosystems as agents of ecological disturbance by comparing the effects of each technique on the area disturbed and the degree of biomass modification, and then ordering them along a disturbance severity gradient. With a strong emphasis on the numerical estimation of the spatial footprint of different disturbances, broadcast burning typically has the highest disturbance severity, followed in order by broadcast herbicide use, mixing, plowing, disc trenching, mounding, scalping, and inverting. The evaluation of disturbance severity of various silvicultural site preparation techniques while using the proposed framework is feasible, in which quantitative assessments of area disturbed and biomass modification could be collected and assessed in most managed forests.

Keywords: boreal mixedwood ecosystems; disturbance footprint; disturbance severity; ecological disturbance; mechanical site preparation

1. Introduction

Silvicultural site preparation techniques are planned anthropogenic disturbances that are designed to aid in the regeneration and establishment of a new stand of trees. The main goal of any silvicultural site preparation is to create a beneficial environment for tree regeneration and growth, which may include favorable germination spots for seeds [1], successful seedling survival and establishment [2,3], and a reduction in realized competition [4]. To accomplish this, many methods can be applied in order improve soil conditions (such as low or excessive moisture, low nutrient availability, and compaction) and control competing vegetation [1,2]. The disturbance of soil and vegetation can create a favorable microclimate for tree establishment [5,6] and it can influence the species composition and the structure of the future forest [7]. Silvicultural site preparation has the potential to mitigate the effects of climate change on early plantation growth in boreal forests by promoting tree establishment and growth, as well as promoting microclimate conditions for vegetation establishment [4].

The efficacy of site preparation depends on appropriately aligning potential methods with the specific site conditions, tree species, competing vegetation, climate, and the desired silviculture objectives. Improper site preparation can negatively affect the chemical, physical, hydrological, and biological properties of the soil, and result in a failure to achieve management objectives [3]. Site preparation may also create unfavorable growing conditions when applied incorrectly or on an inappropriate site or ecosystem [1,3]. Thus, it is important to select the appropriate site preparation method that will minimize any undesired impacts on the ecosystem.

Although widely recognized as an anthropogenic disturbance, silvicultural site preparation techniques are commonly not examined through the lens of disturbance ecology. Understanding the possible effects of silvicultural site preparation on the ecosystem and recognizing site preparation as an ecological disturbance can help to select the best option for different forest management goals. For example, a forester managing a forest for ecological restoration may opt to apply a site preparation technique with little area disturbed and/or little biomass modification. Moreover, the study of site preparation techniques from a disturbance ecology perspective can help direct attention to the biological and ecological legacies [8] that are associated with site preparation activity.

Disturbances are typically characterized by their frequency, intensity, and severity [9]. Frequency is the temporal scale at which disturbance occurs, whereas intensity is a measure of the force of the disturbance [9]. Severity can be defined as the amount of biomass loss or the degree of change in ecosystem structure or function caused by a disturbance [9]. The quantification of disturbance severity provides a framework for evaluating, comparing, and predicting the results of any natural or anthropogenic disturbance [10]. Roberts [11] evaluated clear-cuts and partial cuts with scarification while using a disturbance severity model [10] but did not explore different kinds of silvicultural site preparation. An assessment of the disturbance severity of a range of silvicultural site preparation methods can provide insight into the suitability of site preparation options for various land management goals and their underlying ecological impacts.

The objective of this paper is to explore silvicultural site preparation techniques as agents of ecological disturbance. We focus on eight types of site preparation treatments—prescribed burning and herbicides, plus six types of mechanical site preparation (MSP), namely mixing, plowing, disc trenching, mounding, scalping, and inverting—that are commonly used today in boreal mixedwood ecosystems [1,4]. Mixed forests of white spruce (*Picea glauca* [Moench] Voss) and trembling aspen (*Populus tremuloides* Michx.) in northwestern Canada are typically clear-cut logged, followed by site preparation to offset the effects of cold wet soils and vigorous competition from species, such as trembling aspen and bluejoint (*Calamagrostis canadensis* [Michx.] Beauv.), in order to promote establishment of nursery-grown seedlings of the preferred crop tree, white spruce. Other site preparation techniques, such as motorized-manual screefing, are practiced elsewhere in the world, but they are not treated here. Specifically, we (1) review the general effects of the selected site preparation techniques on soil conditions and seedling performance, (2) propose a framework for evaluating site preparation disturbance severity at the cutblock/stand level, (3) analyze site preparation techniques while using our proposed framework with the objective of ordering them along a disturbance severity gradient, and (4) identify research needs and recommend future work.

2. Silvicultural Site Preparation Techniques

2.1. Prescribed Burning

Prescribed burning can be used to prepare sites prior to planting in order to remove organic materials from the area [12,13]. However, prescribed burning requires planning during the harvesting, as it requires a continuous cover of slash to carry the fire, it is limited seasonally by weather and fuel conditions, and may intensify competition problems when undesired or invasive vegetation is adapted to fire or invades after fire [14,15]. In addition, the use of prescribed burning has declined in response to concerns regarding air quality [12]. Even though the use of prescribed burning has decreased in some areas, it is still used in boreal mixedwood ecosystems, especially in remote locations for the establishment of conifer plantations.

2.2. Herbicide

When employing the chemical site preparation method, which involves the use of systemic herbicides, the two main objectives are to control competing vegetation by reducing its abundance before planting or seeding and to control competing vegetation during the early stages of tree seedling establishment [16,17]. Where the objective of site preparation is not vegetation suppression alone, herbicides may be used in combination with MSP and burning methods [12,18–20]. A broad-spectrum herbicide is sometimes used in order to kill and dehydrate the vegetation prior to the use of broadcast burning, in a technique that is known as "brown and burn". The selection of the chemical treatment involves determining which registered herbicide (e.g., those with glyphosate or 2, 4-D amine as active ingredients) would be the most effective on the identified target species [16]. The use of herbicides is greatly constrained by concerns regarding impacts on non-target organisms (especially non-target plants [21], aquatic animals, cyanobacteria [22], and human health [22]). Although sometimes applied in spot treatments for crop tree release, herbicide applications for site preparation are almost always applied in a uniform broadcast manner, usually by aircraft.

2.3. Mixing

Mixing consists of incorporating the surface organic layer into the underlying mineral soil, which leaves the nutrients of the organic layers immediately available to germinants or planted seedlings. Mixing methods can be divided in three categories: coarse mixing, fine mixing, and spot mixing [1,23]. Coarse mixing is accomplished while using disc implements that heap clods of surface organic and mineral soil layers into a bed. This method provides slight control of competing vegetation, but it provides benefits where low soil temperatures and/or high soil water tables inhibit seedling growth [1]. Fine mixing is usually used on sites with high potential for competing vegetation, so a high-speed rotating implement is required to chop propagating plant parts small enough to control resprouting [1,6,23]. Spot mixing is usually recommended for sites where mixing is biologically suitable, but where stumps, slash, or other obstacles impede the use of strip mixing implements [23]. This method is also prescribed on sites where minimal soil disturbance is required [1,6].

2.4. Disc Trenching

Disc trenching exposes mineral soil in continuous strips. Disc trenching can produce three different planting positions: the trench position for dry sites, hinge position for medium-moist sites, and berm position for moist sites [1]. The trench profile can be adjusted depending on the objective by changing the disc angle. A disc angle that runs more parallel to the direction of travel produces a deeper, narrower trench, and a disc angle that runs more perpendicular to the direction of travel produces a wider trench [1,23].

2.5. Mounding

The main purposes of mounding are to create elevated planting spots [2,23]. Spot mounding is an intermittent form of mounding, which removes vegetation and the surface organic layer and results in inverted or mixed soil on top of an organic surface or bare center. Mounds may be composed of mixed surface organic matter and mineral soil, or of mineral soil capping on top of a largely undisturbed organic layer. Deep planting is usually applied in order to protect roots from being exposed to weathering of the mound surface subsequent to planting [1,23].

2.6. Plowing (Ploughing)

Plowing consists of creating continuous elevated berms [1,2]. In some regions it is listed as continuous mounding because it creates elevated planting spots on ridges [2], while, in other regions, planting is done in the furrows. Ripper plows are modified standard ripper teeth, on the back of a tractor, and they are specifically designed for tilling wet ground when it is frozen [1]. The most common plow design is a double mouldboard, which attaches to the ripper shank of the tractor. The ripper teeth dig into the soil, while the plow attachment displaces soil on either side [1].

2.7. Scalping

Scalping, which is also known as patch scarification, creates patches of mineral soil exposed in a systematic pattern by removing the surface organic layer to expose the underlying mineral soil. Planting spots are usually below the original ground level [2], but different planting spots may be chosen, depending on site conditions (e.g., moisture regime of the site [23]). On most sites, seedlings are recommended to be planted on the shoulder of exposed mineral soil, adjacent to the inverted humus [1]. Seedlings may be planted in the bottom of the scalp on dry sites. The depth of scarification has to be appropriate for the site. Scalping that is too shallow may not be sufficient for controlling vegetation if competing roots are not removed. If scalping is too deep, the planting spot may become waterlogged, especially on a wet site or during wet periods [1,6,23].

2.8. Inverting

Inverting consists of turning the surface organic layer over; the inverted organic layer may be covered with mineral soil (mineral soil cap). Inverting is sometimes listed as one of the mounding techniques, although the planting spot normally is not significantly elevated [2,24]. On some sites, seedlings can have a problem with frost heaving following site preparation by inversion [23,25]. For instance, in Ontario, frost heaving in hand-simulated inverted humus mounds has been severe enough to cause concern regarding the suitability of such planting spots [23,25].

3. Effects of Silvicultural Site Preparation Techniques on Soil Conditions and Seedling Performance

Silvicultural site preparation techniques differ with regard to the type and amounts of soil disturbance that they induce and the concomitant ecosystem impacts. Ballard [26] evaluated the effects of forest management on forest soils and found that there can be both positive and negative impacts. The sensitivity of the site to short- and long-term degradation must be considered when selecting from among silvicultural site preparation options [1]. Site preparation techniques, such as herbicide application and prescribed burning, tend to cause less disturbance to the soil than mechanical practices [17], and they could be an option where MSP would pose a risk to sensitive soils or to water quality.

Many boreal forest ecosystems have a highly developed organic layer that overlies mineral soil [27]. Cold temperatures, short growing seasons, and high soil moisture decrease the decomposition rates of organic matter, resulting in the accumulation of a thick surface organic layer [28], which insulates the undisturbed soil beneath. In unprepared soil (i.e., with no site preparation), the humus layer (H horizon) may remain cool on warm days as the litter and fermentation layers (L and F horizons) act as insulation [27]. Exposed mineral soils warm faster than the undisturbed soil beneath the organic layers [23]. Therefore, MSP methods increase soil temperature by exposing mineral soils in a manner that varies with soil color [29]. In addition, removing shading from vegetation by herbicide, prescribed burning, and MSP methods also promotes the warming of the soil surface.

The increase in soil temperature as a result of site preparation treatments can provide benefits to tree seedlings, especially at high latitudes and high elevations [30,31]. Increasing soil temperature is, in many cases, beneficial for plants in general, because roots grow faster in warm soil than in cold soil [23]. Spruce (*Picea* spp.) seedlings, for example, show minimal growth at soil temperatures under 10 °C [32]. Warm soil can also facilitate water uptake, and it can decrease frost damage to seedlings through re-radiation of heat energy [23,33]. For example, in scalped soils, seedling roots that are stimulated by the increased soil temperature may quickly grow beyond the scalped area, reaching the nutrients in the surrounding undisturbed layers [2,23]. In inverted soils, roots reaching the mineral soil beneath the inverted layers have access to an enhanced supply of moisture [6,23]. Because transpiration from the soil horizons under organic layers is reduced when vegetation is controlled, moisture accumulates beneath these organic layers [23].

Soil disturbance that is caused by MSP techniques can also enhance seed survival and germination during natural regeneration or direct seeding of a number of tree species [34,35]. Seeds and small seedlings may suffer desiccation in the organic layer, thus their contact with mineral soil can facilitate water uptake [2] and enhance seed and seedling survival [36]. MSP may also increase biodiversity because it tends to improve conditions for seed germination of a number of woody and herbaceous species [2].

Silvicultural site preparation techniques have the potential to significantly affect nutrient dynamics [37,38], either positively or negatively. For example, the mixing technique may risk releasing too much of the nutrient reserves of a site in the first few years after treatment, resulting in the long-term depletion of nutrients. Nutrients not taken up by plants may be lost to the ecosystem through leaching or erosion [23]. Prescribed burning can increase short-term (<5 years) availability of nutrients in the soil [13], but it may cause longer-term nutrient deficiencies [39]. Unlike the MSP techniques, herbicides can indirectly affect soil properties. Vegetation reduction by herbicide treatment tends to help retain available soil moisture [40], and there may be significant nutrient losses by leaching, increased daytime temperatures, decreased nighttime temperatures, and increased microbial activity [41].

Mechanical site preparation has been shown to cause reductions in available phosphorus (P) as well as reduced nitrogen (N) and carbon (C) in surface soils [38] if the forest floor is displaced. A study that was conducted in Alberta indicated that fifteen months after different MSP methods, treated sites have either reduced or unchanged concentrations of total N and C, available P, and mineralizable N in the surface mineral soil, as compared with harvested sites with no site preparation [42]. Although many studies indicate that nutrient pools may be affected by MSP [42,43], few studies have shown that MSP has negatively affected the long-term productivity of crop trees [2]. Piatek and Allen [44], for example, indicated that the nutrients lost during MSP had no noticeable effect on foliage production when measured 15 years after planting.

Silvicultural site preparation can lead to deep soil disturbances, depending on the intensity [2]. Most MSP techniques can cause soil erosion if not carefully implemented and adapted for the specific climate and site characteristics [45]. Excessive soil disturbance can result in further soil erosion, compaction, removal of nutrients beyond root systems, or even landslides [1]. The risk of soil erosion is increased when mineral soil is exposed by MSP (especially on slopes), and nutrients can be exported from a site in solution or by mass transport [46]. Alcázar et al. [45] showed elevated levels of soil erosion on areas where MSP treatments exposed the mineral soils. At sites where there is a risk of soil erosion, low fertility, and low soil organic matter, harvesting and MSP in combination with wet weather can cause soil compaction and erosion, with effects on site productivity (i.e., tree survival and growth) [47]. Disc trenching has the risk of accelerating erosion if running across the contours of a slope. Beasley et al. [48] compared the storm flow and sediment losses from mechanically and chemically prepared sites in southwest Arkansas over a four-year period. In contrast to MSP effects, they found that chemical site preparation using herbicides had no significant effect on sediment loss. Prescribed burning can have a major effect on surface soil erosion if it removes much of the forest floor over large areas of the cutblock [13].

Most research to date on silvicultural site preparation methods and plant performance has been conducted utilizing a few conifer tree species with the goal of increased productivity in traditional plantation forestry or during restoration [12,24]. In general, studies indicated improved seedling performance on sites with site preparation as compared to untreated sites. A study conducted in Sweden showed that, after 18 growing seasons, higher levels of Norway spruce (*Picea abies* [L.] Karst.) seedling survival were found following soil inversion (77% for normal intensity and 76% for high intensity) in comparison with mounding (67%) and control (i.e., no scarification, 57%) [34]. The difference in the seedling height between these scarification treatments and control corresponded to a time gain of approximately four years of growth after 18 years [34]. Powelson et al. [12] in British Columbia indicated a greater white spruce growth response to MSP and herbicide treatments when compared to untreated sites over 30 years.

Studies indicate that herbicide site preparation results in better crop tree growth when compared to MSP, particularly on mixedwood sites where hardwoods and shrubs compete with planted conifers [49,50]. For example, a study on black spruce (*Picea mariana* [Mill.] BSP) establishment indicated that the largest stem volume with the lowest vegetation indices was recorded under chemical site preparation as compared to the scarification site preparation methods [49]. Moreover, chemical site preparation can be more effective in improving early conifer performance than when chemical brushing is used after planting [25]. Chen et al. [51] showed that planting trees after herbicide treatments resulted in increased crop tree and stand volume for both container-grown jack pine (*Pinus banksiana* Lamb.) and black spruce, but not bareroot black spruce.

Seedling responses to silvicultural site preparation vary among tree species [36] due to their physiological differences (e.g., tolerance to drought and flooding). Therefore, the choice of silvicultural site preparation methods needs to take the desired species or species composition and its expected responses into consideration. For example, seedlings of species that are intolerant to flooding [e.g., whitebark pine (*Pinus albicaulis* Engelm.), ponderosa pine (*Pinus ponderosa* Dougl. ex P. & C. Laws.), and Douglas fir (*Pseudotsuga menziesii* (Mirb.) Franco)] are expected to have better performance when placed on the elevated spot, rather than the pit resulting from mounding.

4. Assessment of Area Disturbed and Biomass Modification of Silvicultural Site Preparation Techniques

The assessments of the percentage of total disturbed area were based on the typical application of each site preparation technique, whether it follows a systematic pattern of spot treatments, a continuous pattern, or the entire cutblock/stand [1,52] (Table 1). The analysis of the degree of biomass modification was based on which pre-existing organic materials are consumed, dislocated, or otherwise disrupted by each technique [1,13,16,23,52] (Table 1). The biomass considered includes living materials (e.g., understory vegetation), any organic materials left by previous vegetation (e.g., woody material, roots and rhizomes, and dead foliage), and forest floor (i.e., surface organic layers, LFH).

Biomass modification was inferred from general principles, as quantitative measurements are largely absent from the literature. Consequently, a simple ordinal score was assigned in order to represent the degree of biomass modification that is associated with each site preparation technique, ranging from 0 to 6, where 6 is the highest degree of biomass modification. We assigned a disturbance severity index from 0 (non-existent) or 1 (low) to 3 (high) for modifications to the residual biomass (e.g., living materials and organic materials left by previous vegetation) and the forest floor (Table 1). The biomass modification ordinal score corresponds to the sum of the disturbance index of the residual biomass and forest floor. For example, the justification for broadcast burning being assigned an ordinal score of 6 is that the organic material that was left by previous vegetation and the forest floor were both 3, while the herbicide treatment was assigned an ordinal score of 4.5 that was based on a high impact on the residual biomass (3), but a lower impact on the forest floor (1.5).

Table 1. Attributes of eight site preparation techniques commonly used in boreal mixedwood ecosystems.

Site Preparation Technique	Total Area Disturbed	Principal Effect	Biomass Modification (Effect/Disturbance Index)		Summed Score (0 to 6)
			Living Material and Organic Materials Left by Previous Vegetation (Scored 0 to 3)	Forest Floor (Scored 0 to 3)	
Broadcast burning	Whole area 90%–100%	Kills all aboveground vegetation, variable consumption of organic matter	Consumed, charred, or scorched, some incorporated into soil organic matter/3	Consumed or scorched, much incorporated to mineral soil/3	6
Herbicide (broadcast)	Whole area 90%–100%	Kills most of the dominant vegetation	Not removed from area treated, but understory vegetation (shrubs and herbs) mostly killed/3	Not directly affected, but a prolonged pulse of litter from killed vegetation is added/1.5	4.5
Mixing (coarse or fine)	Continuous pattern 50%–75%	Incorporation of the forest floor into the underlying mineral soil	Vegetation and fine/medium woody materials are chopped and mixed with soil/2.5	Mixed with mineral soil and organic materials derived from previous vegetation /2.5	5
Plowing	Continuous pattern 30%–65%	Exposure of mineral soil, organic material buried	Dislocated or buried in area treated/2	Dislocated or buried in area treated/ 2.5	4.5
Disc trenching	Continuous pattern 25%–50%	Exposure of mineral soil planting area	Dislocated within area treated/2	Dislocated within area treated/2	4
Mounding	Systematic pattern (of plantable spots) 15%–35%	Exposure of mineral soil planting spots	Dislocated within spots treated/2	Dislocated within spots treated/2	4
Scalping	Systematic pattern (of plantable spots) 10%–30%	Exposure of mineral soil planting spots	Dislocated from spots treated/1.5	Dislocated from spots treated/1.5	3
Inverting	Systematic pattern (of plantable spots) 10%–30%	Exposure of mineral soil planting spots by turning the forest floor upside down	Not removed from spots treated but moved from above ground to below ground as soil organic layers are turned upside down/1	Not removed or dislocated, but turned upside down/1	2

For drawings of the different MSP techniques and machines involved, refer to Von der Gönna [1] and McMinn and Hedin [23]. Von der Gönna [1] outlines the range of surface areas disturbed by different MSP techniques.

5. Disturbance Severity Analyses

Many studies have discussed disturbance severity [53,54], but with a focus on its effects on the dominant tree layer. Roberts [10] proposed a three-axis model of disturbance severity as an extension of a single-axis model of Oliver and Larson [53], with each of the three major vertical layers in the forest ecosystem (forest canopy, understory vegetation, forest floor, and soil) on a separate axis. The model has been used in order to evaluate a wide range of natural disturbance types [10] and evaluate clear-cuts and partial cuts with scarification [11].

Building on the work of Roberts [10], we have also adapted Oliver and Larson's [53] definition of disturbance severity to apply to the eight types of site preparation at the cutblock/stand level (i.e., the area harvested and treated by silvicultural site preparation). It is presented in the form of a simple two-dimensional framework, with one axis being the percentage of the total area disturbed and the other axis being the degree of biomass modification (Figure 1).

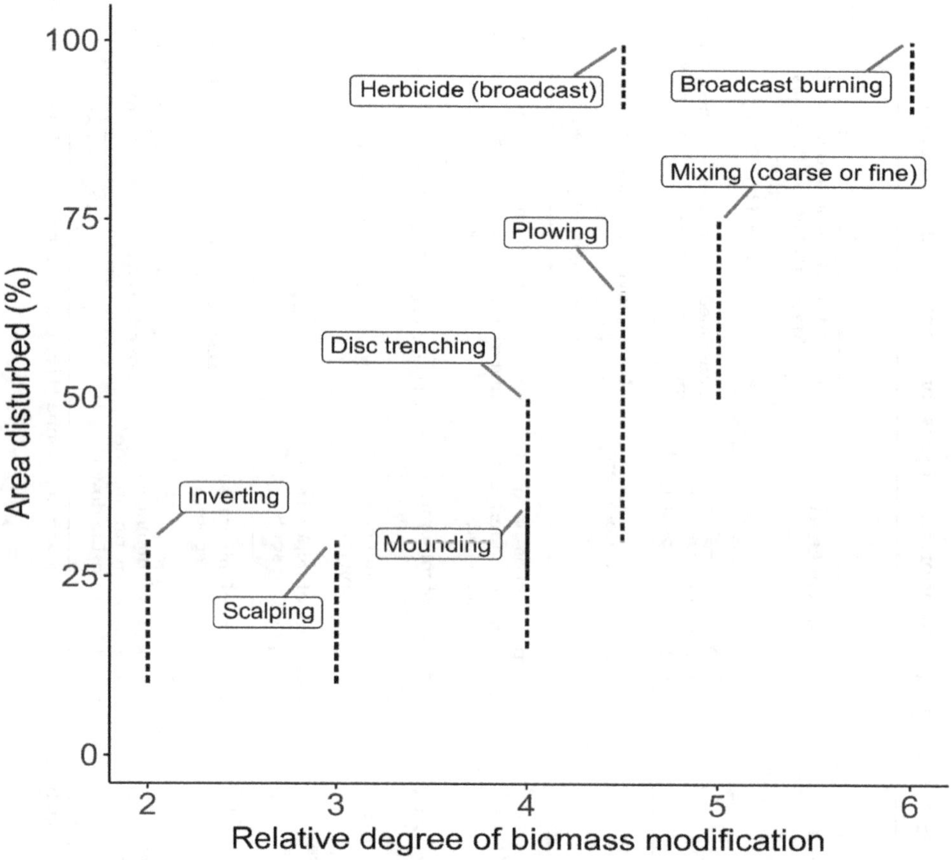

Figure 1. A two-dimensional framework for evaluation of site preparation disturbance-severity at the cutblock/stand level applied to eight site preparation techniques commonly used in boreal mixedwood ecosystems. The vertical axis is the percentage of the total area disturbed, with vertical bars representing the typical ranges of disturbed areas reported, and the horizontal axis is the degree of biomass modification (as per Table 1).

Broadcast burning is interpreted to have the highest disturbance severity, followed in order by herbicide, mixing, plowing, disc trenching, mounding, scalping, and inverting, based on the percentage of total area disturbed and degree of biomass modification of each technique (i.e., the two-dimensional framework) (Figure 1). Because broadcast burning and herbicide site preparation are designed to cover the whole area, not just plantable spots, they have a greater percentage of total area disturbed and typically result in higher degree of biomass modification when compared to MSP techniques. The organic materials of the entire cutblock are consumed or modified by broadcast burning; thus, the amount of biomass loss by these techniques is higher than the other techniques. Herbicide site preparation also kills most of the dominant vegetation, eventually killing the root systems of plants, but it does not remove the organic material left behind, and may even result in a pulse of new litter as foliage dies and falls to the ground.

Because of the physical impact of the MSP techniques, much of the living understory vegetation is also expected to be killed in the disturbed spots. However, because MSP methods are either applied in a systematic or continuous pattern instead of across the whole cutblock, they result in a lower degree of modification to the cutblock's biomass than broadcast burning or herbicide. Although displaced, rhizomes, stolons, and root stocks of many species eventually resprout [55]. Mixing has the highest percentage of area that is disturbed of mechanical disturbance (50%–75%) and a higher degree of biomass modification than the other MSP techniques, because the organic materials that are left by the harvested trees, understory vegetation, and the organic matter are chopped and incorporated into the mineral soil.

Plowing, disc trenching, scalping, and mounding all remove the organic materials that are left by harvested trees and organic matter from the areas treated, but plowing has the greater percentage of area disturbed (30%–65%), followed by disc trenching, mounding, and scalping. Although both scalping and mounding are applied in a systematic pattern, mounding is expected to have a higher ground area disturbed than scalping, because mounding displaces the organic layer in order to create an adjacent elevated planting spot. The lower severity that is assigned to inverting reflects its low percentage of area disturbed (10%–30%) combined with the fact that it turns the organic layer upside down, leaving the organic materials disrupted but in place in the spots disturbed.

The two-dimensional framework (Figure 1) allows for the evaluation of disturbance severity of each site preparation method at the cutblock/stand level and provides an ordination of the techniques along a severity gradient. This ordering of site preparation techniques gives strong emphasis to the numerical estimation of the spatial footprint of these anthropogenic disturbances. Depending on land management goals or the concerns of particular natural resource management disciplines (e.g., carbon sequestration, biodiversity protection, and soil conservation), alternative degrees of emphasis might be placed on the fate of residual biomass (e.g., as long-lasting black carbon [56]) or on the widespread maintenance of soil health [57].

Taking these ideas forward from a conceptual ordering to an empirically based ordination that builds on our conceptual framework of disturbance severity is feasible. This framework could then incorporate quantitative assessments of area disturbed, biomass loss and alteration, and forest floor/soil disruption, which could be evaluated across a variety of boreal mixedwood cutblocks that have just undergone site preparation, and that could also be applied to forest operations in other parts of the world. Statistical multivariate ordination of disturbance severity using proportional area affected, biomass disrupted, and other factors, such as carbon emissions and wildlife habitat value, could also be undertaken.

6. Conclusions

Silvicultural site preparation effects on forest ecosystems can vary according to site conditions and management objectives. Selecting a suitable silvicultural site preparation method requires an understanding of the possible effects on forest ecosystems, on planting spot microclimates, and the potential for enhanced tree growth and carbon sequestration. Many studies in boreal ecosystems have concentrated on evaluating the effectiveness of silvicultural site preparation techniques in promoting successful tree regeneration, but there is still a need to study the effects on a wide range of additional species, under different site conditions, and their potential for ameliorating the effects of timber harvesting, site degradation, and climate change. To this end, there is value in looking at site preparation techniques through the lens of disturbance ecology. A preliminary conceptual assessment indicates that broadcast burning appears to have the highest severity, followed in order by herbicide, mixing, plowing, disc trenching, mounding, scalping, and inverting, but other perspectives and statistical analyses are also possible. Future studies could apply the proposed two-dimensional disturbance severity framework and further elaborations of this approach to empirically evaluate additional site preparation techniques and operational forest management effects in forests around the world.

Author Contributions: J.C.C. wrote the original draft preparation; P.J.B. conceived the approach, contributed to the manuscript and provided editorial suggestions; C.M.E. contributed to the manuscript and provided editorial suggestions. All authors have read and agreed to the published version of the manuscript.

Acknowledgments: This paper would not have been possible without the knowledge acquired from the graduate course in Disturbance Ecology offered by the University of Northern British Columbia. We thank anonymous reviewers for their constructive comments and suggestions.

References

1. Von der Gönna, M. *Fundamentals of Mechanical Site Preparation*; Forestry Canada, B.C. Ministry of Forests: Victoria, BC, Canada, 1992; p. 28.
2. Löf, M.; Dey, D.C.; Navarro, R.M.; Jacobs, D.F. Mechanical site preparation for forest restoration. *New For.* **2012**, *43*, 825–848. [CrossRef]
3. Otchere-Boateng, J.; Herring, L.J. Site preparation: Introduction. In *Regenerating British Columbia's Forests, Section Three*; Lavender, D.P., Parish, R., Johnson, C.M., Montgomery, G., Vyse, A., Willis, R.A., Winston, D., Eds.; Government of Canada: Victoria, BC, Canada, 1990; p. 130.
4. Cortini, F.; Comeau, P.G.; Boateng, J.O.; Bedford, L.; McClarnon, J.; Powelson, A. Effects of climate on growth of lodgepole pine and white spruce following site preparation and its implications in a changing climate. *Can. J. For. Res.* **2010**, *41*, 180–194. [CrossRef]
5. Outcalt, K.W. *Mechanical Site Preparation Improves Growth of Genetically Improved and Unimproved Slash Pine on a Florida Flatwoods Site Pinus Elliottii*; United States Department of Agriculture (USDA): Washington, DC, USA, 1983; pp. 11–13.
6. Örlander, G.; Gemmel, P.; Hunt, J. *Site Preparation: A Swedish Overview*; FRDA Report; The Canada-British Columbia Forest Resource Development Agreement (FRDA): Victoria, BC, Canada, 1990.
7. Haeussler, S.; Bartemucci, P.; Bedford, L. Succession and resilience in boreal mixedwood plant communities 15–16 years after silvicultural site preparation. *For. Ecol. Manag.* **2004**, *199*, 349–370. [CrossRef]
8. Franklin, J.F.; Lindenmayer, D.; MacMahon, J.A.; McKee, A.; Magnuson, J.; Perry, D.A.; Waide, R.; Foster, D. Threads of continuity. *Conserv. Pract.* **2000**, *1*, 8–17. [CrossRef]
9. Walker, L.R. *The Biology of Disturbed Habitats*; Oxford University Press (OUP): Oxford, UK, 2012.
10. Roberts, M.R. Response of the herbaceous layer to natural disturbance in North American forests. *Can. J. Bot.* **2004**, *82*, 1273–1283. [CrossRef]
11. Roberts, M.R. A conceptual model to characterize disturbance severity in forest harvests. *For. Ecol. Manag.* **2007**, *242*, 58–64. [CrossRef]
12. Powelson, A.; Heineman, J.; Bedford, L.; McClarnon, J.; Nemec, A.F.L.; Otchere-Boateng, J. *White Spruce Responses to Mechanical Site Preparation, Chemical Site Preparation, and Post-Planting Vegetation Control over Three Decades in the Boreal White and Black Spruce Zone of British Columbia*; Crown Publications: Victoria, BC, Canada, 2016; pp. 1–65.
13. Hawkes, B.C.; Feller, M.C.; Meehan, D. Site preparation: Fire. In *Regenerating British Columbia's Forests, Section Three*; Lavender, D.P., Parish, R., Johnson, C.M., Montgomery, G., Vyse, A., Willis, R.A., Winston, D., Eds.; Government of Canada: Victoria, BC, Canada, 1990; pp. 131–149.
14. Iverson, L.R.; Hutchinson, T.F.; Prasad, A.M.; Peters, M.P. Thinning, fire, and oak regeneration across a heterogeneous landscape in the eastern USA: 7-year results. *For. Ecol. Manag.* **2008**, *255*, 3035–3050. [CrossRef]
15. Rebbeck, J. Fire management and woody invasive plants in oak ecosystems. *USDA For. Serv. North. Res. Stn.* **2012**, *p-102*, 142–155.
16. Otchere-Boateng, J.; Herring, L.J. Site preparation: Chemical. In *Regenerating British Columbia's Forests, Section Three*; Lavender, D.P., Parish, R., Johnson, C.M., Montgomery, G., Vyse, A., Willis, R.A., Winston, D., Eds.; Government of Canada: Victoria, BC, Canada, 1990; pp. 164–178.

17. Lowery, R.F.; Gjerstad, D.H. Chemical and mechanical site preparation. In *Forest Regeneration Manual*; Duryea, M.L., Dougherty, P.M., Eds.; Springer: Dordrecht, The Netherlands, 1991; pp. 251–261. ISBN 978-94-011-3800-0.
18. Ross, D.W.; Scott, W.; Heninger, R.L.; Walstad, J.D. Effects of site preparation on ponderosa pine (*Pinus ponderosa*), associated vegetation, and soil properties in south central Oregon. *Can. J. For. Res.* **1986**, *16*, 612–618. [CrossRef]
19. Bolte, A.; Löf, M. Root spatial distribution and biomass partitioning in *Quercus robur* L. seedlings: The effects of mounding site preparation in oak plantations. *Eur. J. For. Res.* **2010**, *129*, 603–612. [CrossRef]
20. Löf, M.; Rydberg, D.; Bolte, A. Mounding site preparation for forest restoration: Survival and short term growth response in *Quercus robur* L. seedlings. *For. Ecol. Manag.* **2006**, *232*, 19–25. [CrossRef]
21. Wood, L.J. The presence of glyphosate in forest plants with different life strategies one year after application. *Can. J. For. Res.* **2019**, *49*, 586–594. [CrossRef]
22. Rolando, C.; Baillie, B.; Thompson, D.; Little, K. The risks associated with glyphosate-based herbicide use in planted forests. *Forests* **2017**, *8*, 208. [CrossRef]
23. McMinn, R.G.; Hedin, I.K. Site preparation: Mechanical and manual. In *Regenerating British Columbia's Forests, Section Three*; Lavender, D.P., Parish, R., Johnson, C.M., Montgomery, G., Vyse, A., Willis, R.A., Winston, D., Eds.; Government of Canada: Victoria, BC, Canada, 1990; pp. 150–163.
24. Sutton, R.F. Mounding site preparation: A review of European and North American experience. *New For.* **1993**, *7*, 151–192. [CrossRef]
25. Wood, J.E.; von Althen, F.W. Establishment of white spruce and black spruce in Boreal Ontario: Effects of chemical site preparation and post-planting weed control. *For. Chron.* **1993**, *69*, 554–560. [CrossRef]
26. Ballard, T.M. Impacts of forest management on northern forest soils. *For. Ecol. Manag.* **2000**, *133*, 37–42. [CrossRef]
27. Khomik, M.; Arain, M.A.; McCaughey, J.H. Temporal and spatial variability of soil respiration in a boreal mixedwood forest. *Agric. For. Meteorol.* **2006**, *140*, 244–256. [CrossRef]
28. Harden, J.W.; O'Neill, K.P.; Trumbore, S.E.; Veldhuis, H.; Stocks, B.J. Moss and soil contributions to the annual net carbon flux of a maturing boreal forest. *J. Geophys. Res. Atmos.* **1997**, *102*, 28805–28816. [CrossRef]
29. Sewerniak, P.; Stelter, P. Wpływ sposobu przygotowania gleby na dynamikę jej temperatury na wydmach Kotliny Toruńskiej (Effect of site preparation method on dynamics of soil temperature on inland dunes of the Toruń Basin). *Sylwan* **2016**, *160*, 923–932.
30. Dobbs, R.C.; McMinn, R.G. The effects of site preparation on summer soil temperatures in spruce-fir cutovers in the British Columbia interior. *Bi Mon. Res. Notes* **1973**, *29*, 6–7.
31. Mallik, A.U.; Hu, D. Soil respiration following site preparation treatments in boreal mixedwood forest. *For. Ecol. Manag.* **1997**, *97*, 265–275. [CrossRef]
32. Grossnickle, S.C. *Ecophysiology of Northern Spruce Species*; NRC Research Press: Ottawa, ON, Canada, 2000; ISBN 978-0-660-17959-9.
33. Lahti, M.; Aphalo, P.J.; Finér, L.; Ryyppö, A.; Lehto, T.; Mannerkoski, H. Effects of soil temperature on shoot and root growth and nutrient uptake of 5-year-old Norway spruce seedlings. *Tree Physiol.* **2005**, *25*, 115–122. [CrossRef] [PubMed]
34. Johansson, K.; Nilsson, U.; Örlander, G. A comparison of long-term effects of scarification methods on the establishment of Norway spruce. *For. Int. J. For. Res.* **2013**, *86*, 91–98. [CrossRef]
35. Löf, M.; Birkedal, M. Direct seeding of *Quercus robur* L. for reforestation: The influence of mechanical site preparation and sowing date on early growth of seedlings. *For. Ecol. Manag.* **2009**, *258*, 704–711. [CrossRef]
36. Prévost, M. Effets du scarifiage sur les propriétés du sol, la croissance des semis et la compétition: Revue des connaissances actuelles et perspectives de recherches au Québec. *Ann. Sci. For.* **1992**, *49*, 277–296. [CrossRef]
37. Reyes, J.; Thiers, O.; Gerding, V.; Donoso, P. Effect of scarification on soil change and establishment of and artificial forest regeneration under *Nothofagus* spp. in Southern Chile. *J. Soil Sci. Plant Nutr.* **2014**, *14*, 115–127. [CrossRef]
38. Munson, A.D.; Margolis, H.A.; Brand, D.G. Intensive silvicultural treatment: Impacts on soil fertility and planted conifer response. *Soil Sci. Soc. Am. J.* **1993**, *57*, 246–255. [CrossRef]
39. Carter, M.C.; Darwin Foster, C. Prescribed burning and productivity in southern pine forests: A review. *For. Ecol. Manag.* **2004**, *191*, 93–109. [CrossRef]
40. Cole, E.; Lindsay, A.; Newton, M.; Bailey, J.D. Vegetation control and soil moisture depletion related to herbicide treatments on forest plantations in northeastern Oregon. *Weed Technol.* **2018**, *32*, 461–474. [CrossRef]

41. Gregory, J.M. Impact of forest weed control on soils. In *Weed Control in Forest Management*; Holt, H.A., Fischer, B.C., Eds.; Purdue University: West Lafayette, IN, USA, 1981; pp. 231–236.
42. Schmidt, M.G.; Macdonald, S.E.; Rothwell, R.L. Impacts of harvesting and mechanical site preparation on soil chemical properties of mixed-wood boreal forest sites in Alberta. *Can. J. Soil Sci.* **1996**, *76*, 531–540. [CrossRef]
43. Yildiz, O.; Eşen, D.; Sarginci, M. Long-term site productivity effects of different *Rhododendron* control methods in eastern beech (*Fagus orientalis* Lipsky) ecosystems in the Western Black Sea region of Turkey. *Soil Use Manag.* **2009**, *25*, 28–33. [CrossRef]
44. Piatek, K.B.; Allen, H.L. Site preparation effects on foliar N and P use, retranslocation, and transfer to litter in 15-years old *Pinus taeda*. *For. Ecol. Manag.* **2000**, *129*, 143–152. [CrossRef]
45. Alcázar, J.; Woodard, P.M.; Rothwell, R.L. Soil disturbance and the potential for erosion after mechanical site preparation. *North. J. Appl. For.* **2002**, *19*, 5–13. [CrossRef]
46. Edeso, J.M.; Merino, A.; González, M.J.; Marauri, P. Soil erosion under different harvesting managements in steep forestlands from northern Spain. *Land Degrad. Dev.* **1999**, *10*, 79–88. [CrossRef]
47. Aust, W.M.; Miwa, M.; Burger, J.A.; Patterson, S.C.; Carter, E.A. Wet-weather timber harvesting and site preparation effects on Coastal Plain sites: A review. *South. J. Appl. For.* **2004**, *28*, 137–151. [CrossRef]
48. Beasley, R.S.; Granillo, A.B.; Zillmer, V. Sediment losses from forest management: Mechanical vs. chemical site preparation after clearcutting. *J. Environ. Qual.* **1986**, *15*, 413–416. [CrossRef]
49. Sutherland, B.; Foreman, F.F. Black spruce and vegetation response to chemical and mechanical site preparation on a boreal mixedwood site. *Can. J. For. Res.* **2000**, *30*, 1561–1570. [CrossRef]
50. Heineman, J.L.; Simard, S.W.; Sachs, D.L.; Mather, W.J. Chemical, grazing, and manual cutting treatments in mixed herb–shrub communities have no effect on interior spruce survival or growth in southern interior British Columbia. *For. Ecol. Manag.* **2005**, *205*, 359–374. [CrossRef]
51. Chen, H.Y.H.; Byman, W.J.; Bell, F.W. Chemical site preparation influences productivity, composition, and structure of boreal mixedwoods in Ontario, Canada. *For. Ecol. Manag.* **2006**, *229*, 145–154. [CrossRef]
52. Glen, L.M. *Drag Scarification in British Columbia*; Information Services Branch, Ministry of Forests: Victoria, BC, Canada, 1979; ISBN 978-0-7719-8143-2.
53. Oliver, C.D.; Larson, B.A. *Forest Stand Dynamics, Update Edition*; John Wiley & Sons: Hoboken, NJ, USA, 1996.
54. Frelich, L.E. *Forest Dynamics and Disturbance Regimes: Studies from Temperate Evergreen–Deciduous Forests*; Cambridge University Press: Cambridge, UK, 2002.
55. Haeussler, S.; Bedford, L.; Leduc, A.; Bergeron, Y.; Kranabetter, J.M. Silvicultural disturbance severity and plant communities of the southern Canadian boreal forest. *Silva Fenn.* **2002**, *36*, 307–327. [CrossRef]
56. DeLuca, T.H.; Aplet, G.H. Charcoal and carbon storage in forest soils of the Rocky Mountain West. *Front. Ecol. Environ.* **2008**, *6*, 18–24. [CrossRef]
57. Dollinger, J.; Jose, S. Agroforestry for soil health. *Agrofor. Syst.* **2018**, *92*, 213–219. [CrossRef]

2

The Impact of Forest usage and Accessibility on the Perceptions of its Users and Surrounding Residents

Gloria C. Blaise [1], Jeffrey A. Brown [2,†], Rebecca C. Jordan [3,*,†] and Amanda E. Sorensen [3,†]

1 Department of Natural Resources and the Environment, Cornell University, Ithaca, NY 14850, USA; gcb83@cornell.edu

2 Global Institute of Sustainability, Arizona State University, Tempe, AZ 85281, USA; jeffrey.alexander.brown@asu.edu

3 Department of Community Sustainability, Michigan State University, East Lansing, MI 48824, USA; soren109@msu.edu

* Correspondence: jordanre@msu.edu

† Authors are listed alphabetically after first author.

Abstract: Forests provide valuable ecosystem services to individuals that live near them and visit them. However, many forests, especially in highly developed areas, are specifically managed for resource conservation purposes. A common practice for conservation is restricting access to people to ensure minimal human-driven harm. While the restriction of human access to a forest increases its biological ecosystem services, it limits its cultural services and may reduce the public's perceived value of the forest. To investigate how access influences the perceived value of a forest, two forests in the state of New Jersey, USA, were compared. The forests, the Rutgers Ecological Preserve (RUEP) and Hutcheson Memorial Forest (HMF), represent accessible and inaccessible forests respectively. The study reported here evaluated the effects of visitation and accessibility on public perception of both sites. Residents near the RUEP and HMF, and visitors of both sites, were surveyed (n = 48). The results of the surveys demonstrated that the frequency of visitation and community location can have noticeable influences on a number of variables, including one's belief that forests provide important services and the thought that protected areas take away from an individual's freedom to use the area. A more positive view of the forests was present in those who visited the public forest most often.

Keywords: protected areas; conservation; public perceptions

1. Introduction

Urbanization is rapidly changing the global landscape with the continuing growth of cities separating people from nature [1,2]. With most of the world's population living in urban settings, there is a growing disconnect between people and their natural environment [3,4]. This separation of humans from the environment is a possible driver of poor human health outcomes [5]. These health outcomes may be further exacerbated in cities, given the frequently prominent legacy of cultural and socioeconomic histories that result in a disproportionate impact of environmental injustices on minority communities in urban areas [6,7]. Forests can provide a variety of benefits to individuals living near them, including improving air quality, reducing atmospheric carbon dioxide (CO_2), moderating local microclimates, lowering rainfall runoff and flooding and reducing noise pollution [8–10]. However, in many developed and developing areas, forests are targeted for conservation not for their benefits to human health but because of their importance in conserving biodiversity [11,12]. The difference in drivers behind forest conservation may cause potential conflicts as, often, protected areas and green spaces lose conservation value as they become more accessible to humans [13,14]

Sustainable management of forests has various definitions in the literature but broadly encompasses the core goal of maintaining ecological functionality (i.e., biodiversity, productivity) while promoting community benefits (i.e., economic, social) [15,16]. To maintain ecological functionality, some forests across the Northeast of the United States have been managed and classified as protected areas. Protected areas are clearly defined as those where development and human activity is limited in order to achieve long term conservation of nature and the associated ecosystem [17]. While the initial goal of protected areas was to preserve habitat and biodiversity, in recent years they have also been used to preserve spaces of cultural importance and to provide communities with access to nature [18] Protected areas are classified into six categories according to their management objectives (see [17].org "Protected Areas Categories"). As a result of protected area classification, forests across the Northeast vary in their level of public access. Some classified protected forests entirely restrict public access, whereas others exist on a continuum of public access and use. Regardless of the accessibility or classification of a forests, the presence of these forests provides ecosystem services to the communities around them.

Prior research has found that physical engagement with the natural world increases understanding and knowledge of biodiversity and thus willingness to protect it [19]. Regular engagement in nature has been shown to be particularly important in childhood for promoting life-long appreciation and affinity for nature [20–25]. Additionally, urban green spaces have been shown to be an important source of social and psychological benefits for those communities surrounding them [26]. However, recent work has found that the accessibility of forests and green spaces may influence individuals' valuation of these spaces. One study found that individuals were willing to pay larger amounts for homes near parks and that this behavior increased when parks were not heavily used [27]. Conversely, near parks where a lot of recreational activities were present, property values were lower. In less developed areas, property values were dependent on proximity to a forest or open area [27]. Other work has similarly found that house prices, and in turn property taxes, increased when in close proximity to protected forests [28]. From this work, it is clear that proximity to green space benefits those surrounding it, but this benefit is reduced with increased human use.

Although limiting use of protected forests may have an ecological benefit, this benefit may not strongly influence individuals' perceptions of protected forests. Instead, studies have found that visiting parks and natural areas and elements of individual identity may be more important in influencing perceptions of protected natural areas. A recent study of public views of parks found that there was a strong association between visitation frequency, age, gender and the perceived benefits of green spaces [29]. This work is supported by other studies, which found that parks and natural areas facilitate social interactions that establish a sense of community, pride, identity [30] and involvement [31]. As some protected forest areas are managed for public use, people may see more value in these forests for the social opportunities provided rather than the ecosystem service benefits, or be entirely unaware of the ecosystem services these protected forests provide. However, communities surrounding private-access protected forests may not receive the social benefits of public use forests and may be unaware of or not value the ecosystem service benefits provided by the forest. This lack of access may lead to decreased support for maintaining and protecting these forests, which play an important role in environmental health. Recent work has acknowledged that there is limited research that investigates the relationship between visitation/access to green spaces and perceptions of their personal and community-wide benefits [29]. It is, therefore, important to understand how access to protected forests influences people's perceptions of them, particularly considering the role of the public in protected area conservation.

In this study, we asked what factors, including environmental orientation, perceptions and use of public green spaces and sociodemographics, influence how individuals perceive the value of a forest. Additionally, we asked if any of these factors impact whether people believe that "protecting" forests limits people's "freedom to enjoy" the forests. Further, by comparing the responses of individuals who live near a protected forest with restricted access and those of individuals who live near a forest open to the public, we further investigated whether physical proximity to a protected forest shapes perceptions of the value of the forest and the influence of protected status on use. This research sought to understand the relationship between visitation opportunities of protected forests and public attitudes towards these forests. Understanding these relationships can guide management strategies to increase public support for the creation and conservation of protected forests.

2. Materials and Methods

2.1. Study Location

The Northeast of the United States has experienced rapid urbanization over the last 100 years [32]. Although much of the region has been developed, in areas such as New Jersey, conservation efforts dating back to the 1950s have set aside many patches of forest and other natural landscapes and developed an extensive network of protected areas [33]. As such, the Middlesex and Somerset counties of New Jersey contain a mixture of developed landscapes and forested areas with 20% and 39% of the counties made up of forests, respectively. However, the management of the forests within the counties is not uniform, with some forests established as strict conservation areas, such as private easements which are inaccessible to the public, while other forests are open to the public and feature hiking trails and other amenities. For this study, we focus on two forests, Hutcheson Memorial Forest (HMF) in Somerset County and the Rutgers Ecological Preserve (RUEP) in Middlesex County. The two green spaces are about 10 miles apart (Figure 1) and while HMF is private (strict accessibility, in this context meaning only accessible to the public on select days with research staff-led tours), the RUEP is a multi-use forest open to the public, which also serves as a teaching area. This research was conducted during the summer months of 2017 in the neighborhoods surrounding the two forests.

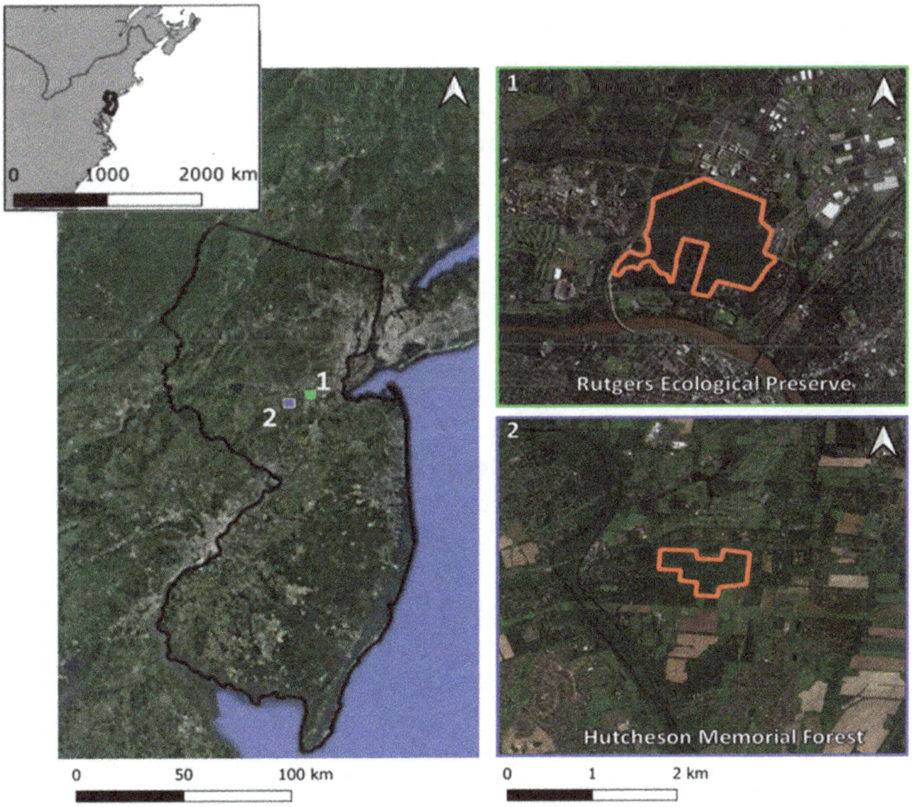

Figure 1. Location of the Rutgers University Ecological Preserve (RUEP, 1) and Hutcheson Memorial Forest (HMF, 2) within New Jersey on the East Coast of the United States. Imagery shows the differences in landscape surrounding these two areas, with the RUEP surrounded primarily by Rutgers University and residential land while HMF is surrounded primarily by agricultural and residential land. Images from Google Earth (2017).

2.2. Hutcheson Memorial Forest

Hutcheson Memorial Forest (HMF) is a 96-ha patchwork of abandoned agricultural fields and uncut old growth deciduous forest located in Somerset County, New Jersey, USA (40°29′55.7′′ N 74°33′46.6′′ W, Figure 1). The forest is surrounded primarily by undeveloped land (46%) with a mixture of developed land (28%) and agricultural land (24%). The core of HMF was established in 1711 by Dutch settlers and, as such, the 26-ha core is dominated by hickory and oak trees averaging 200 years of age [34]. Surrounding the old growth section of the forest, former agricultural fields have been undergoing succession over the past 60 years, resulting in young closed-canopy forest composed primarily of juniper. In 1955, Rutgers University acquired HMF and since then the forest has been a center for research. The forest is closed to the public and signs are posted around the forest perimeter declaring it private property and prohibiting trespassing. Additionally, in 2015, a 10-foot deer fence was built around the perimeter of the forest creating a physical barrier. For those interested in learning about the research done in the forest and the forest's history, tours are available one to two times a month during the spring and summer and once or twice during the winter. Tours are always guided by a researcher and range in size from 5–20 individuals. Generally, the audience of the tours tends to be older and the Hutcheson Memorial Research Center has paired with senior living centers to provide tours.

2.3. Rutgers University Ecological Preserve

The Rutgers University Ecological Preserve is a 128-ha tract of land that abuts Rutgers University's Livingston Campus in Piscataway, New Jersey (40°31′02.5″ N 74°26′27.0″ W, Figure 1) and is surrounded primarily by developed land (81%, Supplemental Figure S1). Almost all of the area has been used for agricultural purposes since the 1800s, with a few areas released from cultivation in the mid-1800s. All agricultural practices were stopped in the 1940s and Rutgers University acquired the land between 1945 and 1963. The Rutgers Ecological Preserve was officially established in 1976 [35]. The preserve is dominated by maple and oak trees with interspersed cherry and ash. There are six established trails through the Ecological Preserve and various events are hosted there to encourage people to use the area. The RUEP is also used by several courses at Rutgers as an outdoor classroom and many class projects revolve around creating signs and displays throughout the preserve to educate the public about the natural history of the area.

2.4. Survey Design and Implementation

We surveyed residences within a specified 300-m band area surrounding each of the study sites. The bordering residential areas were selected because individuals in these areas were more likely to be aware of and potentially use nearby forests. A sample of 100 residential households per neighborhood was selected using Google Maps. To the best of our ability, steps were taken to ensure that multihousehold residents were accounted for with the use of WhitePages. In the summer months (May through July) of 2017, two survey mail-outs were conducted about a month apart. When completed, participants received an Amazon gift card compensation with a monetary value of $10.00.

2.5. Questionnaire Design

The questionnaire focused on assessing individuals' knowledge and perception of the local forest, as well as their environmental orientation and demographic information.

2.6. Forest Perception Variables

Eight questions investigated individuals' perceptions and knowledge of the forest. There were five Likert-type scale questions (1. "The quality of the forest is excellent", 2. " The forest is valuable to the community and the environment", 3. "The forest has a diverse number of species", 4. "The forest has a variety of birds" and 5. "Most people in the community are not aware of this forest"), with 1 representing strong disagreement and 5 representing strong agreement. A sixth question asked if respondents were aware of the forest's history (note: this was coded as yes or no). A seventh question asked how long the respondent lived adjacent to the forest and was scored between 1–4 with 1 representing 1–5 years, 2 representing 6–10 years, 3 representing 11–15 years and 4 representing 15 years or more. The final question asked how often a respondent visited the forest.

2.7. *Environmental Orientation*

Three questions assessed individual's attitudes toward protected areas. These five-point Likert-type questions were "People in my community believe that a sense of community is important", "I consider myself an environmentalist" and "Protecting lands take away from our freedom to enjoy them."

2.8. *Sociodemographic Variables*

Three questions investigated sociodemographic factors of the respondents. Age was clustered on a scale from 1 to 5 including the categories: 18–24, 25–32, 33–40, 40–50 and 50+. Individuals were only given two options for gender, so it was used as a binary variable. Ethnicity was treated as a binary variable so that it could be included in the analysis and due to the lack of non-white respondents. Individuals were categorized as either white or non-white.

2.9. *Analysis*

We investigated whether people's perceptions of protected forests differed between individuals who lived near a strictly protected forest (HMF) and those who lived near a more openly protected forest (RUEP) by comparing the average responses to the questions "Protecting lands take away from our freedom to enjoy them" and "The forest is valuable to the community and the environment". Comparisons between the responses for the two groups were conducted via a two-tailed t-test with uneven variance, as the sample size between the two groups were different.

We also ran a generalized linear mixed model to investigate whether other factors led to differences in perceptions of how protecting lands may limit individuals "freedom to enjoy them". In this model, the dependent variable was respondents' responses to the question above while independent variables included the three sociodemographic variables, the eight forest perception variables and the two additional environmental orientation variables as fixed effects and the location where the respondents lived (near the RUEP or HMF) as a random effect. We also investigated what factors influenced the perceived value of a forest to the community. For this analysis, we followed the same methods as the previous model, except that we used the response to the question "The forest is valuable to the community and the environment" as the dependent variable. Before running our models, a Pearson's correlation analysis was conducted on all independent variables. The questions "The forest has a diverse number of species" and "The forest has a variety of birds" were correlated above 0.7 so the second question was dropped from our models. No other variables were correlated at a level above 0.7. All analysis was run in R version 3.6.1.

3. Results

A total of 54 surveys were received from participants, providing a 27% response rate. For our analysis, 48 participants responded to the survey fully enough to be included in the analysis, as incomplete surveys were removed from the data. Of the 48 responses included, 13 responses were from the area surrounding HMF and 35 were from the RUEP (see the supplementary materials for full survey).

3.1. Demographics

Overall demographics were similar between the respondents from around HMF versus those from around the RUEP (Table 1). A total of 68% of respondents to the survey sample were in the 40–50 and 50+ age intervals. with no significant different between groups (p = 0.19). Twenty-two of the twenty-seven respondents who provided racial identification identified as white, with no significant difference between the two groups, although the number of non-white respondents was higher among those from the Rutgers University Ecological Preserve (p = 0.09), with more individuals identifying as non-white. There was an equal representation at a proportion of 50:50 for those that identified as male or female, with no difference between groups (p = 0.71). Overall, the survey population was generally older white individuals but nonsignificant differences in demographics implied that differing opinions on forests were not a result of differences between sample populations around HMF versus the RUEP. The majority (75%) of locals had lived at their residences for more than 15 years, 15% for 1–5 years, 10% for 6–10 years and 2% for 11–15 years. Participation in the survey varied by forest, with 73% of survey respondents living near the RUEP.

Table 1. Sociodemographic information for participants displayed by overall composition, as well as separated for those who lived near HMF and those near the RUEP. Average values are shown along with standard deviation in parenthesis.

Variable	Combine	RUEP	HMF
Age	3.81 (1.48)	3.66 (1.47)	4.23 (1.48)
Caucasian (1 = yes, 2 =no)	1.31 (0.47)	1.36 (0.49)	1.17 (0.39)
Gender (male =1, female =2)	1.5 (0.51)	1.49 (0.51)	1.53 (0.51)
Length of time in neighborhood	3.33 (1.15)	3.31 (1.16)	3.38 (1.19)

3.2. Perceptions of Protected Status and Freedom to Use Land

Results from the t-tests showed that the responses for the HMF versus RUEP residents did not differ with regard to how individuals perceived the protected status influencing the freedom of individuals to use the land (p =0.41). Both groups disagreed with the statement that, in general, protecting forests limits individuals' freedom to enjoy these lands, with RUEP average response of 1.66 and the HMF average response of 1.74 (Table 2).

Table 2. Average responses for those who lived near HMF vs. those who lived near the RUEP to the questions "Protecting lands take away from our freedom to enjoy them" and "The forest is valuable to the community and the environment". Average value and standard deviation are shown in parentheses. For responses, 1 indicated strong disagreement and 5 indicated strong agreement.

Variable	RUEP	HMF
"Protecting lands take away from our freedom to enjoy them"	1.66 (1.25)	1.74 (1.41)
"The forest is valuable to the community and the environment"	4.91 (0.28)	4.62 (0.77)

Our mixed model investigating perceptions of whether protected status limited individuals' freedom to use forests had an R^2 of 0.41. The model revealed that only two significant variables—an individual's perceived sense of community and their self-identification as environmentalist—influenced respondents' perceptions of protected status. The more an individual agreed that a sense of community was important, the more likely they were to also think that protecting forests removes people's freedom to enjoy the forests (beta = 0.59, p = 0.04; Figure 1, Table 3). Additionally, individuals who identified as environmentalists were less likely to see protected status as removing the freedom of individuals using forests (beta = −0.95, p = 0.02; Figure 1, Table 3). No other variables were significant in our model (Table 3).

Table 3. Results from the linear mixed models. Beta estimates for each variable, as well as the p-value for each variable (in parentheses), are shown for our model investigating whether people perceived protected status as limiting freedom to use the forest, as well as for our model investigating how people perceived the value of the forest. Significant beta values are marked with *. R^2 for each model is displayed in parentheses after the question the model investigated as a dependent variable.

Variable	"Protecting Lands Take Away from Our Freedom to Enjoy Them" (R^2 0.41)	"The Forest Is Valuable to the Community and the Environment" (R^2 0.39)
Age	−0.05 (0.84)	−0.05 (0.63)
Caucasian (1 = yes, 2 = no)	−0.23 (0.67)	−0.08 (0.72)
Gender (male = 1, female = 2)	−0.09 (0.85)	0.41 (0.05) *
Length of time in neighborhood	0.47 (0.15)	−0.08 (0.50)
"The quality of the forest is excellent"	0.40 (0.11)	−0.10 (0.33)
"The forest has a diverse number of species"	0.19 (0.56)	0.37 (0.01) *
"Most people in the community are not aware of this forest"	0.34 (0.11)	−0.15 (0.49)
Awareness of forest's history	−0.66 (0.24)	0.08 (0.31)
Frequency of visitation	−0.39 (0.19)	0.03 (0.80)
"People in my community believe that a sense of community is important"	0.58 (0.05) *	0.14 (0.20)
"I consider myself an environmentalist"	−0.94 (0.02) *	0.00 (0.98)

3.3. Perception of Forest's Value to the Community

Between the two groups of individuals surveyed, there was a small but statistically significant difference in how individuals perceived the value of the forest to the community and the environment ($p = 0.05$); however, both groups agreed or strongly agreed that the forest was valuable, with those living near the RUEP averaging a response of 4.91 and those near HMF averaging 4.61 (Table 2). Our mixed model investigating perceptions of the value of the forest had an R^2 of 0.39. The model revealed two significant variables which influenced respondents' perceptions of the value of the forest. The more an individual agreed that the forest contained a diverse number of species, the more likely they were to think that the forest was valuable to the community and the environment (beta = 0.37, $p < 0.01$). Additionally, one demographic factor, gender, influenced the perceived value of the forest, with females more likely to agree that the forest was valuable to the community and the environment (beta = 0.41, $p = 0.04$; Table 3).

4. Discussion

These data suggest that, in general, people may not believe that protected status limits the ability of individuals to enjoy forests. However, this view may be held more strongly by those that are ecologically oriented, as those that identified as environmentalists were less likely to think that protected forests limited the freedom of individuals to enjoy forests. Those that are more anthropocentrically oriented may in fact disagree, as increased emphasis on a sense of community correlated with increased agreement that protecting forests could be limiting to forest use. This may be the case because those with an anthropocentric orientation may feel that resources exist solely for human use.

We also acknowledge that the results of this study might not be universal. Given that HMF is surrounded by a large amount of undeveloped and natural land, its restricted status may not limit community members from accessing nature. However, if the surrounding landscape were more developed, like the land surrounding the RUEP, people might have felt more negatively about HMF's limited access. Additionally, with the RUEP's recent housing development, the community is newer and may be less aware of the access points.

This work aligns with previous literature about perceptions of the value of protected areas and green spaces. In conservation, land is often deemed valuable because of high levels of biodiversity [36]. The respondents in this study also seemed to agree that biodiversity may drive the perceived value of a forest, as perceived increased species diversity was correlated with increased forest value. Although biodiversity was not measured in this study, other work has found a connection between biodiversity, or perceived biodiversity, and increased human wellbeing [37]. Thus, individuals who perceive an area to have higher biodiversity may, in fact, benefit from proximity to these areas. The role of gender and perception of the environment has also been seen in other studies in which females more positively perceived protected areas and their associated ecosystem services [38,39].

This study has some limitations because of the difference in the response rates of HMF participants. It is reasonable to believe that those individuals who have taken time to participate in a tour of HMF would hold different perceptions of the forest than those that never visited HMF. We also considered the implications of using Likert questions, as these questions are ideal for initial exploration but not in-depth analysis. Therefore, more work is needed to explore public perceptions of open-use versus protected forests.

Other scholars have highlighted the need to further explore the value of protected areas for people living outside of them [40]. Prior work has characterized the values derived from protected areas [41], indicating that these spaces influence the lives and wellbeing of visitors and residents. In-depth interviews with a subset of the users of and residents nearby the forests from this study would be particularly useful to help tease apart the relationships between the perceived and real value of protected areas, user-ship and personal identity.

While the restricted access of HMF serves an important purpose for scientific research, our work suggests that there may be broader consequences of these restrictions in terms of community support for maintenance of the forest. However, additional educational and community outreach programming may bolster community awareness and support of the forest. As protected areas are increasingly becoming the main mode by which people can experience nature [40], it is important to maintain these areas for future generations. Perhaps those who might be less ecologically oriented might be affected by appeals to community, which may be a motivation to support protected area preservation. Regardless, this work highlights, in conjunction with prior research, that it is important to prioritize management strategies of protected areas that directly involve the local community, with the explicit goal of generating community interest and buy-in to support these spaces.

Author Contributions: G.C.B., A.E.S. and R.C.J. conceived the idea. G.C.B. and R.C.J. carried out the experiment. J.A.B. performed analysis. All authors discussed the results and contributed to the final manuscript. All authors have read and agreed to the published version of the manuscript.

Acknowledgments: The authors would like to thank the survey respondents who made this work possible, the Rutgers University Center for Remote Sensing and Spatial Analysis (RU CRSSA), and Rutgers Douglass College. All work was done with Rutgers University institutional IRB approval #17-651.

References

1. Turner, W.R.; Nakamura, T.; Dinetti, M. Global Urbanization and the Separation from Nature. *Bioscience* **2004**, *54*, 585–590. [CrossRef]

2. Soga, M.; Evans, M.J.; Yamanoi, T.; Fukano, Y.; Tsuchiya, K.; Koyanagi, T.F.; Kanai, T. How can we mitigate against increasing biophobia among children during the extinction of experience? *Biol. Conserv.* **2020**, *242*, 108420. [CrossRef]
3. Soga, M.; Gaston, K.J. Extinction of experience: The loss of human–nature interactions. *Front. Ecol. Environ.* **2016**, *14*, 94–101. [CrossRef]
4. Cox, D.T.; Shanahan, D.F.; Hudson, H.L.; Fuller, R.A.; Anderson, K.; Hancock, S.; Gaston, K.J. Doses of nearby nature simultaneously associated with multiple health benefits. *Int. J. Environ. Res. Public Health* **2017**, *14*, 172. [CrossRef]
5. Cox, D.T.; Shanahan, D.F.; Hudson, H.L.; Fuller, R.A.; Gaston, K.J. The impact of urbanisation on nature dose and the implications for human health. *Landsc. Urb. Plan.* **2018**, *179*, 72–80. [CrossRef]
6. Keniger, L.E.; Gaston, K.J.; Irvine, K.N.; Fuller, R.A. What are the benefits of interacting with nature? *Int. J. Environ. Res. Public Health* **2013**, *10*, 913–935. [CrossRef]
7. Schell, C.J.; Dyson, K.; Fuentes, T.L.; Des Roches, S.; Harris, N.C.; Miller, D.S.; Woelfle-Erskine, C.A.; Lambert, M.R. The ecological and evolutionary consequences of systemic racism in urban environments. *Science* **2020**, *369*, eaay4497. [CrossRef]
8. Smith, W.H. *Air Pollution and Forests*; Springer Series on Environmental Management; Springer: Berlin/Heidelberg, Germany, 1990.
9. Rosenzweig, C.; Solecki, W.; Slosberg, R. *Mitigating New York City's Heat Island with Urban Forestry, Living Roofs, and Light Surfaces. A Report to the New York State Energy Research and Development Authority*; A report to the New York State Energy Research and Development Authority: Albany, NY, USA, 2006.
10. Nowak, D.J.; Dwyer, J.F. Understanding the benefits and costs of urban forest ecosystems. In *Urban and Community Forestry in the Northeast*; Springer: Dordrecht, The Netherlands, 2007; pp. 25–46. [CrossRef]
11. Endreny, T.; Santagata, R.; Perna, A.; Stefano, C.D.; Rallo, R.F.; Ulgiati, S. Implementing and managing urban forests: A much needed conservation strategy to increase ecosystem services and urban wellbeing. *Ecol. Model.* **2017**, *360*, 328–335. [CrossRef]
12. Amaya-Espinel, J.D.; Hostetler, M.E. The value of small forest fragments and urban tree canopy for Neotropical migrant birds during winter and migration seasons in Latin American countries: A systematic review. *Landsc. Urban Plan.* **2019**, *190*, 103592. [CrossRef]
13. Gardner, T.A.; Caro, T.I.M.; Fitzherbert, E.B.; Banda, T.; Lalbhai, P. Conservation value of multiple-use areas in East Africa. *Conserv. Biol.* **2007**, *21*, 1516–1525. [CrossRef] [PubMed]
14. Joppa, L.N.; Loarie, S.R.; Pimm, S.L. On population growth near protected areas. *PLoS ONE* **2009**, *4*, e4279. [CrossRef] [PubMed]
15. Wiersum, K.F. 200 years of sustainability in forestry: Lessons from history. *Environ. Manag.* **1995**, *19*, 321–329. [CrossRef]
16. Dwyer, J.F.; Nowak, D.J.; Noble, M.H. Sustaining urban forests. *J. Arboric.* **2003**, *29*, 49–52.
17. International Union for Conservation of Nature. 2008. Available online: https://www.iucn.org/theme/protected-areas/about (accessed on 29 August 2018).
18. Phillips, A. Turning ideas on their head: The new paradigm for protected areas. In *The George Wright Forum*; No. 2; George Wright Society: Hancock, MI, USA, 2003; Volume 20, pp. 8–32.
19. Kareiva, P. Ominous trends in nature recreation. *Proc. Natl. Acad. Sci. USA* **2008**, *105*, 2757–2758. [CrossRef] [PubMed]
20. Bunting, T.E.; Cousins, L.R. Environmental dispositions among school-age children: A preliminary investigation. *Environ. Behav.* **1985**, *17*, 725–768. [CrossRef]
21. Chawla, L. Children's concern for the natural environment. *Child. Environ. Q.* **1988**, *5*, 13–20.
22. Wilson, R.A. *Fostering a Sense of Wonder during the Early Childhood Years*; Greyden Press: Columbus, OH, USA, 1993.
23. Pyle, R.M. *The Thunder Tree: Lessons from an Urban Wildland*; Houghton Mifflin: Boston, MA, USA, 1993; Volume 220.
24. Schultz, P.W.; Shriver, C.; Tabanico, J.J.; Khazian, A.M. Implicit connections with nature. *J. Environ. Psychol.* **2004**, *24*, 31–42. [CrossRef]
25. Miller, J.R. Biodiversity conservation and the extinction of experience. *Trends Ecol. Evol.* **2005**, *20*, 430–434. [CrossRef]

26. Chiesura, A. The role of urban parks for the sustainable city. *Landsc. Urban Plan.* **2004**, *68*, 129–138. [CrossRef]
27. Crompton, J.L. The impact of parks on property values: A review of the empirical evidence. *J. Leis. Res.* **2001**, *33*, 1–31. [CrossRef]
28. Kettunen, M.; ten Brink, P. *Social and Economic Benefits of Protected Areas: An Assessment Guide*; Routledge: Abingdon, UK, 2013.
29. Moyle, B.D.; Weiler, B. Revisiting the importance of visitation: Public perceptions of park benefits. *Tour. Hosp. Res.* **2017**, *17*, 91–105. [CrossRef]
30. Jordan, R.; Sorensen, A.; Clark, D. Urban/Suburban Park Use: Links to Personal Identity? *Curr. World Environ.* **2015**, *10*, 355. [CrossRef]
31. Bedimo-Rung, A.L.; Mowen, A.J.; Cohen, D.A. The significance of parks to physical activity and public health: A conceptual model. *Am. J. Prev. Med.* **2005**, *28*, 159–168. [CrossRef] [PubMed]
32. U.S. Census Bureau. Demographic Trends in the 20th Century. Census Special Reports. 2002. Available online: https://www.census.gov/prod/2002pubs/censr-4.pdf (accessed on 29 August 2018).
33. Fenn, K.H. Protected Area Networks in an Urbanizing Landscape: Spatial Characteristics and Land Acquisition Strategies. Ph.D. Dissertation, Rutgers University-School of Graduate Studies, New Brunswick, NJ, USA, 2018. [CrossRef]
34. Buell, M.F. The mature oak forest of Mettler's Woods. William Hutcheson L. *Meml. For. Bull.* **1957**, *1*, 16–19.
35. Rutgers Ecopreserve. (n.d.). Brief History. Available online: https://ecopreserve.rutgers.edu/ruep-info/ruep-histor/ (accessed on 29 August 2018).
36. Bakhtiari, F.; Jacobsen, J.B.; Strange, N.; Helles, F. Revealing lay people's perceptions of forest biodiversity value components and their application in valuation method. *Glob. Ecol. Conserv.* **2014**, *1*, 27–42. [CrossRef]
37. Shanahan, D.; Bush, R.; Gaston, K.; Lin, B.B.; Dean, J.; Barber, E.; Fuller, R.A. Health benefits from nature experiences depend on dose. *Sci. Rep.* **2016**, *6*, 28551–28561. [CrossRef] [PubMed]
38. Mohai, P. Gender differences in the perception of most important environmental problems. *Race Gender Class* **1997**, 153–169.
39. Martino, D. Gender and urban perceptions of nature and protected areas in Bañados del Este Biosphere Reserve. *Environ. Manag.* **2008**, *41*, 654. [CrossRef]
40. West, P.; Igoe, J.; Brockington, D. Parks and peoples: The social impact of protected areas. *Annu. Rev. Anthr.* **2006**, *35*, 251–277. [CrossRef]
41. Harmon, D. Intangible values of protected areas: What are they? Why do they matter. In *The George Wright Forum*; No. 2; George Wright Society: Hancock, MI, USA, 2004; Volume 21, pp. 9–22.

3

The Impact of COVID-19 on the Management of European Protected Areas and Policy Implications

James McGinlay, Vassilis Gkoumas, Jens Holtvoeth, Ruymán Federico Armas Fuertes, Elena Bazhenova, Alessandro Benzoni, Kerstin Botsch, Carmen Cabrera Martel, Cati Carrillo Sánchez, Isabel Cervera, Guillermo Chaminade, Juliana Doerstel, Concepción J. Fagundo García, Angela Jones, Michael Lammertz, Kaja Lotman, Majda Odar, Teresa Pastor, Carol Ritchie, Stefano Santi, Mojca Smolej, Francisco Soriano Rico, Holly Waterman, Tomasz Zwijacz-Kozica, Andreas Kontoleon, Panayiotis G. Dimitrakopoulos and Nikoleta Jones

1 Department of Land Economy, University of Cambridge Conservation Research Institute & Cambridge Centre for Environment, Energy and Natural Resource Governance, University of Cambridge, Cambridge CB2 3QZ, UK; jm2365@cam.ac.uk (J.M.); vg345@cam.ac.uk (V.G.); j.holtvoeth@bristol.ac.uk (J.H.); ak219@cam.ac.uk (A.K.)

2 Grupo Tragsa, Avenida Quinto Centenario, Edifico San José, Local 4, 38800 San Sebastián de La Gomera, La Gomera, Canary Islands, Spain; rarmas1@tragsa.es

3 Kullaberg Nature Reserve, County Administrative Board of Skåne, Italienska vägen 323, 26377 Mölle, Sweden; elena.bazhenova@lansstyrelsen.se

4 Prealpi Giulie Nature Park, Resia, Piazza del Tiglio 3, 33010 Resia, Italy; alessandro.benzoni@parcoprealpigiulie.it (A.B.); stefano.santi@parcoprealpigiulie.it (S.S.)

5 Department of National Park Planning, Regional Development and Tourism, National Park Black Forest, Schwarzwaldhochstraße 2, 77889 Seebach, Germany; Kerstin.botsch@nlp.bwl.de

6 Centro Administrativo del Parque Nacional y Parque Natural de Sierra Nevada, Carretera Antigua de Sierra Nevada km 7, 18191 Pinos Genil, Granada, Spain; mariac.cabrera.martel@juntadeandalucia.es

7 Subdirección General de Patrimonio Natural y Cambio Climático, Dirección General del Medio Natural, Consejería de Agua, Agricultura, Ganadería, Pesca y Medio Ambiente, Región de Murcia, Plaza Juan XXIII, s/n. Desp. C1.10, 30008 Murcia, Spain; cetssierraespuna@carm.es (C.C.S.); francisco.soriano2@carm.es (F.S.R.)

8 Las Batuecas-Sierra de Francia Natural Park, Casa del Parque Natural Las Batuecas-Sierra de Francia, Ctra/Las Batuecas, n° 22, 37624 La Alberca, Salamanca, Spain; isabel.cervera@patrimonionatural.org

9 Söderåsen National Park, County Administrative Board of Skåne, Skäralid 747, 26453 Ljungbyhed, Sweden; Guillermo.chaminade@lansstyrelsen.se

10 Wald und Holz NRW, Nationalparkforstamt Eifel, Fachgebiet Kommunikation und Naturerleben, Urftseestraße 34, 53937 Schleiden-Gemünd, Germany; doerstel@nationalpark-eifel.de (J.D.); lammertz@nationalpark-eifel.de (M.L.)

11 Garajonay National Park, Centro de Visitantes Juego de Bolas, La Palmita C.P., 38330 Agulo, La Gomera, Canary Islands, Spain; cfaggar@gobiernodecanarias.org

12 Snowdonia National Park Authority, Penrhyndeudraeth, Gwynedd LL48 6LF, UK; angela.jones@eryri.llyw.cymru

13 Environmental Board of Estonia, 15172 Tallinn, Estonia; kaja.lotman@keskkonnaamet.ee

14 Triglav National Park, Ljubljanska cesta 27, 4260 Bled, Slovenia; majda.odar@tnp.gov.si (M.O.); mojca.smolej@tnp.gov.si (M.S.)

15 EUROPARC Federation, Waffnergasse 6, 93047 Regensburg, Germany; teresa.pastor@europarc.org (T.P.); carol.ritchie@europarc.org (C.R.)

16 Peak District National Park Authority, Aldern House, Bakewell DE45 1AE, UK; holly.waterman@peakdistrict.gov.uk

17 Tatra National Par, Kuźnice 1, 34-500 Zakopane, Poland; tzwijacz@tpn.pl

18 Department of Environment, University of the Aegean, 81100 Mytilene, Greece

* Correspondence: pdimi@env.aegean.gr (P.G.D.); nj322@cam.ac.uk (N.J.);

Abstract: The COVID-19 pandemic led to many European countries imposing lockdown measures and limiting people's movement during spring 2020. During the summer 2020, these strict lockdown measures were gradually lifted while in autumn 2020, local restrictions started to be re-introduced as a second wave emerged. After initial restrictions on visitors accessing many Nature Protected Areas (PAs) in Europe, management authorities have had to introduce measures so that all users can safely visit these protected landscapes. In this paper, we examine the challenges that emerged due to COVID-19 for PAs and their deeper causes. By considering the impact on and response of 14 popular European National and Nature Parks, we propose tentative longer-term solutions going beyond the current short-term measures that have been implemented. The most important challenges identified in our study were overcrowding, a new profile of visitors, problematic behavior, and conflicts between different user groups. A number of new measures have been introduced to tackle these challenges including information campaigns, traffic management, and establishing one-way systems on trail paths. However, measures to safeguard public health are often in conflict with other PA management measures aiming to minimize disturbance of wildlife and ecosystems. We highlight three areas in which management of PAs can learn from the experience of this pandemic: managing visitor numbers in order to avoid overcrowding through careful spatial planning, introducing educational campaigns, particularly targeting a new profile of visitors, and promoting sustainable tourism models, which do not rely on large visitor numbers.

Keywords: biodiversity conservation; conflict; national parks; management; pandemic; public health; wellbeing

---

1. Introduction

Nature Protected Areas (PAs) are important because of their high biodiversity value and the socio-economic benefits they provide for people [1]. In addition to their crucial role in biodiversity conservation, most PAs in Europe are recognized as multifunctional landscapes providing multiple benefits and ecosystem services. These may range from provisioning services, for example in working farmed landscapes, to regulating services, from water and air quality regulation to carbon storage, but also a very wide range of cultural ecosystem service benefits [2] ranging from psychological restoration [3–5] and improved physiological health [6–8] to better social relations [9–12], and spiritual development [13,14]. Whilst, to a certain extent, many of these benefits may be available to people from a local urban and peri-urban green space, countries designate a suite of high-quality nature protected areas that are of exceptional quality for biodiversity conservation, and provision of the above provides a wide range of benefits to people, and high profile sites such as National Parks are typically extremely popular with visitors. These protected landscapes, therefore, have a crucial role in improving physical and mental health [15], assisting in the improvement of people's wellbeing [16,17], and protecting local social and cultural values [18].

Europe is the region with the largest number of PAs internationally [19]. European PAs are of various sizes with overlapping designations such as the Ramsar Convention, the NATURA 2000 network, the Emerald network, and nationally designated parks [20]. PAs are a significant source of income for local communities living inside or near their boundaries. This is in part due to the high number of visitors they attract [21]. In Europe, Schägner et al. [22] estimated that 449 national parks attract over 2 billion visitors with a total value of €14.5 billion annually. These estimates represent only a fraction of the actual value of tourism in European Protected Areas considering that Europe has over 100,000 Protected Areas [23].

The COVID-19 pandemic led many European countries to impose lockdown measures to limit people's movement [24]. These measures were aimed at reducing the spread of the virus but also decreased significantly the number of people visiting outdoor spaces [25], including PAs, particularly those located in more remote areas, as is often the case for many larger PAs. In European countries, where strict lockdown restrictions were imposed, a reduction in visitor numbers was initially observed (e.g., [26,27]). Likewise, the number of visitors increased rapidly as soon as these restrictions were eased [27–29].

These changes in visitor numbers are expected to have posed significant management issues for PAs in Europe. As we write this paper (October 2020), lockdown measures that had been lifted during the past summer are gradually being re-introduced in several European countries. Central and regional government authorities are looking into ways of containing the virus' transmission, focusing significantly on measures that are enforced locally. During the lockdown, most people only had access to green spaces near where they lived and access to more remote sites was limited to those living nearby. In the case of popular and high-profile PAs, such as national and regional parks, coming out of strict lockdown in late spring and summer 2020 resulted in a steep increase in visitor numbers. Therefore, the management authorities in these areas needed to introduce new measures to enable all users to visit them safely. However, a key issue for popular nature Protected Areas is that the requirements of nature conservation and public health safety may, at times, be difficult to balance.

In normal times, PA staff manage to channel visitors in a way that minimizes disturbance of more sensitive species, such as ground-nesting birds in the nesting season, or fragile ecosystems, such as high altitude montane habitats. This normally means encouraging visitors to spend most of their time in less sensitive locations, often leading to the creation of busier 'honeypot sites', where crowds may occur. However, social distancing regulations that have been implemented in response to COVID-19 involve avoiding crowded locations, so visitors spread out more evenly across the PA, thus increasing the likelihood of human disturbance of species and habitats. In addition to the challenges presented solely by the increase in visitor numbers, visitor behavior can also conflict with landscape and nature conservation.

In this paper, we present an early analysis of how COVID-19 has impacted European PAs so far. We focus on two key issues: (a) the challenges that COVID-19 presented to the management of PAs in Europe, especially on the ability of PAs to perform their functions of conserving nature and providing nature-based benefits to visitors, and (b) review indicative measures that have been implemented across different parks in the region to tackle these challenges. In the discussion section we analyze the problems and their deeper causes, consider what lessons can be learned from the COVID-19 pandemic, and propose tentative longer-term solutions going beyond the current short-term measures.

2. Materials and Methods

In order to capture the key challenges faced by PAs due to COVID-19 and the measures applied to address them, we followed a two-stage approach: we initially surveyed the existing limited academic and grey literature and websites of park authorities to identify measures that had been introduced by park authorities across Europe during the first months of the pandemic. We then organized two workshops with key informants from the management bodies of a selected sample of European nature PAs.

Our study sample was selected based on three broad criteria: (a) high profile sites were targeted which are popular with visitors (including sites which have been awarded the EUROPARC Federation Charter of Sustainable Tourism) as they were more likely to prove very popular after the lockdown was eased and so particularly challenging to manage regarding the trade-off between visitor management and nature conservation; (b) geographical spread across a diverse range of European countries from West to East, North to South; (c) varying national regulations on COVID-19.

For the selection of the final participant PAs (Table 1), we decided to narrow down our research to eight countries which reflected a range of government responses to the pandemic from strict lockdown to softer measures (criterion c). The eight countries were the UK, Spain, Italy, Estonia, Germany, Poland, Slovenia, and Sweden. During the first months of the pandemic, the UK, Spain, and Italy were three of the most badly affected countries in Europe by the virus and strict lockdown restrictions were imposed restricting significantly people's movement [30–32]. Germany also had a large number of cases but the death rate was lower compared to other countries and restrictions were less severe compared to the UK, Spain, and Italy [33,34]. Estonia, Poland, and Slovenia had a lower fatality rate [35] and restrictions on movement were also imposed [36–39] but were eased earlier compared to other countries. Finally, Sweden was the only country in our study which did not impose strict lockdown restrictions compared to the rest of the European countries. Although no strict measures were imposed in Sweden, it is considered that a large majority of people followed social distancing recommendations while the most vulnerable self-isolated voluntarily [40].

After determining the sample of eight countries, a call was announced inviting park authorities to participate in workshops focusing on the impacts of COVID-19. Two workshops were organized (co-hosted with the EUROPARC Federation) three months apart (1 July and 6 October 2020) with invited representatives of the management of 14 selected sites in order to explore with them the challenges that parks face due to the pandemic. All sites were included in the study, as they are all popular visitor destinations, they spread across multiple European countries, and they face new challenges due to COVID-19. The 14 sites also represent two different types of designations according to the IUCN categories (II and V). (IUCN Categories are described in detail on the IUCN website: https://www.iucn.org/theme/protected-areas/about/protected-area-categories). Details of the participating parks are given in Table 1 below. Fourteen attendees participated in the first workshop representing 13 parks and 16 attendees in the second workshop also representing 13 parks.

The first workshop took place during the first COVID-19 peak when a potential 6-month pandemic was considered likely. At the time of the second workshop, the second peak had begun and it appeared that a 12-month crisis or longer was probable.

Table 1. Nature Protected Areas (PAs) participating in the workshops.

Name	Country & Region	IUCN Protection Level
Matsalu National Park	Estonia	II
Eifel National Park	Germany (Nord-Rhein Westphalen)	II
Black Forest (Schwarzwald) National Park	Germany (Baden-Württemberg)	II
Prealpi Giulie Natural Park	Italy (Friuli Venezia Giulia)	V
Tatra National Park *	Poland (Carpathians)	II
Triglav National Park	Slovenia (Upper Carniola)	II & V
Sierra Espuña Regional Park	Spain (Murcia)	V
Sierra Nevada National Park	Spain (Andalusia)	II
Las Batuecas-Sierra de Francia Natural Park **	Spain (Castilla y León)	V
Garajonay National Park	Spain (Canary Islands)	II
Kullaberg Nature Reserve	Sweden (Skåne Province)	V
Söderåsen National Park	Sweden (Skåne Province)	II
Peak District National Park	UK (England)	V
Snowdonia National Park *	UK (Wales)	V

* attended first workshop only, ** attended second workshop only.

The first workshop focused on the main positive and negative impacts of the PAs on their local communities prior to the COVID-19 pandemic, and on the main challenges that COVID-19 had presented to them, with a particular focus on changes in visitor numbers and the impact on local people. Participants were then asked about the measures they had implemented to cope with the impact of COVID-19 on the PA and local people, and about their plans to reduce or manage the negative impacts in the future. The second workshop then focused more on the tensions and conflicts between stakeholder groups created by the PA and the COVID-19 pandemic. Participants were asked whether

COVID-19 had caused new or increased existing tensions and conflicts in their PA (e.g., between local people, between visitors, or between visitors and local people). Finally, it covered any research done by the PA management in the past that could explain why tensions or conflicts had emerged (for example, a social impact assessment).

Both workshops were recorded and facilitated by experienced colleagues who took notes during the workshop. Online polls were also conducted during the workshops to obtain quantitative data using the Zoom video-conferencing platform, and further qualitative comments were collected regarding the above discussion topics using Mentimeter.

The notes and the recordings were then analyzed by researchers experienced in qualitative data analysis to identify key emergent themes focusing on two main broad topic areas: (a) the challenges in the management of the PAs due to the pandemic and (b) actions to overcome these challenges. The findings of both qualitative and quantitative data analysis are presented below in Section 3.

3. Results

3.1. *Challenges Due to COVID-19 in European Nature Parks*

A range of challenges caused by COVID-19 was identified by participants (Table 2). In all 14 parks, an increase in visitors was observed especially during the summer compared to the same period the previous year. An increase in weekday visitors was also noted in certain parks. The increase in visitors led to overcrowding incidents and park authorities had to introduce very quickly new social distancing measures and recommendations that would ensure that all users were able to safely enjoy the area. In countries with strict lockdown restrictions (UK, Italy, Spain), the initial low visitation numbers (due to strict travel regulations) were followed by a significant increase in visitors during the summer. In the case of Swedish sites, an increase in visitors was noticed at the beginning of the pandemic (as no restrictions on movement were imposed), which continued throughout the summer months. Similarly, at the German sites, where the movement of people within specific regions was not significantly restricted, a gradual increase in visitors was observed from the beginning of the health emergency in the country. It should be noted that some participant parks experienced an almost 100% increase in visitors on certain days relative to expectations for that time of year. A possible explanation for this increase suggested by the participants was that people felt safer in outdoor and more remote locations, such as the ones protected by National Parks and nature reserves across Europe, compared to indoor and urban spaces. Furthermore, in some countries, the weather was relatively mild in spring 2020 (UK, Germany, and Sweden), which may also have resulted in a significant increase in visitors where people's movement was allowed.

A second important challenge was that the increase in visitors was often combined with incidents of problematic behavior, which is defined here as behavior that conflicts with either the conservation aims of the PA or the widely accepted social norms of behavior within the local communities around the PA. Although irresponsible behavior does occur in PAs, our analysis revealed that such issues became more frequent during the first months of the pandemic.

During the second workshop, park authorities were asked to specify which behaviors had become more frequent at their sites during the pandemic period (Figure 1). Parking and road congestion was the most frequently mentioned issues, which was partly due to the greatly increased number of tourists visiting the parks and a tendency to drive rather than commute by public transport or join organized groups with coaches to reach the parks. Illegal parking became a common problem in several PAs, with rangers having to routinely monitor for illegal behavior. A second important problem mentioned by respondents was linked to waste management issues, such as littering. Workshop participants noted that the observed increase in problematic behavior may be linked to a different profile of people visiting the areas due to COVID-19, who were unaware of the main regulations that are in place and of widely accepted norms of behavior in conservation areas. This was noted across most sites irrespective of the geographical region. Finally, the mountainous parks in the sample expressed concern over increased visits from inexperienced hikers in the winter and the anticipated increased demand for first-aid provisions and search and rescue missions. This is expected both due to a decrease in or ban on guided tours due to social distancing regulations and due to more visitors attempting mountain hiking as an alternative to traveling abroad or elsewhere domestically.

Table 2. Challenges in managing European PAs due to COVID-19.

Challenge	Key Issue
Overcrowding	A significant increase in (mainly domestic) visitors was reported especially as soon as people were allowed to travel further from their home
Problematic behavior by PA users	An increase in problematic behavior when using the PA was observed. This referred to a range of issues including waste management and disturbance e.g., littering and dog/human waste, noise nuisance, illegal/unauthorized activities e.g., camping
Parking and traffic issues	Incidents of irresponsible parking were reported including parking in non-designated areas and parking in a way that disturbed people and nature. Traffic was increased as people accessed parks with their own car avoiding public transport and organized groups
Social distancing	Changes were needed in how certain activities were being run, such as guided tours and visitor centers, to ensure social distancing
Conflicts	Conflicts between local people, and between visitors and locals emerged because of overcrowding and the fear of virus transmission, and over behavioral issues
Cancellation of educational and cultural activities	Several activities which are organized regularly in the parks, such as guided tours and festivals, had to be canceled or limited to a very low number of participants

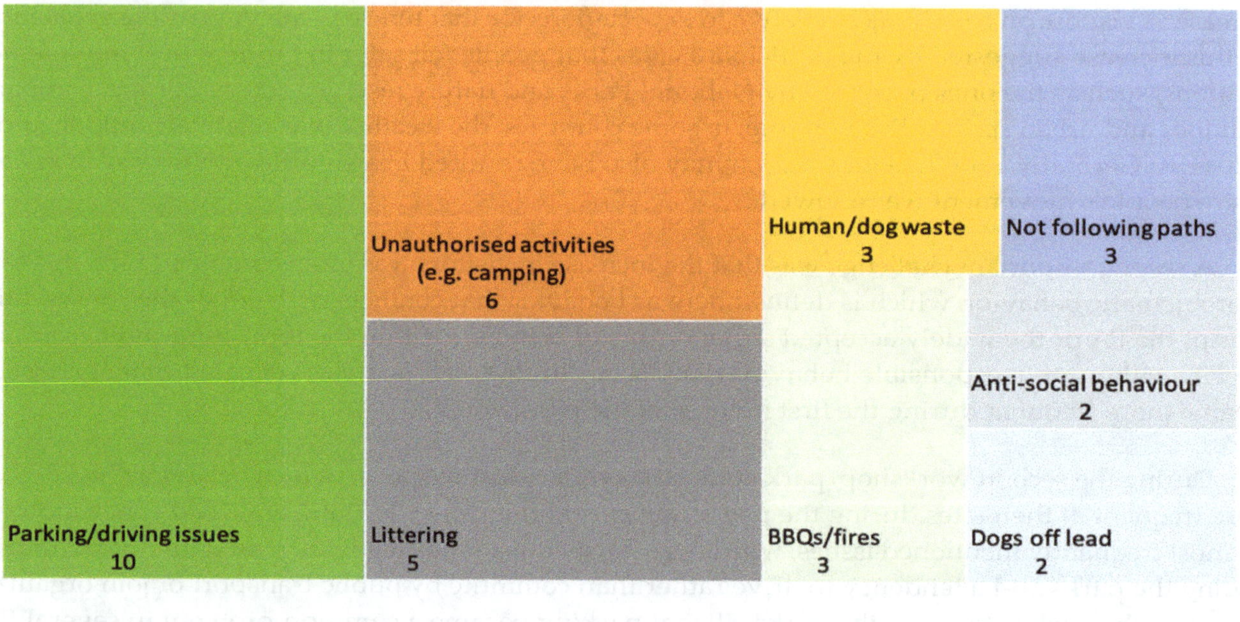

Figure 1. Frequency of problematic behavior during the first seven months of the pandemic mentioned by park authorities in the workshops.

Overcrowding incidents and irresponsible behavior also led to conflicts between local residents, and between locals and visitors at several sites. Conflicts between residents tended to arise in cases where behavior, which otherwise would be considered unremarkable or unproblematic, was perceived to contravene emergency restrictions and social distancing recommendations, or where attempts to modify behavior to conform to the new situation led to new conflicts and tension that did not normally arise. Examples of this included local people going for a walk or cycling in their local area and either being criticized for this or not being recognized as local.

Conflicts between visitors and locals arose over either local norms of behavior between locals and visitors, such as inconsiderate parking or littering or in some areas, because visitors were considered 'transmitters' of the virus and locals would prefer the government to have restricted access to the park until they felt safe. Incidents of vandalism and placement of signs on the road stating that visitors were not welcome were also reported.

Another important issue that was mentioned by several parks was the cancellation of or changes to educational and cultural activities, such as school visits, guided tours, and festivals. Although demand for such activities was high during the summer, due to new social distancing rules, several park authorities decided to limit the number of people participating in these activities or cancel them entirely to limit virus transmission. Apart from the wider social and economic impact of having reduced environmental education and cultural activities at the different sites, a related issue was people choosing to go on their own in the park instead of with a guided tour. This also led to a higher number of cars trying to access the PAs (instead of groups using coaches or public transportation).

Collectively, the coincidence of large increases in visitor numbers, attempts to social distance and avoid crowds, lack of availability of organized tours, as well as new types of visitors less aware of the susceptibility to disturbance of many natural systems, all contributed to increasing the risk and extent of disturbance in remoter more sensitive areas of PAs, as well as more general threats to the tranquility, quality, and integrity of the protected landscapes.

3.2. COVID-19 Measures in Order to Overcome the New Challenges

Different measures were introduced in the parks in order to address the new challenges (Table 3). To address overcrowding, several PAs proceeded with measures limiting access for visitors with different levels of restrictions depending on the virus transmission rates. These measures were often guided by the restrictions imposed at a higher level of administration, either by a regional or central government. In Sweden for example, no restrictions were imposed as the national guidelines did not limit people's movement. On the contrary, at the UK sites, these measures tended to be stricter compared to Sweden (e.g., complete closure of facilities and parking areas) but as lockdown measures were eased, a larger number of visitors were able to access these protected landscapes.

As visitor numbers increased, a key task for all park authorities was to introduce new measures in order to ensure social distancing. A variety of tools were introduced including a one-way system on popular and narrow paths; restricting the number of participants on guided tours; restricting the number of people allowed within facilities (e.g., restaurants, visitor centers, restrooms), and counting the number of visitors entering the area. New measures to maintain good hygiene and limit the spread of the virus were also introduced in PAs. These included enhanced cleaning and waste disposal measures, such as placing hand sanitizers in key locations, cleaning toilet facilities and frequently touched surfaces regularly, and banning cash payments (allowing only contactless payments). Protection of staff was also a key priority with the provision of PPE equipment and installing plexiglass barriers in customer-facing facilities such as restaurants and visitors' centers. A mobile application was also used at one site, which assisted in people having an overview of how many users were on a trail at the same time and reminded them of the current recommendations due to COVID-19.

Table 3. Measures to tackle new challenges due to COVID-19.

Challenge	Measures
Overcrowding	Closure of major facilities, closure of parking areas, cease advertising/promoting the PA to visitors, temporary closure of specific honeypot sites, online updates on car park capacity and overcrowding incidents, replace guided tours and school visits with online educational programs
Irresponsible users	Information campaigns including signs on local notice boards, key entrance points, information on websites, and social media (e.g., Twitter and Facebook). Use of social media to promote appropriate pro-environmental behavior in the PA. Increased number of rangers or increased presence of local police. Fix damaged or vanadalized signage as soon as possible.
Parking and traffic issues	Information campaigns letting people know when a car park is full and also about responsible parking. Increased number of rangers. Introduce new regulations. Towing vehicles away
Conflicts	Information campaigns including signs, information on websites and social media (e.g., Twitter and Facebook). Increased number of rangers.
Social distancing	Banning social gatherings, restricted number of people on guided tours, restrictions on the number of people within facilities (e.g., restaurants, visitor centers, restrooms), establishing one-way system on popular paths, rigorous and enhanced cleaning regimes and waste collection e.g., at visitor centers, placing hand sanitizers in key locations, regular cleaning of toilet facilities, and banning of cash payment (allowing only contactless payments). Protection of staff was also a key priority with the provision of PPE equipment and installing plexiglass in key facilities such as restaurants and visitor centers.
Cancellation of educational and cultural activities	Online learning, a limit on the number of people who are able to attend guided tours

Regarding the increase in problematic behavior at some sites, which was partly attributed to the different profiles of new users, most park authorities recognized that it was necessary to inform and educate people on permissible activity within the PA and on responsible behavior. Several information campaigns were initiated by the park authorities. These included leaflets informing people of key regulations at key entrance points and also clear signage promoting the dispersal of visitors from the main car parks. Information about regulations was also widely promoted via the websites of the PA authorities and also via social media such as Facebook and Twitter.

The number of rangers patrolling the PAs was also increased in several parks whilst one PA noted that rangers preferred to patrol in pairs as they anticipated more hostile responses from visitors, who were asked to comply with social distancing regulations. Another PA liaised with the local police to enhance their presence in the PA area. Similar measures were introduced to tackle illegal parking, insufficient car parking capacity, and traffic incidents. Several park authorities also informed people of parking and traffic issues via social media. In one park, the parking fee was increased in order to discourage visitors while in other parks, the possibility of introducing a parking fee charge is currently being considered in order to reduce obstructive and illegal parking and manage traffic.

Another key challenge that emerged from our analysis refers to conflicts between in- and out-of-area users in certain PAs. This was mainly because several lockdown restrictions had a geographical component to them with people only allowed to travel up to a certain distance (such as a 5 mile (8 km) limit in Wales, UK) or allowed to travel only within specific regions (Germany and Spain). Although no measures were recorded to tackle this specific challenge, the combined measures mentioned above aimed to reduce conflicts as management authorities tried to reduce issues of overcrowding and problematic behavior, which tended to cause local communities to complain. Regarding the broader social conflicts among residents of the different PAs or the economic impacts of discouraging visitors, no measures were noted at the local level, as such, issues would normally be beyond the remit of park authorities, falling rather to state entities such as the police or government.

As far as the reduction of educational activities is concerned, this is regarded as one of the most important challenges for park authorities. One park mentioned that they have reverted to online learning instead of face-to-face educational activities, while in most parks, some activities have resumed but with a reduced number of participants.

Finally, at this stage of the pandemic, it should be noted that PA management authorities were largely focused on coping with the short-term impacts of the pandemic, with few comments made on its long-term implications and its impacts on the management of the PAs, which are still largely unclear.

4. Discussion

Although strict restrictions for COVID-19 were eased during summer 2020 across Europe it is clear that the pandemic is not over at the time of writing this paper. On the contrary, in September 2020, Europe entered a second wave of the pandemic and indeed such pandemics might become more frequent in the future [41]. Thus, similar to other parts of the world, it is important to reflect on what has happened in these first months of the pandemic and propose ways that will facilitate the long-term management of such PAs in times of public health crises and associated restrictions and uncertainty [42]. There are two broad categories of management challenges: visitor number management and visitor behavior management.

COVID-19 so far has had significant impacts on the management of PAs across the world. In the United States, an increase in visitors was observed in outdoor spaces creating a number of challenges similar to the ones identified in this paper for Europe [42]. Conversely, in other parts of the world, different concerns have been raised with African PAs seeing a significant reduction of tourism in wildlife reserves [43] leading to reduced financial resources for park authorities and raising concerns about illegal practices.

The increase in visitor numbers to European PAs during the pandemic comes as awareness of PAs has been increasing over time [44] and people are increasingly visiting areas of natural beauty in order to improve their wellbeing [16,17,45]. PAs benefit physical and mental health [45–47] by providing people with the opportunity to come closer to nature [48–51]. In addition, a significant increase in users of outdoor spaces [52] has also been documented during the pandemic that appears to be motivated by people trying to find relatively remote places where they felt safe from the virus.

Indeed, Nature Parks are promoted as a national and regional asset, and so it is not therefore irrational or unreasonable for people to choose to visit such locations when advised to avoid crowded and indoor spaces by the Government. Additionally, travel restrictions have reduced alternative options for people to travel to, such as urban areas or destinations abroad. However, the potential conflict between the rights of visitors to access a national asset versus the right of local residents to be safe in their local area during a pandemic adds an additional dimension to existing tensions between visitors and local residents over access to and competition for local resources.

As noted, this increase in visitors has led to the emergence of new conflicts or exacerbation of existing tensions due to overcrowding incidents and problematic behavior by visitors in several European PAs. Park authorities across Europe had to react quickly to these challenges and introduced several measures. These tools aimed to manage the number of visitors in PAs whilst also accommodating new social distancing measures. Our review revealed that the severity and extent of these measures varied across locations but overall a significant effort has been invested by management authorities to face the challenges brought by the pandemic.

Regarding conflicts, the designation of PAs in Europe has often resulted in conflicts of interest between diverse users and local residents [53,54], especially as competition for space has intensified [55] due to increased tourism [56,57]. Thus, democratizing access to PAs and minimizing disturbance to natural systems was a major challenge for European PAs before the pandemic. Overcrowding in PAs [58–60] often resulted in increased noise levels and disturbance of tranquility [61] causing disruptions in the life of local communities and distortion of human ties [55,62] as well as disturbing wildlife and ecosystems. As a result, the pandemic intensified tensions between locals and visitors in many cases. People living inside or in close proximity to PAs felt under threat from growing crowds of tourists both because of the potential transmission of the virus [44] but also because of overcrowding incidents interrupting how people enjoy nature.

In the past, this has meant focusing people's attention on honeypot sites with high visitor capacity and low sensitivity to disturbance and creating buffer zones around PAs. COVID-19 and the need for social distancing make this solution problematic under the current circumstances. Indeed, well-established measures to minimize disturbance to wildlife by clustering visitors in parks, and social distancing measures to keep people apart, appear to be in conflict and need to be balanced against each other to establish a satisfactory equilibrium between nature protection and public health protection. The pandemic has in effect reduced safe limits in terms of visitor densities at popular sites thereby increasing pressure in less busy locations, but with other risks such as increased disturbance of wildlife and ecosystems, degradation of the quality of natural spaces for other visitors and local residents, and possibly also greater public safety risks, as people visit less closely managed locations where accident risks are greater.

This brings us to our first policy recommendation for the long-term management of European PAs. Future solutions will require careful spatial planning which also takes into consideration issues of social equity in accessing PAs [45]. Incidents of overcrowding can be controlled by the careful distribution of visitors within a PA (both temporal and spatial). This solution would minimize the need to reduce the number of visitors (thus having also a minimum impact on the local economy). We should note, however, that this approach may be difficult to apply in certain areas where a significant part of the land is privately owned and so access rights are limited. Furthermore, some areas of habitat, ecosystems, and wildlife are more sensitive to human disturbance than others (such as areas of ground-nesting birds). Issues of public safety, protecting disturbance-sensitive environments, and land access rights, therefore need to be balanced.

Many management authorities are heavily constrained and have limited space to distribute visitors within the territory. As well as local management strategies, this problem can be alleviated in the future by increasing provision through the designation of additional PAs which are established on land which is publicly accessible. Such a solution would be in accordance with the new EU Biodiversity Strategy, where it has been announced that 30% of land and 30% of waterways will be protected by 2030 [19].

Another important issue we highlight in our study is that the increased number of visitors was also accompanied in certain cases by an increase in contentious or problematic behavior by PA users. Our study revealed that this is probably at least in part linked to a new profile of visitors coming to these areas during the pandemic. A possible explanation is that people coming from more urban areas having a limited sense of nature connectedness and may not necessarily have developed norms of environmental behavior and are unaware of the recommendations for responsible use of the park. Consequently, a new approach may be needed, targeting new groups visiting parks around Europe, while also managing conflicts between different users. Cultural sensitization and education of new visitors are necessary, possibly supported by enforcement of regulations. Although changing behavior is an extremely challenging task, it could be an opportunity to identify which tools are the most efficient in terms of altering people's behavior, when they come to visit a PA. Indeed, the arrival of a new profile of visitors could provide a window of opportunity for PA management to reach new audiences. Whilst some new visitors may only value PAs temporarily for the access to outdoor space they provide during the pandemic and then return to their former practices, a proportion of them may well be open to new experiences in natural spaces and be open to developing a greater sense of connection to nature.

As documented in our study, park authorities have invested significant time in introducing measures to manage overcrowding and visitor behavior during the COVID-19 pandemic. However, a lot of these authorities have limited powers to enforce certain measures, such as fines for illegal parking. Also, in many cases, government guidelines on COVID-19 may overrule regulations introduced by management authorities of PAs. Therefore, a second recommendation is that closer collaboration is needed between park authorities and more centralized institutions in order to propose measures, which ensure that PAs continue to benefit the wellbeing of local communities and visitors.

Tackling overcrowding and resolving conflicts in PAs necessitates also finding a balance between local economic development and the wellbeing of locals. Recreational activities within PAs are a major source of income for locals [63]. Sustainable tourism has been at the core of management plans for many PAs in Europe [64] allowing local communities to maintain an income [65,66], which often compensates for economic losses due to the PA designation. However, several PAs currently function at a maximum visitor capacity during peak periods. Reducing visitor demand to manage COVID-19 outbreaks would also imply a reduction in income for local communities, especially for those working in the hospitality and recreation sectors and, therefore, may necessitate the development of lower-impact and higher quality tourism experiences for visitors, combined with a more holistic rural development policy to reduce reliance solely on tourism.

Our third policy recommendation is, therefore, that local economies in PAs cannot rely on models with a maximum visitor capacity in order to be sustainable and must avoid scenarios of 'over-tourism'. The number of visitors needs to be managed to a level where economic benefits continue to flow but the well-being of locals is safeguarded while visitors' experience remains satisfactory. There are various tools to conduct carrying capacity studies and model alternative scenarios of visitor numbers. Furthermore, overcrowding can be an issue both for locals and visitors and thus studies should be exploring the views of these different users when determining capacity levels [58–60]. A wide application of such assessments in PAs, taking also into consideration social distancing measures, would allow management authorities to specify the optimum number of visitors in a PA but also the distribution of these visitors within the PA [67].

The challenges presented by the pandemic will persist possibly for several years as there are multiple impacts of this pandemic that will be experienced in the future, and indeed future pandemics remain a possibility. Consequently, it is important that park authorities carefully consider these challenges in order to manage PAs in a sustainable and resilient way. As noted, at this stage in the pandemic, PA management authorities were largely focused on coping with the short-term impacts of the pandemic, with few comments made about the longer-term implications of the pandemic and its impacts on the management of the PAs, which still are largely unclear. These long-term impacts will depend on a wide range of factors such as the medium to long-term impact on the economy and the future actions of governance actors.

5. Conclusions

This study has explored the impacts and challenges that the COVID-19 pandemic presented to a range of European nature PAs and their local communities, and the measures they implemented to mitigate those impacts and associated conflicts. We have also considered the lessons that might be learned from the pandemic experience to inform longer-term PA management, particularly where the pandemic exacerbated existing tensions, such as between local people and visitors.

The pandemic crisis has made the job of PA management more complex and shifted the balance of priorities in trying to achieve a balance between on the one hand nature and landscape conservation, and on the other maintaining accessibility for the visiting public. Indeed the large influxes of visitors during the periods when lockdown regulations were more relaxed demonstrates the importance of such landscapes to people and their well-being, all the more so during this period of the health crisis. Focusing people's attention on outdoor recreation and its many benefits makes logical sense whilst social distancing measures are needed but presents challenges to PA managers in managing disturbance. In consequence, PA management bodies and local communities will need support during such crises where their income becomes more variable and unpredictable, social relations become strained, and nature conservation more difficult to manage.

Overcrowding incidents and an increase in problematic behavior and use of the PAs by visitors were the most significant challenges and also led to an increase in conflicts between locals and visitors. As a response to these challenges, park authorities were quick to respond and find ways to tackle them. Through the use of social media, education campaigns, and a number of other tools, they have tried to keep the PAs open while keeping visitors and locals safe, assisting also in the recovery of the local economy. However, as European communities are now experiencing a second wave of the pandemic, it is important that longer-term solutions are introduced by management authorities. In consequence, careful management of the spatial distribution of visitors in PAs might be necessary for the future along with educational campaigns targeting groups with a new profile of visitors which has emerged during the pandemic. Thus, although COVID-19 has introduced many challenges for PAs in Europe, it can also be seen as an opportunity to promote new and more sustainable ways to manage protected landscapes.

The pandemic has proved longer-lasting than anticipated, and enhanced global mobility may mean that such a health crisis may become more frequent in the future and so lessons from this pandemic may be worth learning for the longer term, even if the situation normalizes somewhat as the pandemic subsides. Indeed some of the conflicts such as between locals and visitors were pre-existing and exacerbated by the crisis, and the impacts of greater visitor numbers and new types of visitors may constitute a warning at a time when the popularity of nature Protected Areas is increasing and governments actively seek to encourage people to visit them. The advent of new types of visitors to the PAs studied, whilst presenting problems, also presents an opportunity to engage new audiences and foster a sense of connectedness to nature among a broader spectrum of the public.

Author Contributions: Conceptualization, V.G., J.M., N.J., J.H., A.K. and P.G.D.; Methodology, J.M., P.G.D., N.J., V.G., and J.H.; Formal analysis, V.G., J.M., N.J.; Investigation, V.G., J.M., N.J., A.K., J.H., P.G.D., R.F.A.F., E.B., A.B., S.S., K.B., C.C.M., C.C.S., F.S.R., G.C., I.C., J.D., M.L., C.J.F.G., A.J., K.L., M.S., T.P., M.O., C.R., H.W., T.Z.-K.; Writing—original draft preparation, V.G., J.M., and N.J.; Writing—review and editing, V.G., J.M., N.J., A.K., J.H., P.G.D., R.F.A.F., E.B., A.B., S.S., K.B., C.C.M., C.C.S., F.S.R., G.C., I.C., J.D., M.L., C.J.F.G., A.J., K.L., M.S., T.P., M.O., C.R., H.W., T.Z.-K.; Visualization N.J.; Supervision N.J., and P.G.D.; Project administration, N.J.; Funding acquisition, N.J. All authors have read and agreed to the published version of the manuscript.

References

1. Millennium Ecosystem Assessment (MEA). *Ecosystems and Human Well-Being: Synthesis*; Island Press: Washington, DC, USA, 2005; p. 160.
2. McGinlay, J.; Parsons, D.J.; Morris, J.; Graves, A.; Hubatova, M.; Bradbury, R.B.; Bullock, J.M. Leisure activities and social factors influence the generation of cultural ecosystem service benefits. *Ecosyst. Serv.* **2018**, *31*, 468–480. [CrossRef]
3. Kaplan, S. The restorative benefits of nature: Toward an integrative framework. *J. Environ. Psychol.* **1995**, *15*, 169–182. [CrossRef]
4. Hartig, T.; Evans, G.W.; Jamner, L.D.; Davis, D.S.; Gärling, T. Tracking restoration in natural and urban field settings. *J. Environ. Psychol.* **2003**, *23*, 109–123. [CrossRef]
5. White, M.P.; Pahl, S.; Ashbullby, K.J.; Herbert, S.; Depledge, M.H. Feelings of restoration from recent nature visits. *J. Environ. Psychol.* **2013**, *35*, 40–51. [CrossRef]
6. English, J.; Wilson, K.; Keller-Olaman, S. Health, healing and recovery: Therapeutic landscapes and the everyday lives of breast cancer survivors. *Soc. Sci. Med.* **2008**, *67*, 68–78. [CrossRef]
7. Jordan, M. Back to nature. *Therapy Today* **2009**, *20*, 26–28. Available online: http://about.brighton.ac.uk/staff/profiles/jordan/therapy-today.pdf (accessed on 24 September 2020).
8. Hanski, I.; Von Hertzen, L.; Fyhrquist, N.; Koskinen, K.; Torppa, K.; Laatikainen, T.; Karisola, P.; Auvinen, P.; Paulin, L.; Mäkelä, M.J.; et al. Environmental biodiversity, human microbiota, and allergy are interrelated. *Proc. Natl. Acad. Sci. USA* **2012**, *109*, 8334–8339. [CrossRef]
9. Kuo, F.E.; Sullivan, W.C. Aggression and Violence in the Inner City. *Environ. Behav.* **2001**, *33*, 543–571. [CrossRef]
10. O'Brien, L.; Murray, R. *A marvellous Opportunity for Children to Learn: A Participatory Evaluation of Forest School in England and Wales*; Forest Research: Surrey, UK, 2006; p. 52.
11. Morris, J.; Urry, J. *Growing Places: A study of Social Change in The National Forest*; Forest Research: Surrey, UK, 2006; p. 48.

12. Weinstein, N.; Balmford, A.; DeHaan, C.R.; Gladwell, V.; Bradbury, R.B.; Amano, T. Seeing Community for the Trees: The Links among Contact with Natural Environments, Community Cohesion, and Crime. *Bioscience* **2015**, *65*, 1141–1153. [CrossRef]
13. Bhagwat, S.A. Ecosystem Services and Sacred Natural Sites: Reconciling Material and Non-material Values in Nature Conservation. *Environ. Values* **2009**, *18*, 417–427. [CrossRef]
14. Lewicka, M. Place attachment: How far have we come in the last 40 years? *J. Environ. Psychol.* **2011**, *31*, 207–230. [CrossRef]
15. Buckley, R.; Brough, P.; Hague, L.; Chauvenet, A.; Fleming, C.M.; Roche, E.; Sofija, E.; Harris, N. Economic value of protected areas via visitor mental health. *Nat. Commun.* **2019**, *10*, 5005–5010. [CrossRef] [PubMed]
16. Naidoo, R.; Gerkey, D.; Hole, D.; Pfaff, A.; Ellis, A.M.; Golden, C.D.; Herrera, D.; Johnson, K.; Mulligan, M.; Ricketts, T.H.; et al. Evaluating the impacts of protected areas on human well-being across the developing world. *Sci. Adv.* **2019**, *5*, eaav3006. [CrossRef] [PubMed]
17. Jones, N.; Malesios, C.; Kantartzis, A.; Dimitrakopoulos, P. The role of location and social impacts of Protected Areas on subjective wellbeing. *Environ. Res. Lett.* **2020**, *15*, 114030. [CrossRef]
18. Stolton, S.; Hockings, M.; Dudley, N.; MacKinnon, K.; Whitten, T. *Reporting Progress in Protected Areas: A Site-Level Management Effectiveness Tracking Tool*, 2nd ed.; World Bank/WWF Alliance by WWF International: Gland, Switzerland, 2007; p. 15.
19. European Commission. *EU Biodiversity Strategy for 2030. Bringing Nature Back into Our Lives*; Communication from the Commission to the European Parliament, the Council, the European Economic and Social Committee and the Committee of the Regions: Brussels, Belgium, 2020; p. 25. Available online: https://ec.europa.eu/info/sites/info/files/communication-annex-eu-biodiversity-strategy-2030_en.pdf (accessed on 21 September 2020).
20. EEA (European Environment Agency). *Protected Areas in European Overview*; Report no. 5; EEA: Copenhagen, Denmark, 2012; p. 136. Available online: https://www.eea.europa.eu/publications/protected-areas-in-europe-2012 (accessed on 21 September 2020).
21. Balmford, A.; Green, J.M.H.; Anderson, M.; Beresford, J.; Huang, C.; Naidoo, R.; Walpole, M.; Manica, A. Walk on the Wild Side: Estimating the Global Magnitude of Visits to Protected Areas. *PLoS Biol.* **2015**, *13*, e1002074. [CrossRef] [PubMed]
22. Schägner, J.P.; Brander, L.; Maes, J.; Paracchini, M.L.; Hartje, V. Mapping recreational visits and values of European National Parks by combining statistical modelling and unit value transfer. *J. Nat. Conserv.* **2016**, *31*, 71–84. [CrossRef]
23. EEA (European Environment Agency). Nationally Designated Protected Areas, Indicator Assessment. 2017. Available online: https://www.eea.europa.eu/data-and-maps/indicators/nationally-designated-protected-areas-10/assessment (accessed on 28 September 2020).
24. European Commission Joint Research Centre. ECML Covid Measures Database. 2020. Available online: https://covid-statistics.jrc.ec.europa.eu/Measure/DashboardMeasures?view=1 (accessed on 28 September 2020).
25. Our World in Data. Google Mobility Trends. How Has the Pandemic Changed the Movement of People Around the World? 2 June 2020. Available online: https://ourworldindata.org/covid-mobility-trends (accessed on 28 September 2020).
26. Manenti, R.; Mori, E.; Di Canio, V.; Mercurio, S.; Picone, M.; Caffi, M.; Brambilla, M.; Ficetola, G.F.; Rubolini, D. The good, the bad and the ugly of COVID-19 lockdown effects on wildlife conservation: Insights from the first European locked down country. *Biol. Conserv.* **2020**, *249*, 108728. [CrossRef]
27. Jones, N.; McGinlay, J.; Holtvoeth, J.; Gkoumas, V.; Malesios, C.; Kontoleon, A. *Snowdonia National Park: Exploring Views of Local Communities Regarding the Social Impacts of the National Park, Changes Due to COVID-19 on Everyday Life and Potential Management Options during the Pandemic*; University of Cambridge/Project FIDELIO: Cambridge, UK, 2020; p. 13. Available online: https://www.fidelio.landecon.cam.ac.uk/publications (accessed on 19 October 2020).
28. Derks, J.; Giessen, L.; Winkel, G. COVID-19-induced visitor boom reveals the importance of forests as critical infrastructure. *For. Policy Econ.* **2020**, *118*, 102253. [CrossRef]
29. Jones, N.; McGinlay, J. *The Impact of COVID-19 Restrictions on Local Communities of Peak District National Park and Management Options during the Pandemic*; University of Cambridge/Project FIDELIO: Cambridge, UK, 2020; p. 20. Available online: https://www.fidelio.landecon.cam.ac.uk/publications (accessed on 19 October 2020).

30. Remuzzi, A.; Remuzzi, G. COVID-19 and Italy: What next? *Lancet* **2020**, *395*, 1225–1228. [CrossRef]
31. Pepe, E.; Bajardi, P.; Gauvin, L.; Privitera, F.; Lake, B.; Cattuto, C.; Tizzoni, M. COVID-19 outbreak response, a dataset to assess mobility changes in Italy following national lockdown. *Sci. Data* **2020**, *7*, 230. [CrossRef]
32. Hunter, D.J. Covid-19 and the Stiff Upper Lip—The Pandemic Response in the United Kingdom. *N. Engl. J. Med.* **2020**, *382*, e31. [CrossRef]
33. Armbruster, S.; Klotzbücher, V. *Lost in Lockdown? COVID-19, Social Distancing and Mental Health in Germany*. Diskussionsbeiträge, No. 2020-04; Albert-Ludwigs-Universität Freiburg, Wilfried-Guth-Stiftungsprofessur für Ordnungs- und Wettbewerbspolitik: Freiburg im Breisgau, Germany, 2020. Available online: https://www.econstor.eu/bitstream/10419/218885/1/1698957106.pdf (accessed on 28 September 2020).
34. Mutz, M.; Gerke, M. Sport and exercise in times of self-quarantine: How Germans changed their behaviour at the beginning of the Covid-19 pandemic. *Int. Rev. Sociol. Sport* **2020**. [CrossRef]
35. World Health Organization (WHO). WHO Coronavirus Disease (COVID-19) Dashboard. WHO, 2020. Available online: https://covid19.who.int/?gclid=EAIaIQobChMIntvg3cWL7AIVj-ntCh1QGQ7_EAAYASAAEgJoPfD_BwE (accessed on 28 September 2020).
36. Bojanowska, A.; Kaczmarek, L.D.; Kościelniak, M.; Urbańska, B. Values and well-being change amidst the COVID-19 pandemic in Poland. *PsyArXiv* **2020**. [CrossRef]
37. Sutrop, M.; Simm, K. Developing guidelines for the distribution of scarce medical resources during the COVID-19 pandemic. The Estonian case. *Trames. J. Humanit. Soc. Sci.* **2020**, *24*, 251–268. [CrossRef]
38. Estonian Government. Special Notice: As of Tomorrow, Movement Restrictions between the Islands and the Mainland Estonian Will Be Lifted. *News*, 7 May 2020. Available online: https://www.kriis.ee/en/news/special-notice-tomorrow-movement-restrictions-between-islands-and-mainland-estonia-will-be (accessed on 28 September 2020).
39. Republic of Slovenia. Coronavirus Disease COVID-19. Available online: https://www.gov.si/en/topics/coronavirus-disease-covid-19/ (accessed on 28 September 2020).
40. Kamerlin, S.C.L.; Kasson, P.M. Managing Coronavirus Disease 2019 Spread with Voluntary Public Health Measures: Sweden as a Case Study for Pandemic Control. *Clin. Infect. Dis.* **2020**, 864. [CrossRef] [PubMed]
41. Griffin, D.; Denholm, J. This Isn't the First Global Pandemic and It Won't be the Last. Here's What We've Learned from 4 Others Throughout History. *The Conversation*, 16 April 2020. Available online: https://theconversation.com/this-isnt-the-first-global-pandemic-and-it-wont-be-the-last-heres-what-weve-learned-from-4-others-throughout-history-136231 (accessed on 7 September 2020).
42. Jacobs, L.A.; Blacketer, M.P.; Peterson, B.A.; Levithan, E.; Russell, Z.A.; Brunson, M. Responding to COVID-19 and future times of uncertaintly: Challenges and opportunities associated with visitor use, management and research in parks and protected areas. *Parks Steward. Forum* **2020**, *36*, 483–488. [CrossRef]
43. Newsome, D. The collapse of tourism and its impact on wildlife tourism destinations. *J. Tour. Future* **2020**. [CrossRef]
44. Nürnberg, M.; Edrmann, K.-H. Naturbewusstsein 2019: Bevölkerungsumfrage zu Natur und biologischer Vielfalt. Bundersministrerium für Umwelt, Naturschutz und nukleare Sicherheit. 2019. Available online: https://www.bmu.de/fileadmin/Daten_BMU/Pools/Broschueren/naturbewusstseinsstudie_2019_bf.pdf (accessed on 3 November 2020).
45. Jones, N.; Graziano, M.; Dimitrakopoulos, P.G. Social impacts of European Protected Areas and policy recommendations. *Environ. Sci. Policy* **2020**, *112*, 134–140. [CrossRef]
46. Romagosa, F. Physical health in green spaces: Visitors' perceptions and activities in protected areas around Barcelona. *J. Outdoor Recreat. Tour.* **2018**, *23*, 26–32. [CrossRef]
47. Burdon, D.; Potts, T.; McKinley, E.; Lew, S.; Shilland, R.; Gormley, K.; Thomson, S.; Forster, R. Expanding the role of participatory mapping to assess ecosystem service provision in local coastal environments. *Ecosyst. Serv.* **2019**, *39*, 101009. [CrossRef]
48. Lopes, R.; Videira, N. A Collaborative Approach for Scoping Ecosystem Services with Stakeholders: The Case of Arrábida Natural Park. *Environ. Manag.* **2016**, *58*, 323–342. [CrossRef] [PubMed]
49. Lopes, R.; Videira, N. How to articulate the multiple value dimensions of ecosystem services? Insights from implementing the PArticulatES framework in a coastal social-ecological system in Portugal. *Ecosyst. Serv.* **2019**, *38*, 100955. [CrossRef]

50. Kenter, J.O.; Bryce, R.; Davies, A.; Jobstvogt, N.; Watson, V.; Ranger, S.; Solandt, J.-L.; Duncan, C.; Christie, M.; Crump, H.; et al. *The Value of Potential Marine Protected Areas in the UK to Divers and Sea Anglers*; UNEP-WCMC: Cambridge, UK, 2013; p. 125.
51. Bennett, N.J.; Di Franco, A.; Calò, A.; Nethery, E.; Niccolini, F.; Milazzo, M.; Guidetti, P. Local support for conservation is associated with perceptions of good governance, social impacts, and ecological effectiveness. *Conserv. Lett.* **2019**, *12*, 12640. [CrossRef]
52. Venter, Z.S.; Barton, D.N.; Gundersen, V.; Figari, H.; Nowell, M. Urban nature in a time of crisis: Recreational use of green space increases during the COVID-19 outbreak in Oslo, Norway. *Environ. Res. Lett.* **2020**, *15*, 104075. [CrossRef]
53. Jentoft, S.; Pascual-Fernandez, J.J.; Modino, R.D.L.C.; Ramallal, M.E.G.; Chuenpagdee, R. What Stakeholders Think About Marine Protected Areas: Case Studies from Spain. *Hum. Ecol.* **2012**, *40*, 185–197. [CrossRef]
54. Gallo, M.; Špela, P.M.; Laktić, T.; De Meo, I.; Paletto, A. Collaboration and conflicts between stakeholders in drafting the Natura 2000 Management Programme (2015–2020) in Slovenia. *J. Nat. Conserv.* **2018**, *42*, 36–44. [CrossRef]
55. Hogg, K.; Gray, T.; Noguera-Méndez, P.; Semitiel-García, M.; Young, S. Interpretations of MPA winners and losers: A case study of the Cabo De Palos- Islas Hormigas Fisheries Reserve. *Marit. Stud.* **2019**, *18*, 159–171. [CrossRef]
56. Hattam, C.; Mangi, S.C.; Gall, S.C.; Rodwell, L.D. Social impacts of a temperate fisheries closure: Understanding stakeholders' views. *Mar. Policy* **2014**, *45*, 269–278. [CrossRef]
57. Povilanskas, R.; Armaitienė, A.; Dyack, B.; Jurkus, E. Islands of prescription and islands of negotiation. *J. Destin. Mark. Manag.* **2016**, *5*, 260–274. [CrossRef]
58. Leung, Y.-F.; Spenceley, A.; Hvenegaard, G.; Buckley, R. (Eds.) *Tourism and Visitor Management in Protected Areas: Guidelines for Sustainability*; Best Practice Protected Area Guidelines Series No. 27; IUCN: Gland, Switzerland, 2018; p. 120.
59. Davis, D.; Tisdell, C. Recreational scuba-diving and carrying capacity in marine protected areas. *Ocean Coast. Manag.* **1995**, *26*, 19–40. [CrossRef]
60. Santana-Jiménez, Y.; Hernandez, J.M. Estimating the effect of overcrowding on tourist attraction: The case of Canary Islands. *Tour. Manag.* **2011**, *32*, 415–425. [CrossRef]
61. Scholtz, M.; Saayman, M. Diving into the consequences of stakeholders unheard. *Eur. J. Tour. Res.* **2018**, *20*, 105–124.
62. Trivourea, M. People and the Mediterranean Monk Seal (Monachus monachus): A Study of the Socioeconomic Impacts of the National Marine Park of Alonissos, Northern Sporades, Greece. *Aquat. Mamm.* **2011**, *37*, 305–318. [CrossRef]
63. Pham, T.T.T. Tourism in marine protected areas: Can it be considered as an alternative livelihood for local communities? *Mar. Policy* **2020**, *115*, 103891. [CrossRef]
64. EUROPARC Federation. *European Charter for Sustainable Tourism in Protected Areas*; EUROPARC Federation: Regensburg, Germany, 2020. Available online: https://www.europarc.org/library/europarc-events-and-programmes/european-charter-for-sustainable-tourism/ (accessed on 7 September 2020).
65. Dang, X.; Gao, S.; Tao, R.; Liu, G.; Xia, Z.; Fan, L.; Bi, W. Do environmental conservation programs contribute to sustainable livelihoods? Evidence from China's grain-for-green program in northern Shaanxi province. *Sci. Total. Environ.* **2020**, *719*, 137436. [CrossRef] [PubMed]
66. Katikiro, R.E. Improving alternative livelihood interventions in marine protected areas: A case study in Tanzania. *Mar. Policy* **2016**, *70*, 22–29. [CrossRef]
67. Kostopoulou, S.; Kyritsis, I. A Tourism Carrying Capacity Indicator for Protected Areas. *Anatolia* **2006**, *17*, 5–24. [CrossRef]

4

A Practical Index to Estimate Mangrove Conservation Status: The Forests from La Paz Bay, Mexico

Giovanni Ávila-Flores [1], Judith Juárez-Mancilla [2], Gustavo Hinojosa-Arango [3,*], Plácido Cruz-Chávez [2], Juan Manuel López-Vivas [1] and Oscar Arizpe-Covarrubias [1]

1 Academic Department of Marine and Coastal Sciences, Autonomous University of Baja California Sur, Baja 23037 California Sur, Mexico; giovanniavila@hotmail.com (G.Á.-F.); jmlopez@uabcs.mx (J.M.L.-V.); oarizpe@uabcs.mx (O.A.-C.)

2 Academic Department of Economics, Autonomous University of Baja California Sur, 23037 Baja California Sur, Mexico; juarez@uabcs.mx (J.J.-M.); pcruz@uabcs.mx (P.C.-C.)

3 CIIDIR Oaxaca, National Polytechnic Institute, 95060 Santa Cruz, Mexico

* Correspondence: ghinojosaar@conacyt.mx

Abstract: Mangrove cover has declined significantly in recent years in tropical and subtropical areas around the world. Under this scenario, it is necessary to elaborate and implement tools that allow us to make estimations on their conservation status and improve their protection and support decision-making. This study developed an index using qualitative and quantitative data. The criterions used in the index were: (1) Remnant Vegetation Index, (2) Delphi Method Survey, and (3) Rapid Assessment Questionnaire. In turn, the weights of the criterions were defined using the analytical hierarchy process (AHP). Once the values of each criterion were obtained, the index was applied to 17 mangrove communities located in La Paz Bay, Mexico. Finally, according to their score, they were classified based on the IUCN Red List of Ecosystems. The results show that five communities were ranked in the category Minor Concern, eight in Little Threatened, one in Vulnerable, one in Endangered, and two were classified as Deficiency of Data. These results are slightly different from other studies in the region and validate this index as a proper method. Therefore, it could be applied to other sites, especially in areas with little information and/or scarce monetary resources.

Keywords: AHP; Delphi Method; GIS; MCDA; Sustainable Development; Wetlands

---

1. Introduction

1.1. Theoretical Background

Mangroves are recognized worldwide due to various resources and ecosystem services (ES) they provide [1,2]. Some of these ES are provision services (food, timber, medicines, etc.), regulation and maintenance services (protection against hurricanes, erosion control, CO_2 capture, etc.), and cultural services (aesthetic, symbolic, religious, etc.) [3,4]. The presence of coastal wetlands such as mangroves helps contribute to human wellbeing [5,6]. However, mangroves are one of the most threatened coastal ecosystems due to changes in environmental factors and impacts induced by human activities [7]. They are particularly vulnerable to degradation as a result of deforestation, aquaculture, agriculture, tourism, urbanization, and pollution from different sources [8]. Other factors, such as sediment dynamics, exotic species, and alteration of the hydrodynamics, also result in severe mangrove deterioration and habitat loss [9].

According to FAO data, about 3.6 million hectares of mangroves were lost in the 1980–2005 period, approximately 20% of the global mangrove cover [10]. Also, it is estimated that mangrove forests worldwide are disappearing between one to two percent each year, at a higher rate than rainforests or coral reefs [11]. Likewise, it is recognized that only 6.9% of the world's mangroves are in the category of a protected area network [12]. In México, the National Commission for the Use and Conservation of Biodiversity (CONABIO) through the Mexican Mangrove Monitoring System (SMMM) began a full examination of the mangroves in this country. The SMMM observed that in 1970–1980, Mexico presented a cover of 856,405 hectares but by 2015 the number of hectares of mangroves decreased to 775,555 [13]. This loss of mangroves has been observed in urban areas and in zones with a tourist-oriented vision. A well-known example is the destruction of "Tajamar" mangrove, in Cancun city, in 2016 [14].

Another example is the case of mangroves located in La Paz city. The capital of Baja California Sur state, has caused a remarkable anthropic pressure to their mangroves due to the growing urbanization [15]. Also, it is expected that tourism will increase in this city. This is evidenced in the list "52 Places to Go in 2020" of The New York Times, where La Paz was the only Mexican city nominated [16]. In this context, different forums, such as the "I Workshop on Mangroves of the Baja California Peninsula" (2005) and the "I Workshop of RAMSAR Sites of Baja California Sur" (2009) have highlighted the need to know the state of mangroves in La Paz bay. Therefore, the goal of this work was to develop an integrative index to evaluate the conservation status of mangroves in this region.

1.2. Study Area

La Paz Bay is in the southwest of the Gulf of California, at the southern portion of the Baja California Peninsula, in Mexico (Figure 1). This bay has diverse biotic and abiotic characteristics that made it a remarkable site with diverse ecosystems, such as beaches, dunes, seagrass beds, rocky reefs, and mangroves [17]. Its coastal zone harbors 17 mangrove communities, 16 of them within the proximity of La Paz city, and one more at the Espíritu Santo Archipelago. These mangroves are within protected areas such as the Espíritu Santo Archipelago National Park, the Balandra Flora and Fauna Protection Area, and the Islands of the Gulf of California Flora and Fauna Protection Area. They are also under the category of international protection, such as the Ramsar Sites "Humedales Mogote-Ensenada de La Paz" (Ramsar site no. 1816) and "Balandra" (Ramsar site no. 1767).

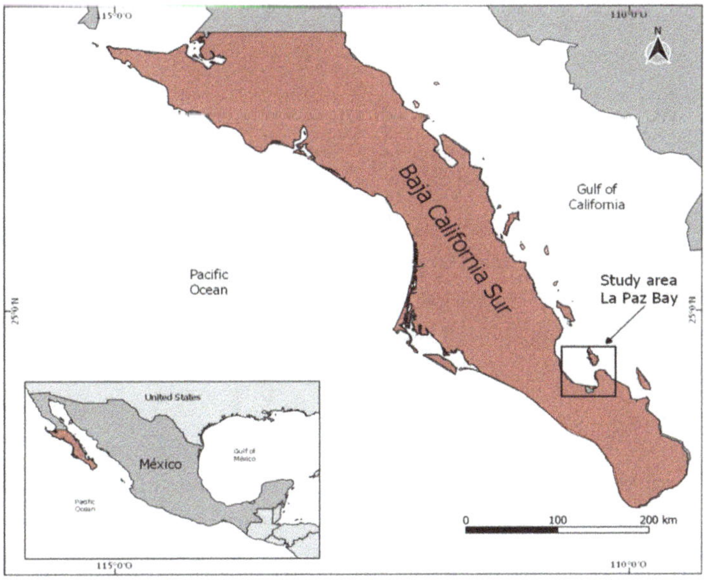

Figure 1. Location of La Paz Bay in the southern end of the Peninsula of Baja California, Mexico.

The 14 communities belonging to the Ramsar site No. 1816 are: Centenario-Chametla, Comitán, El Conchalito, El Mogote, Enfermería, Eréndira, Estero Bahía Falsa, Estero El Gato, La Paz-Aeropuerto, Palmira, Playa Pichilingue-Brujas, Unidad Pichilingue UABCS, Salinas de Pichilingue, and Zacatecas.

Two of these remaining communities belonging to Ramsar site no. 1767 are Balandra and El Merito. Finally, the mangrove Espíritu Santo Archipelago is located within another protected area, it is in the national park category and its name is homonymous to that community (Figure 2).

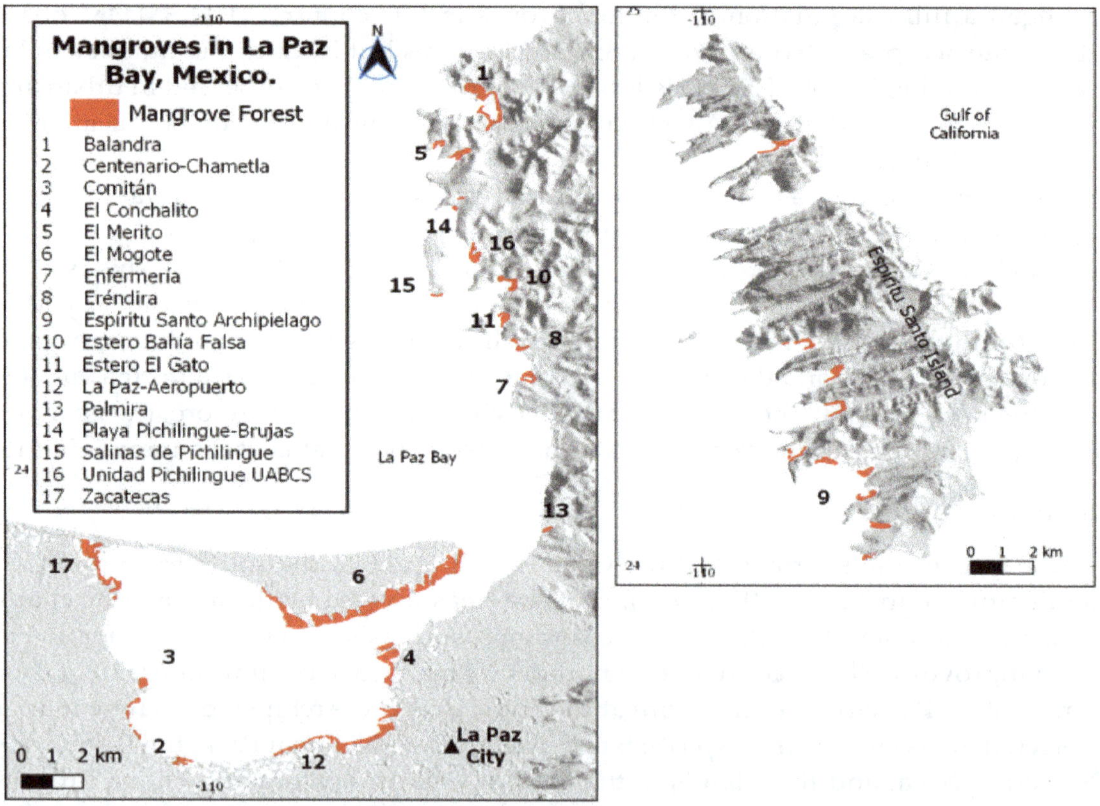

Figure 2. Location of mangroves placed in La Paz Bay, Mexico.

2. Materials and Methods

2.1. Study Design

The development and implementation of the Mangrove Conservation Status Index (MCSI) consisted of five phases carried out between January and September 2019. First, the remnant vegetation index (RVI), Delphi method survey (DMS), and the rapid assessment questionnaire (RAQ) were conducted at the 17 mangrove areas during January–February 2019, followed by the use of the analytical hierarchy process (AHP) to estimate the weights of each one of these indicators. We conducted a spatial analysis on open-source GIS software QGIS (version 3.4.4) to calculate the RVI, designed a questionnaire for the RAQ, and applied surveys to local mangrove experts for the DMS. Finally, we estimated the Mangrove Conservation Status Index for all the mangroves.

2.2. Index Development and AHP

For the construction of the MCSI, three components were selected, as these are widely used in various environmental analyses (RVI, DMS, and RAQ). For example, several studies have carried out comparisons between mangrove cover between different years, globally and nationally and locally. Also, the application of the Delphi method in environmental modeling is considered as a useful tool, with various studies focused on mangroves. Finally, the rapid assessment tool has also been used in forest and wetland analysis. Once the index components were selected, the following formula was generated:

$$\mathbf{MCSI} = (\mathrm{RVI})\mathrm{W1} + (\mathrm{DMS})\mathrm{W2} + (\mathrm{RAQ})\mathrm{W3}.$$

The weights of each component were determined using the analytical hierarchy process (AHP) method developed by Thomas L. Saaty (see Appendix A). For this, paired combinations were made between the three components using a pairwise comparison matrix (Table 1). For example, since the RVI is a quantitative value that reflects the loss or gain of cover in a given period, it was considered of greater relevance than the RAQ and DMS components. In the same way, among these last components, the DMS was considered of greater importance than RAQ. DMS is the result of the opinion of several experts (which includes years of experience) in comparison to RAQ, which takes information from a single field visit.

Table 1. Pairwise comparison matrix (PCM).

Criteria	Sub-Criteria	Number of Comparisons			Total	Weight
		1	2	3		
MCSI	(1) Remnant Vegetation Index		5	5	10	0.62
	(2) Delphi Method Survey.	1		3	4	0.25
	(3) Rapid Assessment Questionnaire.	1	1		2	0.13

2.3. Remaining Vegetation Index (RVI)

The value of RVI of each mangrove community was calculated by considering the vegetation cover obtained in 2018 as the present vegetation area (PVA), divided by the original vegetation area (OVA), which corresponded to the data of the year of 1973. The result was multiplied by 100 to obtain a comparable value on a scale of 0/100. This index was used for the first time in a case study in Colombia [18] following this formula:

$$\mathbf{RVI} = [(\mathrm{PVA})/(\mathrm{OVA})] \times 100.$$

We obtained the vegetation area from scanned aerial photographs and Landsat satellite images. We consulted CONABIO's database. We used the oldest image available from the sources mentioned above for the calculation of the RVI for each mangrove community. In this case, we obtained an aerial photograph from 1974 that captured mangroves, except those at Espíritu Santo Archipelago, in La Paz bay at the Autonomous University of Baja California Sur library's archive. We downloaded Sentinel images (10 m, 20 m, and 60 m, avoiding cloud interference) for May 2018 from the Earth Explorer platform (USGS) to calculate present vegetation cover for the calculation of the RVI.

To digitalize the aerial photographs, we scanned them with the highest available resolution (10,200 × 14,028 pixels). We georeferenced images using geomorphological land references by the control point method. Subsequently, we extracted the sections corresponding to mangroves on QGIS and obtained pixels (1 m × 1 m) by a resampling process. The resampling process did not allow for a higher pixel resolution; nevertheless, it provided better contrast between neighboring pixels. Therefore, observations allowed the precise definition of mangrove areas (Figure 3).

Together, the field data collection and the georeferenced aerial photographs allowed the confirmation of the presence of mangroves and the obtention of the polygons containing mangroves by the use of the on-screen scanning facility in the QGIS software. We transformed the satellite image from geographic coordinates to metrics. For optimal use, we created a composite of bands 4, 5, 3, and panchromatic (Band 8) to increase image resolution and facilitate vegetation recognition [19]. The generation of the base project in the QGIS platform integrated the resulting images in raster format. Once the properties of the images (pixel size, georeference) were validated, we calculated the mangrove cover for each of the 17 sites. We obtained 16 mangrove polygons for 1974 and 17 for 2018 using the manual digitizing technique which has been used by different authors [20–23] and estimated the vegetation cover area. We used the resulting areas to calculate the RVI.

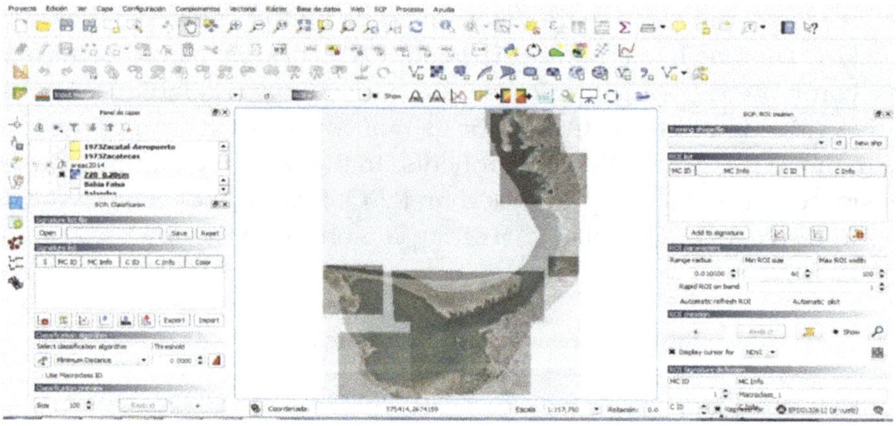

Figure 3. Treatment of spatial images in the QGIS software.

2.4. *Delphi Method Survey*

We applied interviews with regional mangrove experts following the Delphi method, which is a structured way to obtain information and knowledge on a particular topic [24]. This method provides both qualitative and quantitative data (see Appendix B), and it can be adapted for rapid assessments, such as the one implemented during this study [25]. We contacted a total of ten people, but only seven answered the survey. Of these, four were researchers, two worked for government agencies, and one collaborated with non-government organizations. We conducted interviews in person or remotely via electronic media such as Skype or video conference. The interview consisted mostly of open questions, as well as closed questions or fixed-alternatives. Interviewees answered the open question freely with no limit on time. The fixed-alternative questions were formulated to be answered in a scalar way using the Likert measurement tool (Table 2), which consists of obtaining a degree of conformity determined by a range of values. Table 3 shows the main question used in the survey.

Table 2. Remaining Vegetation Index obtained for each mangrove community.

Mangrove Community	Original Vegetation Area (m^2)	Present Vegetation Area (m^2)	Remaining Vegetation Index (Scores)
Balandra	266,044.01	268,577.08	**100.95**
Centenario-Chametla	54,213.98	53,657.47	98.97
Comitán	43,982.36	42,448.27	96.51
El Conchalito	217,785.87	192,957.92	88.59
El Merito	81,397.13	81,373.77	99.97
El Mogote	1,247,826.06	1,254,511.24	**100.53**
Enfermería	56,953.15	37,968.09	66.66
Eréndira	23,627.20	23,969.85	**101.45**
Espíritu Santo Archipelago	-	523,773. 69	N/A
Estero Bahía Falsa	47,744.44	44,573.39	93.35
Estero El Gato	47,916.84	45,429.86	94.80
La Paz-Aeropuerto	160,633.55	360,491.36	**224.41**
Palmira	14,172.06	12,154.07	85.76
Playa Pichilingue-Brujas	11,210.63	3,025.02	26.98
Salinas de Pichilingue	-	2,971.47	N/A
Unidad Pichilingue UABCS	63,284.70	50,329.61	79.52
Zacatecas	227,058.89	259,119.05	**114.11**

Note: Bold scores represent increases on mangrove cover.

Table 3. Key question applied to experts in the Delphi Method Survey component.

Based on your experience, what is the conservation status of the mangroves of the bay of La Paz?		
Mangrove Communities	**Conservation Status**	**I Don't know.**
Balandra	Bad 1 2 3 4 5 Good	

2.5. Rapid Assessment Questionnaire

The rapid evaluation is a reliable and timely estimation method, which allows an approximation of the magnitude and characteristics of a problem. It marks the line to define needs or tasks to consider during a subsequent evaluation [26]. This type of assessment provides complementary information to other sources, in a simple, fast, and flexible way. In the case of the mangroves of La Paz Bay, we visited 17 sites, which were selected according to the management plans of the protected area (Balandra) and Ramsar site (Humedales Mogote-Ensenada de La Paz No. 1816). To evaluate each of the mentioned mangroves, we created a rapid assessment questionnaire (*RAQ*) based on different surveys developed by academics and decision-makers from the region. The *RAQ* considered specific environmental indicators, divided thematically (water, air, soil, flora, fauna, and waste), and used qualitative indicators to assess impacts observed at each mangrove site during the field visits. The values of the *RAQ* ran from 0 to 1; the closer the value to 1, the more impacted the site was. We recorded our observations at the site, and photographic evidence is available from the authors upon request (see Appendix C).

2.6. Application of the Integrative Mangrove Conservation Status Index

We calculated the *MCSI* using the scores of each one of the components of the index, RVI, RAQ, and DMS, and following the formula:

$$\mathbf{MCSI} = (\text{RVI})(0.62) + (\text{RAQ})(0.25) + (\text{DMS})(0.13).$$

We classified mangrove sites depending on their MCSI score following an adapted classification of the IUCN Red List of Ecosystems (see Appendix D). This scale considers eight categories of risk for the earth's ecosystem (Figure 4). Three of them contemplate quantitative thresholds: critically endangered (CR), endangered (EN), and vulnerable (VU)—together, the IUCN describes these ecosystems as threatened. There are several qualitative categories to include: (1) ecosystems that fail to meet the quantitative criteria for the threatened ecosystem categories (NT, near threatened); (2) ecosystems that unambiguously meet none of the quantitative criteria (LC, least concern); (3) ecosystems with poor data (DD, data deficient); and (4) ecosystems that have not been assessed (NE, not evaluated). An additional category (CO, collapse) is assigned to ecosystems that have collapsed throughout their distribution, the analogue of the extinct (EX) category for species [27].

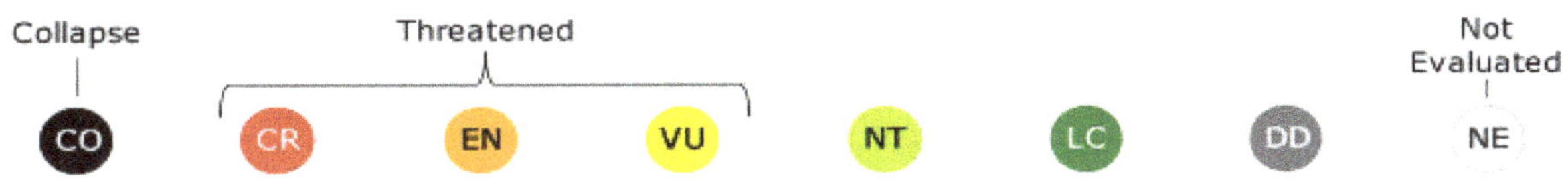

Figure 4. Categories of the IUCN Red List of Ecosystems. Source: IUCN, 2019.

3. Results

To calculate the proposed MCSI, we first estimated each of the components for the 17 mangroves from La Paz Bay. We summarize our findings in the following sections.

3.1. Spatial Analysis and Remaining Vegetation Index (RVI)

According to the RVI analysis, five communities presented an increase in mangrove cover (RVI > 100), ten showed losses of mangrove forest (RVI < 100), and two could not be analyzed because we were not able to obtain an original vegetation area. Playa Pichilingue-Brujas experienced the most significant losses of mangrove vegetation, and only 26.98% of the initial cover remained. Enfermería also presented a significant decrease in mangrove cover (44.44%).

3.2. Delphi Method Survey

We sent a total of 10 surveys to regional mangrove experts; however, just seven people replied. We interviewed them to complement their answers to the surveys (available from the authors upon request). We calculated the median for the score assigned by the experts to each one of the mangrove areas. Only two mangrove communities presented median scores of five, El Merito and Espíritu Santo Archipelago, which presents the best conservation status according to expert opinions (Table 4). Five mangrove areas scored four points for an acceptable conservation level. Moreover the mangroves that scored less than three, meant that they present a major deterioration.

Table 4. Values obtained from the Delphi Method Survey applied to experts.

Mangrove Community	Expert Opinion 1	Expert Opinion 2	Expert Opinion 3	Expert Opinion 4	Expert Opinion 5	Expert Opinion 6	Expert Opinion 7	Median
Balandra	4	3	4	4	5	4	4	4
Centenario-Chametla	2	2	3	2	4	2	2	2
Comitán	2	4	4	4	3	3	4	4
El Conchalito	3	2	3	3	4	2	3	3
El Merito	5	-	5	5	4	5	5	5
El Mogote	3	3	4	4	4	4	4	4
Enfermería	1	-	2	2	3	3	2	2
Eréndira	2	2	2	2	1	4	2	2
Espíritu Santo Archipelago	5	-	5	5	3	-	5	5
Estero Bahía Falsa	3	3	3	2	3	3	3	3
Estero El Gato	4	3	3	4	-	3	4	3
La Paz- Aeropuerto	4	3	4	3	5	4	4	4
Palmira	4	2	2	2	3	2	2	2
Playa Pichilingue-Brujas	2	2	1	1	4	2	2	2
Salinas de Pichilingue	-	-	3	3	-	-	-	-
Unidad Pichilingue UABCS	2	3	3	3	4	3	3	3
Zacatecas	4	-	5	4	4	4	5	4

3.3. Rapid Assessment Questionnaire

The mangroves in La Paz bay presented a varied rank of values for the RAQ. The La Paz-Aeropuerto, Palmira, and Playa Pichilingue–Brujas mangrove communities were the most impacted sites, with RAQ scores between 0.80 and 0.90. Seven mangroves showed medium levels of impact, with scores ranging between 0.50 and 0.70, and another seven sites presented fewer effects, with RAQ values from 0.26 to 0.49 (Table 5).

Table 5. Values obtained from the Rapid Assessment Questionnaire applied in field visits.

Mangrove Community	Rapid Assessment Questionnaires
Balandra	0.4944
Centenario-Chametla	0.6542
Comitán	0.2662
El Conchalito	0.5680
El Merito	0.3184
El Mogote	0.4486
Enfermería	0.5896
Eréndira	0.7260
Espíritu Santo Archipelago	0.2662
Estero Bahía Falsa	0.6382
Estero El Gato	0.5604
La Paz-Aeropuerto	0.8178
Palmira	0.8858
Playa Pichilingue-Brujas	0.8960
Salinas de Pichilingue	0.2884
Unidad Pichilingue UABCS	0.6504
Zacatecas	0.4026

3.4. Application of the Index

We multiplied the scores of the RVI, DMS, and RAQ, as described to calculate the MCSI. The El Merito mangrove community presented the highest MCSI of all the sites (95.54), followed by Balandra, Comitán, El Mogote, Estero El Gato, and Zacatecas, which scored values over 80; therefore, these five communities are of "least concern" in accordance with the IUCN Red List of Ecosystems. The MCSI values of another eight mangroves placed them as "near threatened". Otherwise, Enfermeria was classified as "vulnerable", with an MCSI score of 59.22, and Playa Pichilingue-Brujas was classified as "endangered" (MCSI score of 30.01), these last two mangroves were the worst evaluated. Lastly, two mangroves, Salinas de Pichilingue and Espíritu Santo Archipelago, lacked initial information on mangrove cover; therefore, the MCSI resulted in "data deficient" (Table 6).

Table 6. MCSI values for each mangrove community.

Mangrove Community	RVI Score	DM Score	RA Score	MCSI	IUCN Red List of Ecosystems
Balandra	57	23.2	7.0784	87.2784	Least Concern (LC)
Centenario-Chametla	57	11.6	4.8412	73.4412	Near Threatened (NT)
Comitán	57	17.4	10.2732	84.6732	Least Concern (LC)
Espíritu Santo Archipelago	-	29	10.2732	-	Data Deficient (DD)
El Conchalito	54.6231	17.4	6.048	78.0711	Near Threatened (NT)
El Merito	57	29	9.5424	95.5424	Least Concern (LC)
El Mogote	57	23.2	7.7196	87.9196	Least Concern (LC)
Enfermería	41.8836	11.6	5.7456	59.2292	Vulnerable (VU)
Eréndira	57	11.6	3.836	72.436	Near Threatened (NT)
Estero Bahía Falsa	57	17.4	5.0652	79.4652	Near Threatened (NT)
Estero El Gato	57	17.4	6.1544	80.5544	Near Threatened (NT)
La Paz-Aeropuerto	57	17.4	2.5508	76.9508	Near Threatened (NT)
Palmira	53.8593	11.6	1.5988	67.0581	Near Threatened (NT)
Playa Pichilingue-Brujas	16.9575	11.6	1.456	30.0135	Endangered (EN)
Salinas de Pichilingue	-	-	9.9624	-	Data Deficient (DD)
Pichilingue UABCS	45.5316	17.4	4.8944	67.826	Near Threatened (NT)
Zacatecas	57	23.2	8.3636	88.5636	Least Concern (LC)

4. Discussion

Various scientists have developed evaluation indices that can estimate the conservation status-health of mangroves [28–30]. However, many times the application of some indexes requires financial resources, specialized equipment, and experts, which are not always available. This occurs mainly in developing countries and at the same time in these sites information is required in an expedited manner for decision making. Therefore, we believe that the MCSI could help in the aforementioned scenario, which is so common in various countries in Latin America, Africa, and Asia.

Although this index is easy to apply and requires few financial resources, it is based on a combination of quantitative and qualitative data, which gives adequate support to the decisions generated from the results obtained. It combines mangrove cover (remnant vegetation index) with scientific experts' opinions (Delphi method survey), and perceived conservation status obtained during field visits (rapid assessment questionnaire) to classify mangroves in accordance with the IUCN Red List of Ecosystems. For that purpose, the MCSI used the analytical hierarchy process to define the weight of various indicators following a multicriteria method, as suggested by other authors [31].

Besides this, we can consider that the main finding of the construction and application of the MCSI index was that in general terms the mangrove communities located in the Bay of La Paz have an acceptable state of conservation. On the other hand, these results also indicate that despite being in the same area, mangroves could have very different conservation status.

The MCSI uses mangrove cover as a core indicator, which is also applied by the Mexican Mangrove Monitoring System (SMMM), but also is complemented by experts' opinions and rapid assessments at the mangrove communities. Still, different authors consider that cover is not sufficient to estimate the conservation [32,33], since it does not take into consideration other impacts on the ecosystems or the integrity of the ecological services they provided.

We complemented mangrove cover with the remaining vegetation index (RVI) because it estimates changes on vegetation cover in a specific period. The RVI determined an increase between 1974 and 2018 in the cover area of five mangrove communities in La Paz bay: Balandra, El Mogote, Eréndira, La Paz-Aeropuerto, and Zacatecas; however, the rest of the mangrove areas experienced a decrease in mangrove cover (Table 2). The most affected mangroves, Playa Pichilingue-Brujas and Enfermería, experimented significant human-induced impacts during the last decades, mainly the reduction of their connection to the sea by the construction of roads [34]. Mangroves present in La Paz bay facilitated the use of aerial and satellite imagery for the estimation of the RVI because they are adjacent to desertic areas, which showed high contrast with mangrove species. However, different authors consider that the cover estimations by this method are limited when mangroves are close to other types of forest or wetlands, and additional corrections are necessary [35,36].

The integration of data obtained by different methods has proven difficult [37]; therefore, this study used mixed methods to integrate qualitative and quantitative data, and most importantly, changes in mangrove cover with scientific expert's opinions. This last indicator is essential when the number of experts is limited, but their knowledge in the region is plenty. The information obtained by the DMS contrasted in some cases with the cover estimated by the RVI. In such cases, the scientific experts considered that some mangroves, e.g., El Mogote and Erendira, were in a poor state of conservation (Table 4). Still, those sites presented an increase in cover between 1974 and 2018, according to the RVI (Table 3). For those cases, the RAQ demonstrated the presence of some visible impacts, such as gray water inputs and modification of water circulation, which may be limiting the ecosystem services that mangroves should provide but increased their cover. Some of these impacts change over time; for example, solid waste was observed previously at Comitán but not registered during the RAQ thanks to a cleaning campaign that took place the day before the field visit. Therefore, the MCIS approach highlights the need to integrate expert opinions and field data.

The use of the MCSI scores to classify mangrove communities according to the IUCN Red List of Ecosystems is a novel approach. The analysis for 17 mangroves in La Paz bay resulted in five sites in the category of least concern (LC), and eight considered as near threatened (NT). On the other hand, one of them, the Enfermería mangrove area, is vulnerable (VU), and the Pichilingue-Brujas mangrove is endangered (EN). Finally, two mangroves, Salinas de Pichilingue and Espiritu Santo Archipelago, were classified as data deficient (DD). However, according to partial results, it is currently estimated that Salinas de Pichilingue and Espiritu Santo Archipelago have a good state of conservation, especially in the last case (Figure 5).

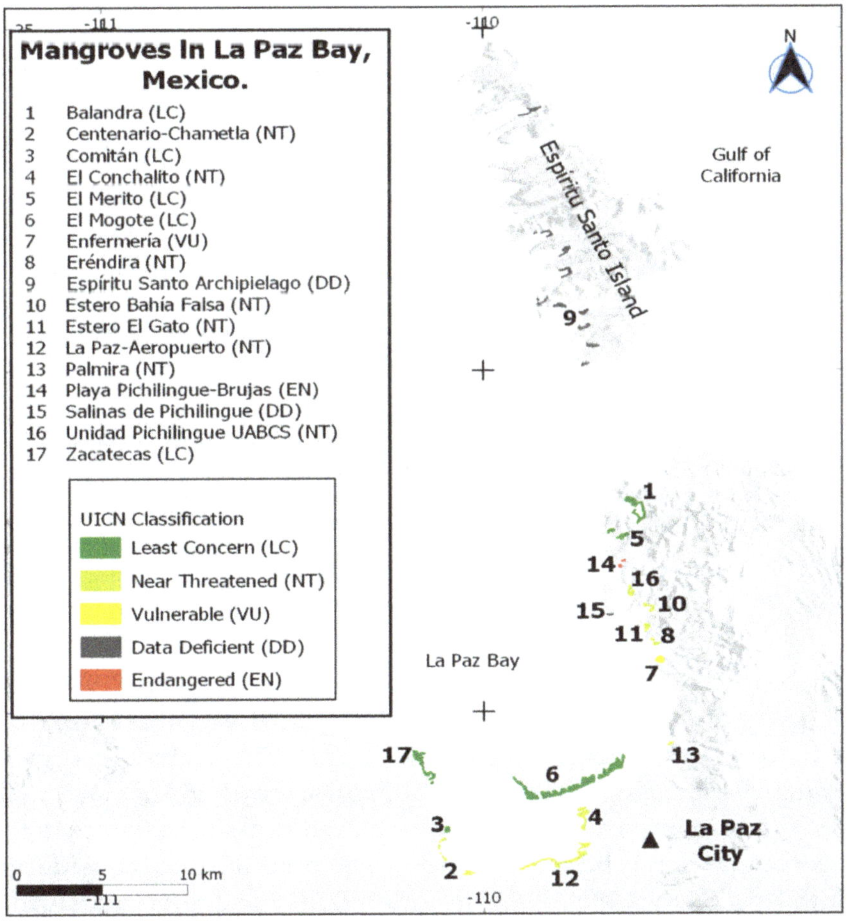

Figure 5. Classification of La Paz Bay mangroves, according to their conservation status.

Accordingly, the MCSI was adequate for the case study and helped to define conservation priorities for the mangroves in the region. The integrative nature of the index allowed for the identification of factors that negatively affect the conservation status of the mangroves, e.g., losses of vegetation, changes in water circulation, or solid waste presence. Then, it can be instrumental for the effective implementation of ecological restoration activities undertaken in areas [38,39]. The application of the MCSI by managers in this region may help to revert the condition of those mangrove areas that have suffered significant deterioration or to address other adverse factors threatening these ecosystems [40]. However, it is estimated that the urbanization caused by the tourism industry will be the most threatening factor for the conservation of mangroves in La Paz Bay, especially if it does not consider the applicable environmental regulations. Although several of the mangrove communities are small, due to their ecosystem services, these must be conserved [41].

Finally, it is concluded that the results on conservation status are more robust than those that include only spatial data and, by their integration into the IUCN Red List of Ecosystems, allow a direct comparison not only with other mangroves but also with different aquatic and terrestrial environments around the world. We recommend the application of the MCSI not only as a decision-making tool but also as an exploratory study; nonetheless, it is advisable to conduct follow up monitoring of quantitative ecological indicators to strengthen and provide feedback to update the MCSI.

5. Conclusions

The results of the present work constitute the first innovative use of the categorization of the Red List of Ecosystems by the IUCN for the mangroves communities. The MCSI integrates not only reliable quantitative indicators but also qualitative indicators to provide a more accurate conservation status for the mangrove communities; however, it is limited by the availability of data for some areas of interest. The multidisciplinary approach of the MCSI allows for its application at data deficient mangrove communities for an initial evaluation and guides future research efforts. The proposed MCSI index is a reliable tool for the management and conservation of mangrove communities; nonetheless, it could be improved as new methods to collect substantial data for additional indicators become available.

Author Contributions: Conceptualization, G.Á.-F., J.J.-M., O.A.-C., and G.H.-A.; Investigation, G.Á.-F.; Methodology, G.Á.-F. and P.C.-C.; Project administration, G.Á.-F. and J.J.-M.; Validation, G.Á.-F.; Writing—original draft, G.Á.-F. and G.H.-A.; Writing—Review and Editing, G.Á.-F., J.J.-M., P.C.-C., O.A.-C., J.M.L.-V. and G.H.-A. All authors have read and agreed to the published version of the manuscript.

Acknowledgments: This paper is based partially on Ávila-Flores, 2016 and other data collected for the doctoral thesis of the first author. We thank Candidate Pablo Hernández-Morales, for his advice on spatial analysis and the M.Sc. Liliana Paredes-Lozano for her support during fieldwork and for proofreading the article. Finally, we would like to thank the mangrove experts who participated in the Delphi Method Survey and the three reviewers whose commentaries and suggestions greatly improved to this paper.

Appendix A

The analytical hierarchy process (AHP) proposed by Thomas L. Saaty (1980) is a method to express the relative importance or dominance of individual element(s) over an assessment, such as criteria or characteristics useful, for example, for decision-making [42]. This technique organizes the evaluation elements into a hierarchy level, assigning numerical values, which gives mathematical support to the organization. The AHP uses pairwise comparison or paired ranking to compare the various elements of the analysis. This comparison is based on a numerical qualification, considering a prime number subjacent scale from 1 to 9 (Saaty scale). Data are entered into a squared matrix to give qualifications; therefore, the level of importance (Table A1).

Table A1. Saaty scale used in paired ranking.

Scale Values	Values Definition	Definition
1	Equal importance	An equal level of importance for both evaluation elements that are being compared
3	Somewhat more or weak importance	One element is slightly more important or relevant than the other
5	Much more or essential importance	One element is more important or essential than the other.
7	Very much or demonstrated importance	One element is much more important or relevant than the other
9	Extreme or absolute importance	One element is definitely more important than the other

Appendix B

The Delphi survey method is a research technique proposed by The RAND Corporation in the 1950s (Table A2). It is a group communication process [43], with reference to a collection of answers in order to obtain and improve information resulting from opinions made by experts in a particular theme [44]. This methodology is based on the application of a series of questions, all of which are submitted once or several times if necessary, with the objective to achieve feedback and enrich opinions [45]. The origin and number of these may come from either the academic sector, industrial, governmental, or civil society organizations. It should be mentioned that the expert selection must be done considering their previous experience and current work on the subject. Using these opinions, this method seeks to develop a previously non-existent or dispersed knowledge useful in the decision-making process in government, science, and industry areas [46].

Table A2. Delphi method process description (based on Mukherjee et al. 2014).

1. FIRST ROUND
2. ⬇
3. PREPARE QUESTIONNAIRE
4. ⬇
5. SELECT EXPERTS IN ORDER TO OBTAIN THEIR OPINION
6. ⬇
7. INVITE THEM TO PARTICIPATE ON DELPHI
8. ⬇
9. COLLECT ANSWERS, ANALYSE, AND PROVIDE FEEDBACK
10. ⬇
11. SECOND ROUND
12. ⬇
13. PREPARE AND SEND QUESTIONNAIRE
14. ⬇
15. COLLECT ANSWERS, ANALYSE, AND PROVIDE FEEDBACK
16. ⬇
17. REPEAT ROUNDS UNTIL CONSENSUS IS REACHED

Appendix C

Rapid assessment or rapid appraisal (RA) is a quick and accurate estimation method to collect in situ information [47]. This technique allows for an immediate approximation to the magnitude and characteristics of a particular situation. It is an approach for developing a preliminary, qualitative understanding [48]. RA marks the line to define needs or tasks to be carried out in posterior assessments (Figure A1); it provides supplemental information to other sources, in a simple, fast, and flexible way [49]. The RA method uses various tools to collect data; for example, the recording of observations in logbooks or through questionnaires, participatory surveys. All these methods have the purpose of giving fast and reliable results. RA's fundamental advantage is that it is itself "a phase or stage in a research process"; therefore, it allows for scientific rigor [50].

Figure A1. Rapid assessment process and elements.

Appendix D

The IUCN Red List of Ecosystems Categories and Criteria is a global standard for how we assess the status of ecosystems, applicable at local, national, regional, and global levels [51,52]. In this work, we artificially established the values to classify mangroves in each of the categories. Each criterion used for each category is described after the definitions mentioned below (Table A3).

Table A3. List of scores for each category of the MCSI Index. (Based on Keith et al., 2015).

IUCN Red List of Ecosystems	Score
Collapsed (CO)	0
Critically endangered (CR)	01–20
Endangered (EN)	21–40
Vulnerable (VU)	41–60
Near Threatened (NT)	61–80
Least Concern (LC)	81–100
Data Deficient (DD)	N/A
Not Evaluated (NE)	N/A

- Collapsed (CO)

"An ecosystem is Collapsed when it is virtually certain that its defining biotic or abiotic features are lost from all occurrences, and the characteristic native biota are no longer sustained. Collapse may occur when most of the diagnostic components of the characteristic native biota are lost from the system, or when functional components (biota that performs key roles in ecosystem organization) are greatly reduced in abundance and lose the ability to recruit" (IUCN, 2019). In this category, the entire vegetation cover of the mangrove community has been destroyed.

- Critically Endangered (CR)

"An ecosystem is Critically Endangered when the best available evidence indicates that it meets any of the criteria A to E for Critically Endangered. It is therefore considered to be at an extremely high risk of collapse" (IUCN, 2019). In this category, the index score obtained by the mangrove community is less than 20 but higher than zero.

- Endangered (EN)

"An ecosystem is Endangered when the best available evidence indicates that it meets any of the criteria A to E for Endangered. It is therefore considered to be at a very high risk of collapse" (IUCN, 2019). In this category, the index score obtained by the mangrove community is less than 40 but higher than 20.

- Vulnerable (VU)

"An ecosystem is Vulnerable when the best available evidence indicates that it meets any of the criteria A to E for Vulnerable. It is therefore considered to be at a high risk of collapse" (IUCN, 2019). In this category, the index score obtained by the mangrove community is less than 60 but higher than 40.

- Near Threatened (NT)

"An ecosystem is Near Threatened when it has been evaluated against the criteria but does not qualify for Critically Endangered, Endangered or Vulnerable now, but is close to qualifying for or is likely to qualify for a threatened category in the near future" (IUCN, 2019). In this category, the index score obtained by the mangrove community is less than 80 but higher than 60.

- Least Concern (LC)

"An ecosystem is Least Concern when it has been evaluated against the criteria and does not qualify for Critically Endangered, Endangered, Vulnerable or Near Threatened. Widely distributed and relatively undegraded ecosystems are included in this category" (IUCN, 2019). In this category, the index score obtained by the mangrove community is equal to or less than 100 but higher than 80.

- Data Deficient (DD)

"An ecosystem is Data Deficient when there is inadequate information to make a direct, or indirect, assessment of its risk of collapse based on decline in distribution, disruption of ecological function or degradation of the physical environment. Data Deficient is not a category of threat, and does not imply any level of collapse risk" (IUCN, 2019). In this category, some of the information considered for the application of the index in a particular mangrove community was not available.

- Not Evaluated (NE)

"An ecosystem is Not Evaluated when it is has not yet been assessed against the criteria" (IUCN, 2019). In this category, it means that it was not evaluated in any way within the three components considered within the index.

References

1. Spalding, M.D.; Kainuma, M.; Collins, L. *World atlas of Mangroves*; Earthscan: London, UK, 2010; pp. 24–25.
2. Van Bochove, J.; Sullivan, E.; Nakamura, E. *The Importance of Mangrove to People: A Call to Action*; United Nations Environment Programme, World Conservation Monitoring Centre: Cambridge, UK, 2014; pp. 43–45.
3. Vo, Q.T.; Kuenzer, C.; Vo, Q.M.; Moder, F.; Oppelt, N. Review of valuation methods for mangrove ecosystem services. *Ecol. Indic.* **2012**, *23*, 431–446. [CrossRef]
4. Atkinson, S.C.; Jupiter, S.D.; Adams, V.M.; Ingram, J.C.; Narayan, S.; Klein, C.J.; Possingham, H.P. Prioritising mangrove ecosystem services results in spatially variable management priorities. *PLoS ONE* **2016**, *11*, e0151992. [CrossRef] [PubMed]
5. Kadykalo, A.N.; López-Rodríguez, M.D.; Ainscough, J.; Droste, N.; Ryu, H.; Ávila-Flores, G.; Harmáčková, Z. Disentangling 'ecosystem services' and 'nature's contributions to people'. *Ecosyst. People* **2019**, *15*, 269–287. [CrossRef]
6. Roldán, V.A.; Galván, D.E.; Lopes, P.F.; López, J.; Bellamy, A.S.; Gallego, F.; Barriga, A.M. Are we seeing the whole picture in land-sea systems? Opportunities and challenges for operationalizing the ES concept. *Ecosyst. Serv.* **2019**, *38*, 100966. [CrossRef]
7. Sandilyan, S.; Kathiresan, K. Mangrove conservation: A global perspective. *Biodivers. Conserv.* **2012**, *21*, 3523–3542. [CrossRef]
8. Romañach, S.S.; DeAngelis, D.L.; Koh, H.L.; Li, Y.; Teh, S.Y.; Barizan, R.S.R.; Zhai, L. Conservation and restoration of mangroves: Global status, perspectives, and prognosis. *Ocean Coast. Manag.* **2018**, *154*, 72–82. [CrossRef]
9. Verdugo, F.F.; Casasola, P.M.; Hernández, C.M.A.; Rosas, H.L.; Pardo, D.B.; Bello, A.C.T. La topografía y el hidroperiodo: Dos factores que condicionan la restauración de los humedales costeros. *Boletín Soc. Botánica México* **2007**, *80*, 33–47.
10. FAO. *The World's Mangroves 1980–2005*; FAO Forestry Paper 153; FAO: Rome, Italy, 2007; p. 77.
11. Duke, N.C.; Meynecke, J.O.; Dittmann, S.; Ellison, A.M.; Anger, K.; Berger, U.; Koedam, N. A world without mangroves? *Science* **2007**, *317*, 41–42. [CrossRef]
12. Giri, C.; Ochieng, E.; Tieszen, L.L.; Zhu, Z.; Singh, A.; Loveland, T.; Duke, N. Status and distribution of mangrove forests of the world using earth observation satellite data. *Glob. Ecol. Biogeogr.* **2011**, *20*, 154–159. [CrossRef]
13. Sitios de Manglar con Relevancia Biológica y Con Necesidades de Rehabilitación Ecológica. Available online: http://www.biodiversidad.gob.mx/ecosistemas/manglares2013/sitioPacNorte.html (accessed on 20 September 2019).
14. Dávalos-Sotelo, R. El papel de la investigación científica en la creación de las áreas naturales protegidas. *Madera y Bosques* **2016**, *22*, 7–13. [CrossRef]
15. Ávila-Flores, G.; Juárez-Mancilla, J.; Hinojosa-Arango, G.O.; Covarrubias, A. The Use of the DPSIR Framework to estimate Impacts of Urbanization on Mangroves: A Case Study from La Paz, Baja California Sur, Mexico. In *WIT Transactions on Ecology and the Environment*; Sustainable City XII; Brebbia, C.A., Sendra, J.J., Eds.; WIT Press: Southampton, UK, 2017; Volume 223, pp. 459–469.
16. Places to Go in 2020. The New York Times. Available online: https://www.nytimes.com/interactive/2020/travel/places-to-visit.html (accessed on 11 January 2020).
17. González-Baheza, A.; Arizpe, O. Vulnerability assessment for supporting sustainable coastal city development: A case study of La Paz, Mexico. *Clim. Dev.* **2018**, *10*, 552–565. [CrossRef]
18. Márquez, G. Colombia: Ambiente, pobreza, violencia. *Fermentum Rev. Venez. Sociol. Antropol.* **2003**, *13*, 25–37.
19. Chauvaud, S.; Bouchon, C.; Manieres, R. Remote sensing techniques adapted to high resolution mapping of tropical coastal marine ecosystems (coral reefs, seagrass beds and mangrove). *Int. J. Remote Sens.* **1998**, *19*, 3625–3639. [CrossRef]
20. Benfield, S.L.; Guzman, H.M.; Mair, J.M. Temporal mangrove dynamics in relation to coastal development in Pacific Panama. *J. Environ. Manag.* **2005**, *76*, 263–276. [CrossRef] [PubMed]
21. Fromard, F.; Vega, C.; Proisy, C. Half a century of dynamic coastal change affecting mangrove shorelines of French Guiana. A case study based on remote sensing data analyses and field surveys. *Mar. Geol.* **2004**, *208*, 265–280. [CrossRef]

22. Murray, M.R.; Zisman, S.A.; Furley, P.A.; Munro, D.M.; Gibson, J.; Ratter, J.; Bridgewater, S.; Minty, C.D.; Place, C.J. The mangroves of Belize. Part 1. Distribution, composition and classification. *For. Ecol. Manag.* **2003**, *174*, 265–279. [CrossRef]
23. Cissell, J.R.; Delgado, A.M.; Sweetman, B.M.; Steinberg, M.K. Monitoring mangrove forest dynamics in Campeche, Mexico, using Landsat satellite data. *Remote Sens. Appl.* **2018**, *9*, 60–68. [CrossRef]
24. Okoli, C.; Pawlowski, S.D. The Delphi method as a research tool: An example, design considerations and applications. *Inf. Manag.* **2004**, *42*, 15–29. [CrossRef]
25. Hasson, F.; Keeney, S.; McKenna, H. Research guidelines for the Delphi survey technique. *J. Adv. Nurs.* **2000**, *32*, 1008–1015.
26. Kelly, C. *Guidelines for Rapid Environmental Impact Assessment in Disasters*; Benfield Hazard Research Centre, University of London: London, UK, 2005.
27. Rodríguez, J.P.; Rodríguez-Clark, K.M.; Baillie, J.E.; Ash, N.; Benson, J.; Boucher, T.; Keith, D.A. Establishing IUCN red list criteria for threatened ecosystems. *Conserv. Biol.* **2011**, *25*, 21–29. [CrossRef]
28. Kovacs, J.M.; Vandenberg, C.V.; Wang, J.; Flores-Verdugo, F. The use of multipolarized spaceborne SAR backscatter for monitoring the health of a degraded mangrove forest. *J. Coast. Res.* **2008**, *24*, 248–254. [CrossRef]
29. Chellamani, P.; Singh, C.P.; Panigrahy, S. Assessment of the health status of Indian mangrove ecosystems using multi temporal remote sensing data. *Trop. Ecol.* **2014**, *55*, 245–253.
30. Faridah-Hanum, I.; Yusoff, F.M.; Fitrianto, A.; Ainuddin, N.A.; Gandaseca, S.; Zaiton, S.; Shamsuddin, I. Development of a comprehensive mangrove quality index (MQI) in Matang Mangrove: Assessing mangrove ecosystem health. *Ecol. Indic.* **2019**, *102*, 103–117. [CrossRef]
31. Wątróbski, J.; Ziemba, E.; Karczmarczyk, A.; Jankowski, J. An index to measure the sustainable information society: The Polish households case. *Sustainability* **2018**, *10*, 3223. [CrossRef]
32. Chowdhury, A.; Maiti, S.K. Assessing the ecological health risk in a conserved mangrove ecosystem due to heavy metal pollution: A case study from Sundarbans Biosphere Reserve, India. *Hum. Ecol. Risk Assess.* **2016**, *22*, 1519–1541. [CrossRef]
33. Aguirre-Rubí, J.R.; Ortiz-Zarragoitia, M.; Izagirre, U., Etxebarria, N.; Espinoza, F.; Marigómez, I. Prospective biomonitor and sentinel bivalve species for pollution monitoring and ecosystem health disturbance assessment in mangrove–lined Nicaraguan coasts. *Sci. Total Environ.* **2019**, *649*, 186–200. [CrossRef] [PubMed]
34. Ávila-Flores, G.; Riosmena-Rodríguez, R. Integrated environmental assessment and scenarios of mangrove community in the Gulf of California case study 'Manglar Enfermería'. In *The Arid Mangrove Forest from Baja California Peninsula 2*; Nova Publishers: New York, NY, USA, 2016; Volume 2, pp. 131–150.
35. Flores-de-Santiago, F.; Kovacs, J.M.; Flores-Verdugo, F. Discrimination of 3 dominant mangrove species from the Pacific coast of Mexico by spectroscopy on intact leaves. *Cienc. Mar.* **2018**, *44*, 185–202. [CrossRef]
36. Pham, T.D.; Yokoya, N.; Bui, D.T.; Yoshino, K.; Friess, D.A. Remote sensing approaches for monitoring mangrove species, structure, and biomass: Opportunities and challenges. *Remote Sens.* **2019**, *11*, 230. [CrossRef]
37. Wątróbski, J.; Jankowski, J.; Ziemba, P.; Karczmarczyk, A.; Ziolo, M. Generalised framework for multi-criteria method selection. *Omega* **2019**, *86*, 107–124. [CrossRef]
38. López-Portillo, J.; Lewis, R.R.; Saenger, P.; Rovai, A.; Koedam, N.; Dahdouh-Guebas, F.; Rivera-Monroy, V.H. Mangrove forest restoration and rehabilitation. In *Mangrove Ecosystems: A Global Biogeographic Perspective*; Springer: Cham, Germany, 2017; pp. 301–345.
39. Reyes, M.A.; Tovilla, C. Restauración de áreas alteradas de manglar con Rhizophora mangle en la Costa de Chiapas. *Madera Bosques* **2002**, *8*, 103–114. [CrossRef]
40. Avila-Flores, G. Diagnosis of the conservation status of mangroves in the city of La Paz. Master's Thesis, Baja California Sur, Mexico, February 2016.
41. Curnick, D.; Pettorelli, N.; Amir, A.; Balke, T.; Barbier, E.; Crooks, S.; Quarto, A. The value of small mangrove patches. *Science* **2019**, *363*, 239. [PubMed]
42. Wind, Y.; Saaty, T.L. Marketing applications of the analytic hierarchy process. *Manag. Sci.* **1980**, *26*, 641–658. [CrossRef]
43. Hsu, C.C.; Sandford, B.A. The Delphi technique: Making sense of consensus. *Pract. Assess. Res. Eval.* **2007**, *12*, 1–8.
44. Gordon, T.J. The Delphi method. *Futures Res. Methodol.* **1994**, *2*, 1–30.

45. Mukherjee, N.; Huge, J.; Sutherland, W.J.; McNeill, J.; Van Opstal, M.; Dahdouh-Guebas, F.; Koedam, N. The Delphi technique in ecology and biological conservation: Applications and guidelines. *Methods Ecol. Evol.* **2015**, *6*, 1097–1109. [CrossRef]
46. Renzi, A.B.; Freitas, S. The Delphi method for future scenarios construction. *Procedia Manuf.* **2015**, *3*, 5785–5791. [CrossRef]
47. Beebe, J. *Basic Concepts and Techniques of Rapid Appraisal*; Human organization: New York, NY, USA, 1995; pp. 2–51.
48. Beebe, J. *Rapid Assessment Process: An Introduction*; AltaMira Press: Lanham, MD, USA, 2001.
49. Hjortsø, C.N.; Christensen, S.M.; Tarp, P. Rapid stakeholder and conflict assessment for natural resource management using cognitive mapping: The case of Damdoi Forest Enterprise, Vietnam. *Agric. Hum. Values* **2005**, *22*, 149–167. [CrossRef]
50. Pedersen, D.; Scrimshaw, N.S.; Gleason, G.R. *Rapid Assessment Procedures: Qualitative Methodologies for Planning and Evaluation of Health Related Programmes*; INFDC: Boston, MA, USA, 1992.
51. Keith, D.A.; Rodríguez, J.P.; Brooks, T.M.; Burgman, M.A.; Barrow, E.G.; Bland, L.; Miller, R.M. The IUCN red list of ecosystems: Motivations, challenges, and applications. *Conserv. Lett.* **2015**, *8*, 214–226. [CrossRef]
52. IUCN. Red List of Ecosystems Web. Available online: https://iucnrle.org/ (accessed on 25 October 2019).

5

Preserving Biodiversity and Ecosystem Services in West African Forest, Watersheds and Wetlands

Oreoluwa Ola and Emmanuel Benjamin

Department of Governance in International Agribusiness and Department of Agricultural Production and Resource Economics, Technical University of Munich (TUM), 85354 Freising, Germany

* Correspondence: oreoluwa.ola@tum.de (O.O.); emmanuel.benjamin@tum.de (E.B.)

Abstract: While biodiversity and ecosystem services derived from the natural environment are the backbones of West African rural livelihood, unsustainable exploitation of natural resources, conflicts, and climate change threaten the continued provision of ecosystem services. This threat creates an urgent need to safeguard the integrity of the environment. Evaluating the effectiveness of environmental conservation projects is central towards designing and scaling-up successful conservation projects. Using secondary literature and project reports, we reviewed ongoing and completed conservation projects in the West African sub-region. Scientific work on incentives for ecosystem services in sub-Saharan Africa typically focuses on Southern and Eastern Africa, leaving Western Africa underserved. This study fills this literature gap by compiling lessons from conservation projects in West Africa to offer region-specific incentives that should inform the design of conservation projects in the region. The study shows that the way forward is a holistic, sustainable development approach that mirrors and meets strategies outlined in Sustainable Development Goals 1, 2, 5, 8, 13, and 17: No Poverty, End Hunger and Promote Sustainable Agriculture, Gender Equality, Decent Work and Economic Growth, Climate Action, and Partnerships for the Goals, respectively.

Keywords: biodiversity; ecosystem services; West Africa; incentives; Sustainable Development Goals

---

1. Introduction

Biodiversity is described as a stock of living resources determined by nature as well as, to some extent, human activity [1,2]. Ecosystem services (ESs), on the other hand, are biodiversity flows beneficial to man. Ecosystem services include provisioning services (food, drinking water, medicine, and hydroelectric power), regulating services (carbon sequestration, climate regulation, and clean air and water), cultural services (tourism and religious and cultural benefits), supporting services (agriculture, soil formation, and flood and erosion control) [3]. In West Africa, watersheds, wetlands, forest, and vegetation are biodiversity hubs and primary providers of ESs. These hubs characterize the environmental landscape in West Africa and are crucial to the livelihood of its citizens.

However, the integrity of these biodiversity hubs is at serious risk. Agricultural expansion, overexploitation of biological resources, population explosion, urbanization, and climate change is destroying biodiversity ecosystems [4]. The occurrence of invasive species as well as industrial and pesticide discharge are threatening the quality and volume of native flora and fauna species residing in watersheds and wetlands across sub-Saharan Africa [4,5]. Consequently, these have degraded and reduced the size of biodiversity hubs in parts of West Africa [3]. Close to 90% of West Africa's original forest has disappeared with small fragments of this natural forest existing in Cote d'Ivoire, Nigeria, and Cameroon [6]. Over-exploitation of forest resources, mining, agricultural expansion, hunting and population growth is fragmenting tropical forests and driving biodiversity losses [7]. These losses threaten both wildlife as well as the current and future livelihoods of those individuals and communities that depend on ESs in West Africa.

In response to these threats, governments, and private sector organizations in developing countries are implementing environmental conservation policies and programs to regulate and preserve biodiversity resources and raise awareness of threats facing biodiversity hubs. For instance, conservation programs such as Reducing Emissions from Deforestation and Forest Degradation (REDD), REDD+, Payments for Ecosystem Services (PES), and Forests and National Park protections laws are increasingly being implemented to protect biodiversity spots and curtail the over-exploitation of forest resources. The + sign adds conservation and sustainable management of forest carbon stocks in developing countries to the existing REDD mandate [8–10]. Reviews of the implemented policies and programs are becoming common because programs and policies must adapt to the changing environmental, institutional, and economic conditions surrounding these programs. Findings from the reviews uncover up-to-date, comprehensive information and steers discussions absent in individual case study research [11]. The results of reviews, therefore, guide policymakers in the design and implementation of new programs as well as in adjusting and scaling-up pilot programs.

In Africa, the few reviews of ecosystem services and conservation programs in Africa concentrate mainly on Eastern and Southern Africa while it should be noted that there are fewer conservation programs in Africa compared to other developing regions of the world [5,11–14]. Sub-Saharan Africa lags behind other developing parts of the world in the implementation of conservation programs [11]. Furthermore, poverty alleviation and equity are dominant themes in conservation (watershed) programs in Africa [5]. Both studies demonstrated that strict regulatory requirements and poor technical knowledge limit opportunities for trading ecosystem services and implementing trading schemes in Africa. Sustainability in Africa depends on the assessment of ecosystem services, especially provisioning ecosystem services, and these assessments are dominant in conservation programs in Eastern and Southern Africa [14].

These studies also lamented the dearth of conservation programs and information on those programs in other parts of Africa. Indeed, an extensive literature search revealed that no study has reviewed conservation programs in West Africa. Given the growing interest in conservation policies and programs in West Africa, it is important that this gap is filled. Assessing the effectiveness of these programs underscores their contribution towards protecting biodiversity, combating poverty, and promoting sustainable agriculture and gender equality, in addition to achieving the Sustainable Development Goals (SDGs).

This study fills that gap by reviewing ongoing (and defunct) policies and programs that protect watersheds, wetlands, and forests in West Africa. We draw attention to the activities of these programs, highlighting their outcomes and challenges to offer region-specific incentives that inform the design of conservation projects in West Africa. Incentives, in this study, refer to mechanisms and tangible benefits that motivate environmental conservation. In a review of conservation programs in Asia, Latin America, and Africa, it was discovered that incentives are more effective environmental conservation measures compared to other instruments [12]. The authors, however, caution that, for incentives to achieve their goals, specific economic, cultural, and institutional conditions must be considered before introducing incentive initiatives. In evaluating conservation programs in West Africa, we tease out those critical cultural, economic, and institutional conditions crucial to successful environmental conservation programs in the region. In different parts of Africa, dependence on ESs differs as socioeconomic conditions, geography, and vegetation changes, creating slight differences in priorities for each region [15]. Our findings should inform the implementation of conservation initiatives in Africa and target potential project sponsors and managers seeking to design and finance projects in West Africa.

The rest of the study is organized as follows. Section 2 introduces the materials and methods used in this study. Section 3 presents an overview of the current state of biodiversity spots in West

Africa. Section 4 summarizes ongoing (and completed) conservation projects in West Africa in Section 4. This section focuses on the objectives of these programs, the ecosystem services they provide, the institutional actors involved in the projects, their outcomes and the challenges they faced. Section 5 discusses various incentives for increasing investments in environmental conservation and projects in West Africa, while Section 6 concludes the paper.

2. Materials and Methods

In putting together this study, a desk review of reports on active and defunct conservation programs in West Africa was conducted. According to [16], qualitative reviews attempts to understand the meaning of and contextualize events and processes. We find this approach suited to the aims and scope of this study. This process unfolded in two stages.

In the first stage, we searched for information on environmental conservation programs implemented in West Africa between September 2017 and January 2018 using the following online social science database: Google Scholar, PubMed, ISI Web of Science, ResearchGate, and ScienceDirect. The search focused on conservation projects that transformed vegetation and forest cover, i.e., deforestation, afforestation, and watershed protection projects. Project implementation and evaluation reports published by the implementing agency in English were consulted to obtain information on these projects. Some of the official reports released by National Governments on some programs in Fracophone West Africa were written in French. The reports were supported by information collected from peer-reviewed articles investigating these projects. The Watershed Markets Database, The Economics of Ecosystem and Biodiversity (TEEB), Forest Trends, Global Environmental Facility (GEF), and the United Nations Framework Convention on Climate Change (UNFCCC) websites were consulted for additional information.

Multi-criteria analysis was employed to evaluate these programs and collect the necessary information [17]. Information on the location of each project, motivation behind its formation, program design, program objectives, ESs, costs, and the actors involved—both implementers and beneficiaries—were gathered. Information on the environmental, economic, and social outcomes of each program as well as the challenges each program faced was also collected.

In the second stage, we consulted reports and documents on ongoing conservation projects in other parts of Africa and developing regions of the world. The information obtained from these projects were combined with information from our case studies in West Africa to structure the discussion on incentives for conservation projects in West Africa and make policy recommendations.

3. West African Wetlands, Watersheds, and Forests

In West Africa, temperature and rainfall operate in the extremes engendering droughts and flooding. The region is characterized by the Sahara Desert to the north and lush green tropical rainforest to the south, which gives way to mangrove swamps connecting the coasts to the mainland. In the middle is the savanna vegetation characterized by grassland and short-to-medium height trees. The diverse vegetation in West Africa supports a vibrant ecosystem and biodiversity hub.

Four of the 17 primary watersheds in Africa (see Figure 1) are located in the Western part of the sub-continent, namely the Senegal River basin, the Volta River basin, the Niger River basin, and the Lake Chad basin [18].

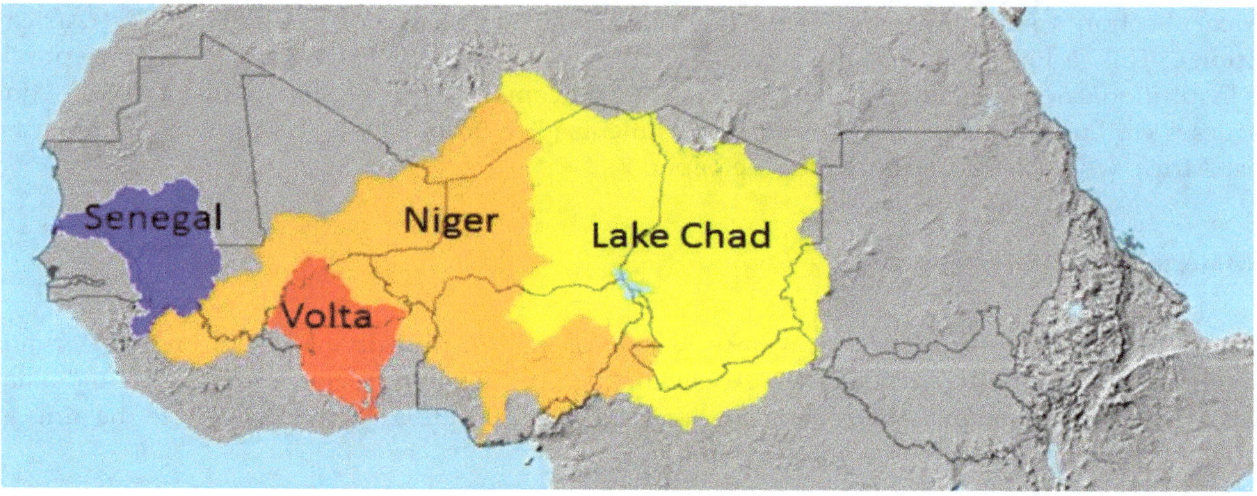

Figure 1. Niger, Lake Chad, Volta, and Senegal basins. Source: [19] Van Der Wijngaart et al., 2019.

The Senegal River basin cuts across four countries: Guinea, Mali, Mauritania, and Senegal. The Volta River basin is shared by six West African countries, namely Mali, Burkina Faso, Benin, Togo, Cote d´Ivoire, and Ghana. The Lake Chad basin spreads to over seven countries, two of which are in the West: Nigeria, Niger, Algeria, Sudan, Central Africa Republic, Chad, and Cameroon. The Niger River Basin spreads through seven West African countries, namely, Guinea, Cote d´Ivoire, Mali, Burkina Faso, Benin, Niger, and Nigeria, as well as Algeria, Chad and Cameroon. Each of the watersheds plays significant roles vital to local economies across West Africa. They are essential sources of water for agriculture, public health, food, clean water, and hydroelectric power. The watershed and wetlands in West Africa are home to 3 million endemic and migratory birds spanning 400 species and support over 2000 known species of indigenous African freshwater fishes [7]. The estimated value of ecosystem services that can be attributed to inland watershed and wetlands ranges between US$5000 and US$100,000 per hectare, while coastal watershed and wetlands were between US$500 and US$1,000,000 per hectare in 2007 [20]. The economic value of Hadejia-Nguru wetlands, which forms part of the Komadougou-Yobe river basin of the Lake Chad basin in Nigeria, was estimated at over US$16 million per year [21].

The Guinean forest, with an estimated area of 553,427 km^2, is also a biodiversity hub. It is a tropical forest in West Africa, "Upper and Lower," that cuts across several Africa countries from Guinea to Sierra Leone as well as beyond the borders of Nigeria (see Table 1) comprising fragmented forest separated by agricultural land, cities, and villages [7].

Table 1. Total area and proportion of hotspots in West African countries.

Country	Total Area (km^2)	Area of Overlap with Guinean Forest Hotspots (km^2)	Percentage of Hotspots in Each Country	Percentage of Countries in Hotspots
Benin	117,650	1462	0.3%	1.2%
Côte d'Ivoire	325,990	150,300	27.2%	46.1%
Equatorial Guinea	28,051	1965	0.4%	7.0%
Ghana	242,178	79,902	14.4%	33.0%
Guinea	249,691	48,488	8.8%	19.4%
Liberia	96,861	95,376	17.2%	98.5%
Nigeria	926,744	127,583	23.1%	13.8%
São Tomé and Príncipe	1001	1001	0.2%	100.0%
Sierra Leone	73,316	47,350	8.6%	64.6%

Source: Adapted from. Carr. et al. (2015) [7].

The forests comprise numerous species of trees providing ecosystem services and various forest products that support human livelihood, e.g., timber, fruits, carbon storage, disease control, tourism, pollination, and water regulation, among other ecosystem services. It also houses over 9000 species of vascular plants, 700 species of birds, 300 species of mammals, 100 reptiles, and 500 freshwater fish [7,22]. Of these species, 1800 plants, 48 birds, 65 mammals, 20 reptiles, and 118 amphibians are native to West Africa [7]. There are approximately 483 plant species, 48 bird species, 65 mammal species, 11 reptile species, 77 amphibian species, and 172 bony fishes under threat of extinction [7]. Among the most threatened mammalian species are indigenous chimpanzees and gorillas, African elephants, pygmy hippopotamus, and primates such as rodents and bats [7].

To the best of our knowledge, few literature has specified a value for forest ecosystem and biodiversity in West Africa; we assume that, for carbon storage, biodiversity conservation, hydrological benefits, and forest products including tourism should be well above US$2,000 per hectare given the combination of primary and secondary forest cover [23,24].

The provisioning services of ES, which improves agricultural productivity, can be achieved through sustainable and environmentally friendly agricultural and pastoral practices, while regulating services are embodied in reforestation and afforestation activities that sequester and reduce carbon emissions. These services reflect the relevance of ESs to agrarian households in West Africa [14,15]. For instance, agriculture accounts for 60% of the labor force but only 35% of the Gross Domestic Product (GDP) in West Africa [25]. The disparity between the GDP and employment figures implies that the majority of West Africans are poor [25]. Indeed, farmers cultivate on poor soils, with limited agricultural technology uptake and adoption, which impedes agricultural development [26]. Climate change affects the poor primarily via agriculture, thus making West African farmers especially vulnerable to climate change [27]. West Africa's population is projected to grow faster than any other region in the world by 2030 [28], putting more pressure on already dwindling natural resource stocks in a region where the majority live in poverty and is vulnerable to climate change. The economic importance of ESs has made forest and watershed (wetland) conservation an absolute priority.

4. Conservation Projects in West Africa

Compared to Latin America and Asia, there are few conservation programs in Africa. Only 3% of Agriculture, Forest, and Land Use projects under the Clean Development Mechanisms (CDMs) are in Africa, while the majority of the Voluntary Carbon projects are in Asia and Latin America [11,29]. Furthermore, about 13% of global watershed protection programs are situated in Africa [13]. West Africa's share of those programs is small because most projects are established in Eastern and Southern Africa.

From the literature search, we identified 14 ongoing and defunct land-use conservation programs in West Africa (see Table 2). We should note that 33 of the 234 CDM projects in Africa are in West Africa, with about 582,000 Certified Emissions Reductions Certificate (CERs) issued, i.e., 5% of total CERs in Africa [29]. Most of these projects, however, are not land-use conservation projects, but thermal, hydro, and renewable energy projects.

Table 2. Conservation Projects in West Africa.

Project	Location	Source (s)
A Sustainable Management Program (SANREM)	Mali	[11,30]
Acacia Community Carbon Plantation	Niger	[31]
Carbon Sequestration and Sustainable Agriculture	Senegal	[32]
Carbon Sequestration Pilot Projects in West African Savannah Optimum	Mali, Benin, and the border between Ghana and Burkina Faso	[32]
Guinean Forest Hotspots	Guinea, Sierra-Leone, Liberia, Ghana, Cote "d" Ivoire, Togo, Benin, Nigeria, Sao Tome and Principe	[7]
Participatory Rehabilitation of Degraded Lands Project	Senegal and Mauritania	[11,33]
Rehabilitation of Degraded Pasture Land Project	Burkina Faso	[34]
Senegal Plantation Project	Mali	[11,31]
Sequestration of Carbon in Soil Organic Matter Program (SOCSOM)	Senegal	[11,25]
Sourou Valley Wetlands Valuation Project	Burkina Faso	[35]
Sustainable Energy Management Project	Burkina Faso	[11]
The Ghana Cocoa Carbon Initiative (GCCI)	Ghana	[36,37]
The Gola REDD Project	Sierra-Leone	[38]
Village-based Management of Woody Savannah and the Establishment of Woodlots for Carbon Sequestration Project	Benin	[11,31,32]
Restoration of Degraded Forest Land Projects	Ghana	[39,40]

4.1. *Types of Project*

Based on their activities, the conservation programs in West Africa were different and fell under various project types. The sustainable management program, sequestration of carbon in soil organic matter program, Sourou Valley Wetlands project, Carbon Sequestration and Sustainable Agriculture program, and Carbon Sequestration projects in Mali, Senegal, Burkina-Faso, Senegal, and Benin, respectively, were experimental or research projects (see Table 2 for full names). Their goal was to assess the economic and ecological potential of various agro-ecological zones to sequestrate carbon. The Gola REDD+ project in Sierra Leone was conceived as a REDD+ project, while the Ghana cocoa carbon initiative is attempting to modify its activities to fit into the REDD+ framework. The regional program to manage the Guinean forest hotspots program combines social, economic development and environmental conservation objectives and bears a strong resemblance to Integrated Conservation Development Projects (ICDPs). The Acacia carbon projects in Niger and Senegal and the plantation project in Mali are funded by the Biocarbon fund managed by the World Bank. Achats Services International sold the carbon credits from the Nigerian project, while the credits from the Malian project remained with the World Bank. The sustainable energy management project in Burkina Faso was implemented under the Activities Implemented Jointly (now Joint Implementation) framework of the Kyoto Protocol, meaning they are Kyoto-compliant. Similarly, the Plan-Vivo project in Burkina Faso was implemented under the Plan Vivo Voluntary Carbon Market requirements, with Plan Vivo managing the carbon credits. The remaining projects, the Participatory Rehabilitation of Degraded Lands in Mauritania and Senegal, and the Village-Based Management of Woody Savannah in Benin have features synonymous with programs that remunerate ES providers for adopting sustainable land-use activities (or PES). However, in this instance, socioeconomic development was a primary objective along with environmental conservation.

The size of land enrolled in these projects varies between 10,000 and 126,000 ha (see supplementary file; Table S1, Column 5). Operating costs range from US$143,000 for the Sustainable Management Project in Mali to US$4–8 million for the Participatory Rehabilitation of Degraded lands project implemented in Senegal and Mauritania. The Guinea Forest Hotspots program initially costs US$ 8.3 million dollars over 10 years, but the program was recently extended to 2021 with an additional US$ 9 million dollars bringing the total cost to US$15.2 million dollars.

4.2. *Institutional Actors*

A consortium of both public and private organizations sponsored and coordinated these programs (see supplementary file; Table S1, Column 7). Similar to findings from [5], the conservation programs in Africa were mainly financed by external bodies. These funds ranged from climate funds, e.g., the Biocarbon climate fund managed by the World Bank, to funding from United Nations agencies and foundations such as McArthur. Funds from foreign government agencies such as the United States Agency for International Development (USAID), governments of Norway, Japan, and Luxembourg and the European Union were used to develop and implement other projects. International research institutes, e.g., National Aeronautics and Space Administration (NASA) and various university departments financed most of these research projects. Programs were also sponsored with funds from international agencies and Non-Governmental Organizations (NGOs) such as Global Environment Facility and Conservation International. In addition to funding, these NGOs also act as intermediaries, playing support roles. They initiate, implement, and monitor programs, as well as sensitize, train, and acquaint participating communities with sustainable land-use practices. These activities helped to reduce the environmental conservation know-how deficit among ES providers and induced the rise of local NGOs and civil society organizations. NGOs were sometimes joined in this supporting role by host governments in each program locale, from the national to the local level.

4.3. *Objectives*

Even though all the programs employed various strategies to achieve their objectives, all those objectives revolved around environmental conservation (see supplementary file; Table S2, Column 4). Most telling, however, is the effort to link economic development and environmental conservation objectives. This reflects the reality in developing regions where conservation programs fundamentally disrupt the livelihood activities of ES providers and where conservation programs without poverty alleviation/socioeconomic development objectives are often deemed unattractive [41].

4.4. *Outcomes*

Information on the outcomes of these programs was only available for five programs (see supplementary file; Table S2, Column 5). The available information showed these programs had some positive effects on the environment, local economies, and institutional units within the participating communities.

The programs mainly promoted reforestation, afforestation, and avoided deforestation. About 842,000 tonnes of CO_2 was sequestered in the Woodlot program in Benin and 3.3 tonnes/ha was sequestered in the Participatory Rehabilitation Program in Senegal and Mauritania [31,33]. The tree planting activities contributed to increased vegetation and forest cover and decline in bush burning. The programs also succeeded in raising awareness of environmental degradation and climate change within the participating communities. The research projects identified deficiencies in local land use practices to ensure that participating communities utilize sustainable agricultural practices that reduce their vulnerability to climate change. For instance, the sustainable management program in Mali introduced a novel system of shifting pastoral activities based on the regenerative abilities of pastures.

The Guinean Hotspots program created a database to store and track the number of flora and fauna species present in the Guinean forest.

The programs also had positive socioeconomic effects. In addition to creating a credit scheme for micro-projects, half of the households participating in the participatory rehabilitation program in Senegal and Mauritania improved their income by 12% [33]. The program, along with the Guinean hotspots program and the Woodlot management program in Benin, introduced activities such as beekeeping that provided alternative sources of income and reduced intensive agricultural production activities.

Within the participating communities, the programs empowered local institutions on conservation activities. For instance, the Woody Lot program in Benin and the sustainable management program in Mali pursued a decentralized decision-making process that actively integrated locals in the program agency. The decentralized process brought together several communities under a single institutional framework that ensured the continued adoption and use of the program activities even after payments had stopped. Furthermore, the programs induced the rise of civil society organizations in West Africa dedicated towards environmental conservation. In Sierra-Leone, the Guinea hotspots program gave rise to other smaller conservation initiatives and multiple NGOs [7]. Overall, the inclusive approach helped to reduce the knowledge gap related to environmental issues that characterize institutions and communities in sub-Saharan Africa. This raised awareness of environmental degradation and the threats posed by climate change.

4.5. Challenges

Unsurprisingly, the leading challenge program developers faced was the poor, technical know-how among local partners (see supplementary file; Table S2, Column 6). The poor, technical know-how manifested in terms of little knowledge of sustainable land-use practices and monitoring, reporting and verifying greenhouse gases. This supports the findings of [10] that, while the basic knowledge of forest management activities is not particularly new in sub-Saharan Africa, locals are unfamiliar with activities such as green accounting, establishing baselines for additionality, monitoring, and verifying carbon offsets. The poor technical know-how often slows down technology uptake by program participants [42]. Organizing training seminars to close this knowledge gap translates to higher costs of operating the programs.

In addition, the legal and institutional framework in West Africa constituted another challenge. Both frameworks in West Africa tend to subordinate environmental conservation, and instead promote economic development [43]. This made introducing conservation activities (e.g., avoided deforestation and reforestation) very difficult in areas that previously favored logging and timber companies. For instance, this was noticeable in Ghana where legal and cultural frameworks supported the exploitation of trees (to promote economic growth) which threatened to undermine the Ghana Cocoa Carbon Initiative program. Managing the conflicts over land rights among communities was another institutional challenge. This situation exacerbates when conflicts among multiple stakeholders with divergent interests arise over access to the resources present on those lands.

Finally, the inability of conservation payments to adequately compensate ES providers and cover opportunity costs was another limiting factor. This highlighted the delicate balance between pursuing environmental conservation goals and meeting the socioeconomic interests of the communities.

5. Discussion

So far, we have provided a synopsis of the conservation programs in West Africa, their activities, outcomes and challenges. In addition to lessons from the programs, we highlight lessons from environmental policies enacted by West African governments, as well as policies and programs in other parts of the developing world to discuss incentives that could encourage the conservation of West African ecosystems. We frame these lessons based on identifying and utilizing incentives that are needed for conservation programs to make headway in West Africa.

We consider incentives from three dimensions: incentive mechanisms, demand-side incentives, and supply-side incentives. Incentive mechanisms are policies, programs, and markets that encourage and promote conservation, e.g., PES, carbon emissions trading markets and REDD+ programs. Demand-side incentives identify potential sources of funding for these programs and conditions that motivate investors to finance biodiversity conservation projects in West Africa. Economic, social, and psychological benefits that induce ES providers in West Africa to adopt environmental conservation activities and mindsets are considered as supply-side incentives.

5.1. Incentive Mechanisms

The development and promulgation of the Kyoto protocol engendered various frameworks that support programs developed solely for environmental conservation. These frameworks, such as Clean Development Mechanisms (CDMs) and Voluntary Carbon Standards (VCS), established markets for ecosystem services and biodiversity protection. Markets operating under CDM and VCS frameworks allow developed countries to channel funds to developing countries to financially support carbon sequestration projects that protect forests and watersheds. Carbon stored in these projects is then sold on market exchanges by developed countries to meet their emission obligations. Carbon credits sold in CDM markets allow developed countries to meet their emission reductions pledges under the Kyoto protocol. On the other hand, credits sold in VCS markets involve entities, both private and public, interested in voluntary emissions reductions. Markets for carbon sequestration are well developed and better funded compared to other ecosystem services, especially since sequestered carbon is a public good whose benefits are widespread [44]. While the benefits of carbon sequestration, focusing on tree planting activities, constitutes a medium to attract funding, the ability of soils to store carbon and adapt to climate change is also gaining attention.

The UNFCCC developed the REDD and, later, the REDD+ frameworks to reduce emissions from forest degradation and promote sustainable forest management. Like CDMs and VCMs, REDD+ facilitates the sale of carbon credits in carbon exchanges or markets. The difference is that the requirements for monitoring, reporting, and verifying carbon credits under these frameworks decrease in rigor from CDMs to VCMs and REDD+. Under these frameworks, PES, Payment for Watershed Services (PWS), and carbon forestry programs, among other programs, have been introduced in developing countries.

Several challenges impeding the implementation of conservation programs in Africa have been highlighted by a number of studies [5,10,45]. These challenges are conflicting legal frameworks defining property rights, limited technical and market information, a lack of institutional experience and adequate business models, and a mistrust of markets for public goods. Clearly defined property rights and access to resource use are crucial to establishing conservation programs, and conflicts between state and customary property rights laws in West Africa undermine this concept [46]. In both the private sector and government agencies, knowledge gaps exist on how to execute and administer processes linked to, and policies that promote, conservation: establishing ES baselines, monitoring, reporting, and verifying ESs, and awareness of sustainable land-use activities [10]. The latter challenge becomes noticeable when the rigorous process of complying with CDMs, VCMs, REDD+, and other markets for public goods requirements becomes necessary. Tradeoffs between environmental goals and developmental goals promulgated by governments to combat food insecurity and reduce poverty hinder conservation. In our case study, the GCCI in Ghana exemplifies this tradeoff. Working around this tradeoff is difficult because three-quarters of poor West Africans live in rural areas and depend on ESs provided by the environment to sustain their livelihood [47]. Finally, corruption that disrupts equitable transfer and distribution of payments from programs to participating communities constitutes another challenge [10]. All these challenges significantly raise the costs of implementing conservation programs in West Africa. Hence, potential project investors are reluctant to invest in conservation projects in West Africa.

However, governments in West Africa are making several policy moves to tackle these problems. These policy moves point to several targets outlined in SDG 13 to alleviate the negative impact of climate change. Specifically, governments at different levels are incorporating conservation measures into development plans and strategies. All the countries in West Africa, except Mauritania, have joined the Forestry Partnership Carbon Facility (FPCF) and UN-REDD REDD+ readiness platforms [48]. A third REDD+ platform, the Forest Investment Program (FIP), is investing in forest management projects in Burkina Faso, Ghana, and Ivory Coast. In addition to the Gola REDD+ project from our case studies, about 10 REDD+ demonstration pilot projects are underway in Liberia, Ghana, and Nigeria [5,8].

The increasing number of REDD+ programs is closing the local knowledge gap regarding conservation practices, green accounting, and raising awareness on climate change. Countries are required to articulate, in concrete terms, strategies and plans for tackling forest degradation, protect watersheds, and reduce carbon emissions to qualify for REDD+ programs. These plans and strategies are changing the institutional landscape in West African countries towards conservation. For instance, the evaluation of the REDD+ strategic framework in Nigeria and Ghana found that previous forest laws that favor logging companies were restructured to push forward responsible forest stewardship [48]. The evaluation also found that various institutions nest together to lay the groundwork for REDD+ and that actors at the local or regional level are actively involved or even drive the process. This decentralized decision-making trend is also consistent with observations from our case studies in Mali and Senegal [30,32]. Burkina Faso and Liberia recently reformed their land tenure laws in 2009 to promote the responsible use of natural resources by recognizing and empowering customary laws above state laws [34,49].

Therefore, what incentive mechanisms are better positioned to take advantage of the evolving environmental conservation landscape in West Africa? There are three general incentives that underpin conservation initiatives: direct payments, compensation-based welfare measures and biodiversity protection [50]. Protecting biodiversity hubs and their ESs entails strict regulations that restrict the ability of ES providers to access ESs supporting their livelihood. This issue becomes more pronounced since governments own most lands and forests that support poor ES providers [49]. Hence, regulations guiding natural resource use are difficult to enforce because they are subject to political pressure [51]. In instances where regulations are enforced, ESs providers often exploit open lands, thereby creating leakages. This implies previous destructive production activities being carried out on lands not enrolled in the program.

On the other hand, direct payments and welfare measures, in the West African socioeconomic context, are more attractive avenues to pursue conservation. This is because they compensate ES providers for forgone alternatives and for switching to new production practices. Examples include compensatory programs such as PES, PWS, and Community-Based Forest Management (CBFM) programs. PES, according to [52], transfers payments from ES users to providers with payments conditional on an agreed process for managing natural resources. The institutional arrangement underpinning PES and PWS is flexible enough to leverage funding from CDMs as well as VCMs and REDD+ frameworks. If implemented under the right circumstances, direct payments and welfare-based programs could be beneficial to both the environment and participating communities [53,54]. In the case of PWS, apart from the Sourou wetland valuation project in Burkina Faso financed by Environment and Agriculture Research Centre (EARC) and Economic and Social Policy Centre (ESPC), no other projects have explicitly focused on watershed and wetland protection in West Africa. There are 15 watershed projects that are currently active or in the works in Africa, with one of the projects currently being developed in West Africa [55]. This raises the question of how to fund these programs.

5.2. Demand-Side Incentives

External funding sources sustained all the programs in our case studies. This is consistent with the view that local markets for ecosystem services in Africa are less developed than their counterparts in Latin America or Asia [56]. However, the increasing development of environmental conservation frameworks by West African governments is providing opportunities for countries to exploit several funding channels.

For countries involved in REDD+, three funding channels currently exist. The first channel allocates funds from the Green Climate Fund, the Forest Investment Program, and the Forest Carbon Partnership Facility to governments participating in REDD+ readiness and demonstration projects [57]. Currently, the Acacia carbon projects in Mali and Niger and sustainable energy management program in Burkina Faso benefit from these carbon funds. The REDD+ pilot projects in Senegal, Malawi, and KawiSafu in East Africa started with funds from the green climate fund [58]. In the second channel, developed countries sign bilateral agreements with developing countries to finance and provide technical support to national governments at different levels of readiness for REDD+. In this model, notable NGOs such as the World Wide Fund for Nature and Global Environmental Facility also provide technical support. From our case study, the Gola REDD+ project in Sierra Leone falls under this category. Another example is the bilateral agreement to support the REDD+ strategic framework between the German Federal Ministry for Economic Cooperation and Development and the Togolese Ministry of Environment and Forest Resources, the governments of Norway and Burkina Faso [11,59]. The third channel establishes VCMs that sell carbon credits generated from REDD+ projects. The projects, however, must fulfill carbon offset requirements set by VCS or by certifiers such as Plan Vivo [60]. To illustrate, Plan Vivo currently sponsors 17 operational projects worldwide: six projects were implemented in East and South Africa and two in Burkina Faso (see Table 2 for the projects).

These funding channels were set up by developed countries to cater mostly to carbon sequestration projects in developing countries. This underscores the first target outlined in SDG 17 to promote global partnerships for sustainable development by encouraging developed countries to assist developing countries financially in achieving sustainable development objectives. Still, developing local markets for ESs is imperative, especially true for watershed and biodiversity protection programs, where location matters both in providing ESs and funding the programs [44,50]. In general, funds for watershed protection come from five sources: hydroelectric power suppliers, large industrial users, municipal water suppliers, irrigation water users, and general tax revenues [61]. Thus, the sound financial health of institutions is an essential prerequisite for PWS schemes, a quality lacking in many public and private African institutions [61].

Still, we can learn from narratives in other parts of the world. Most notable is the Quito Water Trust fund model in Latin America. A consortium of water users in Quito, Ecuador, established a water trust fund in the late 1990s [62]. The fund is managed by a local bank, and returns from investments are used to finance watershed protection projects whose activities are managed by an external environmental NGO. This funding model has since inspired similar models in different parts of the world [63]. The Payments for Watershed Services programs in the Uluguru Mountains in Tanzania and in Lake Naivasha Basin Kenya are similar models that already exist in Africa. In the former program, the local Coca-Cola Company and a public water provider created a trust fund and subsequently entered into contractual agreements with upstream farmers to adopt sustainable land use practices that protected the Ruvu river basin in the Uluguru Mountains [64]. In the latter program, a consortium of water companies, horticultural growers, and tourism industry organizations established a fund from which they compensated farmers cultivating land near the Lake Naivasha to adopt ES generating activities [65].

The trust fund arrangement can be replicated in West Africa, with urban dwellers, public and private utility companies, mining companies, and breweries operating in West Africa as primary funding sources. What incentives can induce these companies to sponsor conservation programs? For instance, declining water quality and quantity that threatened their water source motivated the Quito Trust Fund and subsequent models. Therefore, the first incentive can be tied to service provision. If, for instance, water scarcity threatens water supply for utility companies and breweries, a clear incentive exists to invest in watershed protection and biodiversity conservation programs. Tourist companies and NGOs, both internal and external, are potential funding sources for biodiversity conservation projects. Declining flora and fauna species that attract tourists is a potential incentive for financing conservation programs. We also acknowledge that other viable alternatives such as water treatment and filtration plants that efficiently utilize water exists. Watershed protection programs in cities in North and South America were observed to be more cost-effective, and support a larger population compared to building a treatment plant [66,67].

In instances where the services provided by target companies are not threatened, corporate social responsibility (CSR) could be used to incentivize companies to sponsor conservation programs [68,69]. CSR projects promote the image of the implementing companies and generate public goodwill. In this regard, some multinationals are already active. In Nigeria, oil drilling companies SHELL and Chevron sponsor community development programs in the oil-rich Niger-Delta region and are listed in the Nigerian REDD+ framework as partners and potential buyers of carbon offsets [48]. Furthermore, a local brewery built by Heineken in Sierra Leone and SABMiller in Ghana sponsor projects ranging from water recycling to health care and community development initiatives [70]. See [71] for a review of various CSR projects in Nigeria, and [72] for that of CSR projects in Ghana, Liberia, and Guinea.

A common thread among all these companies is their preference for socioeconomic development programs. Preferences for socioeconomic initiatives align with the priorities in West Africa where environmental conservation is secondary to livelihood improvements. This means that investments in environmental conservation only rise when environmental degradation poses a direct threat to the supply of ecosystem services. Therefore, changing attitudes towards conservation is essential, starting with governments emphasizing environmental conservation alongside the socioeconomic development in their development plans. While we are not advocating prioritizing conservation projects over economic development projects, it would be prudent to link conservation efforts to socioeconomic development. Failure to tackle current environmental degradation decreases the stock of ecosystem services available to future West African ES providers.

5.3. *Supply-Side Incentives*

Incentives on the supply side revolve around inducing behavioral changes conducive to environmental conservation. Successful conservation programs require proper organization, resource tenure, consultation with local stakeholders, and the provision of consistent incentives [52]. The two latter points are pertinent in West Africa because the success of all incentives mechanisms is contingent on how it builds up the socioeconomic interests of the people [73].

However, will poor ES providers in West Africa show any interest in conservation programs? Narratives from studies suggest that adequate payment incentives might stimulate conservation behavior among ES providers. For instance, despite differences in Willingness to Accept (WTA) estimates, it was shown in [74] that opportunity costs largely dictated farmers WTA payments for a watershed program near Lake Naivasha in Kenya. Given sufficient incentives, 74% of ES providers in Liberia demonstrated a willingness to adopt responsible land management practices [75]. Furthermore, among those who expressed opposition to conservation, distrust of government authorities was cited as the main reason. Adequate payments has been highlighted as a critical factor that motivates the uptake of conservation programs [76]. These results imply that, even though conservation programs might intrude on their livelihoods, farmers might be willing to participate in conservation programs provided they are sufficiently compensated.

Thus, the design and magnitude of incentive payments are significant. Payments or incentives must cover opportunity and transaction costs of switching to new livelihood activities or participating in conservation programs [77,78]. Inadequate payments undermine environmental conservation goals by discouraging participation in conservation programs or engendering leakages. What incentives should be considered when designing payment incentives that adequately cover the costs of those participating in conservation programs in West Africa?

One incentive is encouraging diversified livelihood strategies. The UN acknowledged diversification as an SDG 8 target to drive economic growth and provide decent employment. Diversification provides alternative income sources, bolsters income flows, and helps to reduce leakages and pressure on the environment [79,80]. However, diversification activities should be designed to align with the needs of its target group [81]. This should come in various forms, for instance, via business start-ups or upgrading activities. In our case studies, beekeeping was the main diversification activity, but others exist. Alternative environmental activities, such as agroforestry, ecotourism, and poultry-keeping, and informal activities, such as woodcarving and tailoring, are viable diversification activities.

While human capital deficiencies can limit diversification [82], skill and knowledge acquisition training and seminars can reduce these deficiencies and build human capital. Apart from encouraging diversification, skill and knowledge acquisition also helps to reduce knowledge on conservation activities in West African institutions. The reduced knowledge gap becomes significant because future conservation initiatives can reduce initial start-up costs of organizing training seminars to build local capacity. The dissemination of knowledge and training underscores the importance of NGOs (or intermediaries) to the environmental conservation process and activities in West Africa and Africa. This further highlights SDG 17 targets: enhance the flow of information and technology on conservation from developed to developing countries to facilitate capacity building.

Another incentive is introducing conservation activities that build on already existing production systems. This is because ES providers might not be receptive to new production systems, especially if they disrupt everyday livelihood activities. In West Africa, the majority of ES providers engage in agricultural production, with agroforestry often promoted as a viable conservation production system [83,84]. Benefits from agroforestry run from payments for carbon sequestered from afforested trees to water conservation, improved soil fertility, and future returns from the sale of trees, among other benefits. Empirical evidence from TIST and ECOTRUST agroforestry schemes in East Africa reveals that farmers earn payments for carbon sequestration as well as indirect benefits such as access to credit, improved farm productivity, and knowledge of sustainable and efficient farming techniques [31,85,86]. Inherent in agroforestry is the synergy between food security and climate change mitigation (also known as 'Climate Smart Agriculture') [87]. This is why SDG 2 strategies to promote sustainable food production and resilient agricultural practices commonly hinge on agroforestry as its cornerstone, and why opportunities to scale up agroforestry activities should be explored.

Building on existing production systems raises the need to work with local communities or regions. An inclusive decision-making process builds trust and identifies the interests of all stakeholders. The inclusion of all stakeholders in a program's decision-making is central to achieving sustainable forest management [6]. In our case studies, project activities were internalized by ES providers when they were involved in the program decision-making process. Sustainable grazing and land use practices introduced by the Guinea Hotspots and Sequestration of carbon in soil organic matter programs in Senegal continued to be used by the communities after both projects were canceled [30]. Both programs were not only built on pre-existing land management practices but also pursued an inclusive decision-making process that recognized the peculiarity of different participating stakeholders and attempted to reconcile these peculiarities under a single framework [30]. In Ghana, local stakeholders were involved at every stage of the decision-making process of the restoration projects. This approach was singled out as the main driver of success of the projects [40]. Conversely, the impacts of the sustainable management program in Mali was slightly diminished when influential groups hijacked decision marking, which created conflicts over resource use and property rights [30]. Although the REDD+ readiness design are initially inclusionary, an exclusionary approach often emerges when programs are implemented [48].

Therefore, in engaging with participating communities, notions of equity and justice should be strongly considered. Studies that examined the dimensions of justice in REDD+ in West Africa advocated for the inclusion of transparency, equity, and legitimacy as specific elements of justice in conservation instruments [9]. Similarly, equity is crucial to planning and implementing effective policies, and assessing social effects engendered by changes in the value of ESs [88]. In concrete terms, the notion of equity starts with an inclusive decision-making process. A transparent and accountable system that ensures equitable distribution of benefits, costs, and risks is equally urgent. Understanding the cultural, historical, and social background of regions in West Africa is key to achieving this objective. For instance, the immense contribution women make to agriculture in West Africa fails to translate to increased decision-making or ownership of production resources; an area dominated by men. Corrective institutional interventions to correct this imbalance such as equal access to resources and skill acquisition for men and women constitutes a credible pathway to climate smart agriculture [89,90]. Pursuing inclusive decision-making and corrective institutional interventions ensure targets outlined in SDG 5 such as the full participation of women in the leadership process and embarking on reforms aimed at facilitating access to economic resources are met.

The notions of equity also extend to the international community. Western impressions of sustainable development and its manifestation in international conservation frameworks such as REDD+ are, in some instances, incompatible with the prevailing social, economic, and institutional settings in Africa [46]. Expectations regarding environmental conservation must align with realities in West Africa. For instance, understanding that environmental projects can disrupt livelihood activities of rural West Africans and further drive them into poverty. How this plays out would depend on country, regional, and local specifics in West African communities [46].

Efforts at conservation in West Africa holds promises and pitfalls. Current trends suggest that environmental conservation is seeping into the national, regional, and local consciousness in West Africa. Most relevant to environmental conservation are incentives that enhance the environmental, socioeconomic, institutional, and cultural interests of West Africans. From advancing measures that alleviate poverty, improve technical know-how, connect partners from diverse backgrounds, build local institutions, tackle climate change, and promote equity, this incentive approach entails empowering West African ES providers, an approach reminiscent of the ideals of the SDGs and that meets several targets articulated in SDGs 1, 3, 5, 8, 13, and 17 to end poverty, end hunger, promote sustainable agriculture, reduce gender inequalities, push for climate action, and build partnerships for sustainable development.

6. Conclusions

This study reviewed on-going and completed forest and watershed (wetlands) projects in West Africa and their corresponding financing mechanism. From the lessons gleaned from the projects, we outline several incentive measures tailored to address region-specific challenges and inform prospective conservation projects in West Africa. We considered incentive from three dimensions: incentive mechanisms representing conservation programs, demand-side incentives representing the interests of ES users and buyers, and supply-side incentives that induce ES providers to internalize conservation attitudes.

We argue that both environmental conservation and economic objectives should proceed together before meaningful environmental conservation can occur. The current movement towards restructuring existing laws and institutions in West Africa to accommodate conservation programs needs to continue. The ability to finance these mechanisms depends on this movement and signals to potential investors the embrace of sustainable land use activities. Still, poverty alleviation objectives must not be completely ignored. This is because many West Africans rely on resources from forest and watershed (wetlands) to support their livelihoods.

Regarding specific incentives, it is crucial to design and implement incentive mechanisms that build on already existing location systems and platforms. This engenders trust and facilitates uptake of program activities and conservation principles. Intermediaries are also needed to reduce knowledge gaps among potential ES providers in the communities and link them to potential buyers/users of ES who will sponsor and fund programs. Articulating equity goals, fair distribution of costs and benefits, and participatory decision-making should feature prominently in these mechanisms.

Put simply, an ideal conservation incentive mechanism in West Africa should adopt a bottom-up, inclusive, fair, and transparent decision-making process, must combine socioeconomic and conservation objectives, and adequately compensate ES providers.

Author Contributions: O.O. collected the data, performed the analysis, and wrote the paper. E.B. conceived and designed the analysis, contributed data and analysis tools, and wrote the paper.

References

1. Bolt, K.; Cranston, G.; Maddox, T.; McCarthy, D.; Vause, J.; Vira, B.; Pearce-Higgins, J. *Biodiversity at the Heart of Accounting for Natural Capital: The Key to Credibility*; Cambridge Conservation International: Cambridge, UK, 2016.
2. Tisdell, C. *Economics, Ecology and the Environment*; Working Paper No. 41, Sustainability: The Economic Bottom Line; The University of Queensland: Brisbane, Australia, 2000.
3. Aglanu, L. Watersheds and Rehabilitations Measures—A Review. *Resour. Environ.* **2014**, *4*, 104–114.
4. Kabii, T. An overview of African wetlands. In *Wetlands, Biodiversity and the Ramsar Convention: the Role of the Convention on Wetlands in the Conservation and Wise Use of Biodiversity*; Hails, A., Ed.; Ramsar Convention Bureau: Gland, Switzerland, 1997. Available online: https://www.oceandocs.org/bitstream/handle/1834/457/Africa_Wetlands_1.pdf?sequence=1&isAllowed=y (accessed on 6 May 2018).
5. Forest Carbon Partnership Facility (FCPF). REDD+ Countries. 2017. Available online: https://www.forestcarbonpartnership.org/redd-countries-1 (accessed on 21 February 2018).
6. Hillstrom, K.; Hillstrom, L.C. *Africa and the Middle East: A Continental Overview of Environmental Issues*; ABC-CLIO: Denver, CO, USA; Oxford, UK, 2003.
7. Carr, J.; Adeleke, A.; Angu Angu, K.; Belle, E.; Burgess, N.; Carrizo, S.; Choimes, A.; Coulthard, N.; Darwall, W.; Foden, W.; et al. *Ecosystem Profile Guinean Forests of West Africa Biodiversity Hotspot*; Critical Ecosystem Partnership Fund: Arlington, VA, USA, 2015.
8. Cerbu, G.; Swallow, B.; Thompson, D. Locating REDD: A global survey and analysis of REDD readiness and demonstration activities. *Environ. Sci. Policy* **2011**, *14*, 168–180. [CrossRef]
9. Isyaku, U.; Arhin, A.; Asiyanbi, A. Framing justice in REDD+ governance: Centring transparency, equity and legitimacy in readiness implementation in West Africa. *Environ. Conserv.* **2017**, 1–9. [CrossRef]
10. Mbow, M.; Skole, D.; Dieng, M.; Justice, C.; Kwesha, D.; Mane, L.; El Gamri, M.; Von Vordzogbe, V.; Virji, H. *Challenges and Prospects for REDD+ in Africa: Desk Review of REDD+ Implementation in Africa*; GLP Report No. 5. GLP-IPO; GLP International Project Office: Copenhagen, Denmark, 2012.
11. Jindal, R.; Swallow, B.; Kerr, J. Forestry-based carbon sequestration projects in Africa: Potential benefits and challenges. *Nat. Resour. Forum* **2008**, *32*, 116–130. [CrossRef]
12. Bond, I.; Grieg-Gran, M.; Wertz-Kanounikoff Hazlewood, P.; Wunder, S.; Angelsen, A. *Incentives to Sustain Forest Ecosystem Services A Review and Lessons for REDD*; Natural Resouce Issues No. 16; International Institute for Environment and Development: London, UK; CIFOR: Bogor, Indonesia; World Resources Institute: Washington, DC, USA, 2009.
13. IIED. Watershed Markets. 2017. Available online: http://www.watershedmarkets.org/casestudies.html (accessed on 22 July 2017).
14. Wangai, P.; Burkhard, B.; Mueller, F. A review of studies on ecosystem services in Africa. *Int. J. Sustain. Built Environ.* **2016**, *5*, 225–245. [CrossRef]
15. Egoh, B.N.; O'Farrell, P.J.; Charef, A.; Gurney, L.J.; Koellner, T.; Abi, H.N.; Egoh, M.; Willemen, L. An African account of ecosystem service provision: Use, threats and policy options for sustainable livelihoods. *Ecosyst. Serv.* **2012**, *2*, 71–81. [CrossRef]
16. Bryman, A.; Bell, E. *Business Research Methods*; Oxford University Press: Oxford, UK, 2007.
17. Booth, A.; Anthea, S.; Diana, P. *Systematic Approaches to a Successful Literature Review*, 2nd ed.; Sage: London, UK, 2016.

18. UMICH. *Africa, Primary Watersheds*; University of Michigan: Ann Arbor, MI, USA, 2017. Available online: http://www-personal.umich.edu/~{}sarhaus/courses/DirectedStudy/astrid/undpseed/afsheds.html (accessed on 2 December 2017).
19. Van Der Wijngaart, R.; Helming, J.; Jacobs, C.; Garzon Delvaux, P.A.; Hoek, S.; Gomez y Paloma, S. *Irrigation and Irrigated Agriculture Potential in the Sahel: The Case of the Niger River Basin: Prospective Review of the Potential and Constraints in a Changing Climate*; EUR 28828 EN; Publications Office of the European Union: Luxembourg, 2019; ISBN 978-92-79-74275-0. [CrossRef]
20. TEEB. The Economics of Ecosystems and Biodiversity for Water and Wetlands. 2013. Available online: http://www.ieep.eu/assets/1107/TEEB_Water_Wetlands_Executive_Summary.pdf (accessed on 6 May 2017).
21. Schuyt, K. Economic consequences of wetland degradation for local populations in Africa. *Ecol. Econ.* **2005**, *53*, 177–190. [CrossRef]
22. UNESCO. Tiwai Island Wildlife Sanctuary. 2017. Available online: http://whc.unesco.org/en/tentativelists/5742/ (accessed on 8 April 2018).
23. Pearce, D.W. The economic value of forest ecosystems. *Ecosyst. Health* **2001**, *7*, 284–296. [CrossRef]
24. Pimentel, D.; McNair, M.; Buck, L.; Pimentel, M.; Kamil, J. The value of forests to world food security. *Hum. Ecol.* **1997**, *25*, 91–120. [CrossRef]
25. Jalloh, A.; Nelson, G.C.; Thomas, T.S.; Zougmoré, R.; Roy-Macauley, H. *West African Agriculture and Climate Change: A Comprehensive Analysis*; IFPRI Research Monograph; International Food Policy Research Institute: Washington, DC, USA, 2013. [CrossRef]
26. Roudier, P.; Sultan, S.; Quirion, P.; Berg, A. The impact of future climate change on West African crop yields: What does the recent literature say? *Glob. Environ. Chang.* **2011**, *21*, 1073–1083. [CrossRef]
27. Hertel, T.W.; Rosch, S.D. Climate Change, Agriculture, and Poverty. *Appl. Econ. Perspect. Policy* **2010**, *32*, 355–385. [CrossRef]
28. Robison, R.; Findlay Brooks, R. *West Africa: The Climate of Change Climate Change Impacts, Awareness and Preparedness across West Africa*; University of Cambridge Programme for Sustainability Leadership: Cambridge, UK, 2010.
29. ACAD. *Carbon Markets and Africa: A Quick Fact Sheet for Journalists*; United Nations Environmental Programme and RISO Centre: Roskilde, Denmark, 2017. Available online: https://www.afdb.org/fileadmin/uploads/afdb/Documents/Generic-Documents/Carbon%20Market%20Quick%20Facts%20%20ACF%202012.pdf (accessed on 17 August 2018).
30. Roncoli, C.; Jost, C.; Perez, C.; Moore, K.; Ballo, A.; Cisse, S.; Ouattara, K. Carbon sequestration from common property resources: Lessons from community-based sustainable pasture management in north-central Mali. *Agric. Syst.* **2007**, *94*, 97–109. [CrossRef]
31. Masiga, M. Payments for Environmental Services in Sub-Saharan Africa: Taking stock and generating evidence for increased investment and development of PES. In *Payment for Environmental Services Laying the Ground Work*; Mogaka, H., Okeyo-Owuor, J., Kipkoech, A., Eds.; ASARECA: Entebbe, Uganda, 2011; pp. 83–105.
32. FAO. *A Review of Carbon Sequestration Projects*; Food and Agriculture Organisation: Rome, Italy, 2004.
33. Kane, N.; Toure, O.; Quiroga, E. *Conservation of Biodiversity through Participatory Rehabilitation of Degraded Land in Arid and Semi-Arid Cross-Border Zones of Mauritania and Senegal Final Project Evaluation*; United Nations Development Programme and Global Environmental Facility: New York, NY, USA, 2010.
34. Plan Vivo. Rehabilitation and Sustainable Management by AGED of Degraded Pastures in the Sahel region of Burkina Faso. 2017. Available online: http://www.planvivo.org/docs/PDD-Rehabilitation-of-Degraded-Pastures-AGED.pdf (accessed on 19 March 2018).
35. Somda, J.; Nianogo, A.J. TEEB Case: Wetland Valuation Changes Policy Perspectives, Burkina Faso. 2010. Available online: TEEBweb.org (accessed on 13 July 2017).
36. Katoomba. *Sweetening the Deal for Shade-Grown Cocoa: A Preliminary Review of Constraints and Feasibility of 'Cocoa Carbon' in Ghana*; The Katoomba Group: Accra, Ghana, 2009.
37. Asante, W.; Anim, E.; Asare, R. *Institutional Innovations In Africa Smallholder Carbon Projects: Case Study Cocoa Carbon Initiative*; CGIAR Research Program on Climate Change, Agriculture and Food Security (CCAFS): Copenhagen, Denmark, 2012.
38. RSBP. The Gola REDD Project. 2013. Available online: https://www.golarainforest.org/Microsoft%20Word%20-%20VCS%20PD_For%20audit%20Dec%202013.pdf (accessed on 25 November 2018).

39. Appiah, M.; Fagg, M.; Pappinen, A. A review of reforestation approaches in Ghana: Sustainability and genuine local participation lessons for implementing REDD+ Activities. *Eur. J. Sci. Res.* **2015**, *131*, 70–99.
40. International Tropical Timber Organisation (ITTO). Assessing the ITTO Guidelines for the Restoration, Management and Rehabilitation of Degraded and Secondary Tropical Forests: Case studies of Ghana, Indonesia and Mexico. 2015. Available online: https://www.itto.int/direct/topics/topics_pdf_download/topics_id=4632&no=1&file_ext=.pdf (accessed on 5 May 2019).
41. Alix-Garcia, J.; Sims, K.; Yanez-Pagans, P. Only one tree from each seed? Environmental Effectiveness and Poverty Alleviation in Programs for Payment for Ecosystem Services. *AEJ Econ. Policy* **2015**, *7*, 1–40. [CrossRef]
42. Sokona, Y.; Denton, F. Climate change impacts: Can Africa cope with the challenges? *Clim. Policy* **2001**, *1*, 117–123. [CrossRef]
43. UNEP. *Environmental Accounting of National Economic Systems: An Analysis of West African Dryland Countries within a Global Context*; United Nations Environment Programme: Nairobi, Kenya, 2012.
44. Alston, L.; Andersson, K.; Smith, S. Payment for Environmental Services: Hypothesis and Evidence. *Annu. Rev. Resour. Econ.* **2013**, *5*, 139–159. [CrossRef] [PubMed]
45. Unruh, J. Carbon sequestration in Africa: The land tenure problem. *Glob. Environ. Chang.* **2008**, *18*, 700–707. [CrossRef]
46. Mantlana, B. *Readying Africa for REDD+*; COP 17; Heinrich Böll Foundation: Cape Town, South Africa, 2011.
47. GEF. GEF's Programmatic Approach to Biodiversity Conservation in West and Central Africa. 2010. Available online: http://www.thegef.org/gef/sites/thegef.org/files/publication/westafrica-BIO.pdf (accessed on 10 February 2017).
48. Asiyanbi, A.; Arhin, A.; Isyaku, U. REDD+ in West Africa: Politics of Design and Implementation in Ghana and Nigeria. *Forests* **2017**, *8*, 78. [CrossRef]
49. RRI. *Who Owns the Land in Africa? Formal Recognition of Community-Based Land Rights in Sub-Saharan Africa*; Rights and Resources Group: Washington, DC, USA, 2015.
50. Kroeger, T.; Casey, F. An assessment of market-based approaches to providing ecosystem services on agricultural lands. *Ecol. Econ.* **2007**, *64*, 321–332. [CrossRef]
51. Wunder, S.; Engel, S.; Pagiola, S. Taking stock: A comparative analysis of payments for environmental services programs in developed and developing countries. *Ecol. Econ.* **2008**, *65*, 834–852. [CrossRef]
52. Wunder, S. When payments for environmental services will work for conservation. *Conserv. Lett.* **2013**, *6*, 230–237. [CrossRef]
53. Hejnowicz, A.; Raffaelli, D.; Rudd, M.; White, P. Evaluating the outcomes of payments for ecosystem services programmes using a capital asset framework. *Ecosyst. Serv.* **2014**, *9*, 83–97. [CrossRef]
54. Pagiola, S.; Arcenas, A.; Platais, G. Can Payments for Environmental Services Help Reduce Poverty? An Exploration of the Issues and the Evidence to Date from Latin America. *World Dev.* **2005**, *33*, 237–253. [CrossRef]
55. Bennett, G.; Carroll, N. Gaining Depth: State of Watershed Investment 2014. 2014. Available online: www.ecosystemmarketplace.com/reports/sowi2014 (accessed on 10 May 2017).
56. Cisneros, J.A. *Forest Carbon Projects in Africa: A Mapping Study*; Background Report for the 'Political Ecologies of Forest Carbon in Africa' Research Project; STEPS Centre: Sussex, UK, 2012.
57. World Bank Group. World Bank Group Climate Finance: 2017. Available online: http://www.worldbank.org/en/topic/climatefinance (accessed on 10 April 2018).
58. FAO. REDD+ Initiatives, Experience and Challenges in Africa. 2016. Available online: http://www.fao.org/3/a-mp513e.pdf (accessed on 25 June 2018).
59. GIZ. *Support for REDD+ Readiness and Rehabilitation of Forests in Togo (ProREDD)*; German Federal Ministry for Economic Cooperation and Development (BMZ): Eschborn, Germany, 2017. Available online: https://www.giz.de/en/worldwide/31415.html (accessed on 8 June 2017).
60. Joseph, S.; Herold, M.; Sunderlin, W.; Verchot, L. REDDC readiness: Early insights on monitoring, reporting and verification systems of project developers. *Environ. Res. Lett.* **2013**, *8*, 1–15. [CrossRef]
61. Ferraro, P.J. Regional review of payments for watershed services: Sub-Saharan Africa. *J. Sustain. For.* **2009**, *28*, 525–550. [CrossRef]
62. Arias, V.; Benitez, S.; Goldman, R. *Water Fund for Catchment Management in Quito, Ecuador*; The Economics of Ecosystems and Biodiversity (TEEB): Geneva, Switzerland, 2010. Available online: TEEBweb.org (accessed

on 13 April 2017).

63. Goldman, R.; Benitez, S.; Calvache, A.; Ramos, A. *Water Funds: Protecting Watersheds for Nature and People*; The Nature Conservancy: Arlington, VA, USA, 2010.
64. Lopa, D.; Mwanyoka, I.; Jambiya, G.; Massoud, T.; Harrison, P.A.; Ellis-Jones, M.; Blomley, T.; Leimona, B.; van Noordwijk, M.; Burgess, N.D. Towards operational payments for water ecosystem services in Tanzania: A case study from the Uluguru Mountains. *Oryx* **2012**, *46*, 34–44. [CrossRef]
65. Chiramba, T.; Mugoi, S.; Martinez, I.; Jones, T. Payment for Environmental Services pilot project in Lake Naivasha Basin—A viable mechanism for watershed services that delivers sustainable natural resource management and improved livelihoods. In Proceedings of the UN-Water International Conference on Water in the Green Economy in Practice: Towards Rio, Zaragoza, Spain, 3–5 October 2011; UNEP: Nairobi, Kenya, 2011; Volume 20.
66. Appleton, A. How New York City Used an Ecosystem Services Strategy Carried out Through an Urban-Rural Partnership to Preserve the Pristine Quality of Its Drinking Water and Save Billions of Dollars and What Lessons It Teaches about Using Ecosystem Services. In Proceedings of the Katoomba V International Conference, Tokyo, Japan, 5–6 November 2002.
67. Hanlan, J. Watershed protection to secure ecosystem services. *Case Stud. Environ.* **2017**, 1–7. [CrossRef]
68. Benjamin, O.; Maduekwe, E.; Punt, M.; Buchenrieder, G. Corporate sustainability and social responsibility of smallholder farmers: Implications for agriculture financing. In *Handbook of Research on Small Business Social Responsibility: Global Perspectives*; Spence, L., Frynas, J., Muthuri, J., Navare, J., Eds.; Edward Elgar: Cheltenham, UK, 2017.
69. Visser, W. Corporate social responsibility in developing countries. In *Oxford Handbook of Corporate Social Responsibility*; Crane, A., Matten, D., McWilliams, A., Moon, J., Siegel, D.S., Eds.; Oxford University Press: Oxford, UK, 2008.
70. Agterhof, G. *Corporate Social Responsibility in Developing Countries: Comparative Analysis of Breweries in Sierra-Leone and Ghana*; University of Groningen: Groningen, The Netherlands, 2014.
71. Amaeshi, K.M.; Adi, A.B.; Ogbechie, C.; Amao, O.O. *Corporate Social Responsibility in Nigeria: Western Mimicry or Indigenous Influences?* ICCSR Research Paper Series No. 39; ICCSR: Coventry, UK, 2006; SSRN 896500.
72. Forstater, M.; Zadek, S.; Guang, Y.; Yu, K.; Xiao Hang, C.; George, M. *Corporate Responsibility in African Development: Insights from an Emerging Dialogue*; The Joan Shorenstein Center: Boston, MA, USA, 2010.
73. Wollenberg, E.; Springate-Baginski, O. *Incentives + How can REDD Improve Well-Being in Forest Communities?* Center for International Forestry Research (CIFOR): Bogor, Indonesia, 2009.
74. Nyongesa, J.; Bett, H.K.; Lagat, J.K.; Ayuya, O.I. Estimating farmers' stated willingness to accept pay for ecosystem services: Case of Lake Naivasha watershed Payment for Ecosystem Services scheme-Kenya. *Ecol. Process.* **2016**, *5*, 1–15. [CrossRef]
75. Innis, P. *Watershed-Based Payment for Ecosystem Services in Liberia: Examining Prospects and Challenges for Implementation in the St. Paul River Basin*; The International Institute for Industrial Environmental Economics: Lund, Sweden, 2015.
76. Kabii, T.; Horwitz, P. A review of landholder motivations and determinants for participation in conservation covenanting programmes. *Environ. Conserv.* **2006**, *33*, 11–20. [CrossRef]
77. Benjamin, O.; Blum, M. Participation of smallholders in agroforestry agri-environmental scheme: A lesson from the rural mount Kenyan region. *J. Dev. Areas.* **2015**, *49*, 127–143. [CrossRef]
78. Kerr, S.; Pfaff, A.; Lipper, L.; Cavatassi, R.; Davis, B.; Sanchez, A.; Hendy, J. Will Buying Tropical Forest Carbon Benefit The Poor? Evidence from Costa Rica. *Land Use Policy* **2004**, *24*, 600–610.
79. Grieg-Gran, M.; Porras, I.; Wunder, S. How Can Market Mechanisms for Forest Environmental Services Help the Poor? Preliminary Lessons from Latin America. *World Dev.* **2005**, *33*, 1511–1527. [CrossRef]
80. Knoke, T.; Calvas, B.; Aguirre, N.; Román-Cuesta, R.M.; Günter, S.; Stimm, B.; Weber, M.; Mosandl, R. Can tropical farmers reconcile subsistence needs with forest conservation. *Front. Ecol. Environ.* **2009**, *7*, 548–554. [CrossRef]
81. Ngugi, R.K.; Nyariki, D.M. Rural livelihoods in the arid and semi-arid environments of Kenya: Sustainable alternatives and challenges. *Agric. Hum. Values* **2005**, *22*, 65–71. [CrossRef]
82. Jouni, M. Livelihoods, vulnerability and adaptation to climate change in Morogoro, Tanzania. *Environ. Sci. Policy* **2008**, *11*, 642–654.
83. Kiptot, E.; Franzel, S.; Degrande, A. Gender, agroforestry and food security in Africa. *Curr. Opin.*

Environ. Sustain. **2014**, *6*, 104–109. [CrossRef]

84. Minang, P.; Duguma, L.; Bernard, F.; Mertz, O.; van Noordwijk, M. Prospects for agroforestry in REDD+ landscapes in Africa. *Curr. Opin. Environ. Sustain.* **2014**, *6*, 78–82. [CrossRef]
85. Benjamin, O. Improving credit allocation to sustainable agriculture in Sub-Saharan Africa: Review of bio-based economy benefits. *Oida Int. J. Sustain. Dev.* **2012**, *4*, 16–24.
86. Benjamin, O.; Punt, M.; Blum, M. The impact of extension and ecosystem services on smallholder's credit constraint. *J. Dev. Areas* **2016**, *50*, 333–350. [CrossRef]
87. Mbow, C.; Van Noordwijk, M.; Luedeling, E.; Neufeldt, H.; Minang, P.; Kowero, G. Agroforestry solutions to address food security and climate change challenges in Africa. *Curr. Opin. Environ. Sustain.* **2014**, *6*, 61–67. [CrossRef]
88. McDermott, M.; Mahanty, S.; Schreckenberg, K. Examining Equity: A multidimensional framework for assessing equity in payment for ecosystem services. *Environ. Sci. Policy* **2013**, *33*, 416–427. [CrossRef]
89. Farnworth, C.; Fones Sundell, M.; Nzioki, A.; Shivutse, V.; Davis, M.; Kristjanson, P.; Rijke, E. *Transforming Gender Relations in Agriculture in Sub-Saharan Africa*; Swedish International Agricultural Network Initiative (SIANI): Stockholm, Swedish, 2013.
90. Nielsen, J.Ø.; Reenberg, A. Cultural barriers to climate change adaptation: A case study from Northern Burkina Faso. *Glob. Environ. Chang.* **2010**, *20*, 142–152. [CrossRef]

6

Assessment of Small-Scale Ecosystem Conservation in the Brazilian Atlantic Forest

Manoela Sacchis Lopes [1], Bijeesh Kozhikkodan Veettil [2,3,*] and Dejanira Luderitz Saldanha [1]

1 Programa de Pós-graduação em Sensoriamento Remoto, Centro Estadual de Pesquisas em Sensoriamento Remoto e Meteorologia, Universidade Federal do Rio Grande do Sul (UFRGS), Porto Alegre 91501-970, Brazil; manoelasm@gmail.com (M.S.L.); dejanira.saldanha@ufrgs.br (D.L.S.)

2 Department for Management of Science and Technology Development, Ton Duc Thang University, Ho Chi Minh City 758307, Vietnam

3 Faculty of Environment and Labour Safety, Ton Duc Thang University, Ho Chi Minh City, Vietnam

* Correspondence: bijeesh.veettil@tdtu.edu.vn

Abstract: The efficiency of the environmental management of a territory largely depends on previous surveys and systematic studies on the main elements and conditions of the physical environment. We applied remote sensing and digital image processing techniques (Principal Component Analysis and supervised classification) to Landsat imagery for analyzing the spatiotemporal land cover changes occurred in the Rio Canoas State Park in Brazil and its surrounding area from 1990 to 2016. Reforested areas around the park with exotic species is a part of the region's economy and a number of industries depend on it for raw materials. However, it is a matter of concern to avoid contamination with such invasive species, due to the proximity of the Park. From 1990 to 2004, more than 95% of the study area was unchanged and showed minimal distinction in land cover over the 14 years. This was mainly due to the continuous presence of agricultural monocultures around the Park without significant increases (only 3.1% of land cover change during this period). Regarding the interior of the Rio Canoas State Park, from 1990 to 2004, there was no increase in the area of exposed soil. The analysis of the surrounding areas of the park from 2004 to 2016 showed that 5663.78 ha (12.2% of the area) of the land cover has been changed, in most areas, due to reforestation by *Pinus* sp. Notable changes occurred within the park (established in 2004) between 2004 and 2016—there was a partial regeneration of natural species diversity, a small number of invasive species (*Pinus* sp.) and removal of agricultural activities within the park, which contributed a 6.6% (75.45 ha) change in its land cover. We verified that 92.51% (1048.40 ha) of the areas inside the park were unchanged. The results demonstrated that actions were conducted to preserve the natural vegetation cover within the park and to reduce the impacts of anthropogenic activities, including the invasion of exotic species from the surrounding reforested areas into the natural habitat of the park. Given this, our study can aid the environmental management of the Park and its surrounding areas, enabling the monitoring of environmental legislation, the creation of a management plan, and can guide new action plans for the present study area and can be applied to other similar regions.

Keywords: Brazilian Atlantic Forest; ecosystem conservation; Principal Component Analysis; Rio Canoas State Park

1. Introduction

Construction of large hydroelectric dams can induce major habitat loss and degradation in the surrounding areas [1,2]. However, deactivated hydropower landscapes have been used worldwide as an area for tourism development, as well as natural conservation units to compensate the lost ecosystems during the construction of hydroelectric reservoirs since the late 19th century [3]. In recent decades, numerous hydropower landscapes around the world became sites for ecosystem conservation, tourism and are defined as protected areas (PAs). Examples for such models for ecosystem conservation can be found in Brazil [4], Canada [5], Costa Rica [6], and many more.

Brazil has the largest PA system in the world (approximately 220 million ha), even though there has been a reduction in the area of PAs since the late 2000s [7]. The Federal Constitution of Brazil expresses in its article 225 the legal and constitutional duty of transmitting the environmental patrimony in the best of conditions to the future and current generations. Likewise, Law 12,651/2012—the Brazilian Forest Code, in its article 1°-A, item I, confirms Brazil's sovereign compromise with the preservation of its forests and other forms of native vegetation, as well as biodiversity, soil, water resources, and integrity of the climatic system. Based on this fact, it is only with preservation practices and environmental control that we will reach a balance, aiming at the reduction of the direct or indirect degradations caused by anthropogenic activities.

The efficient environmental management of a territory largely depends on previous surveys and systematic studies on the main elements and physical conditions. The inappropriate occupation of space and improper use of natural resources can lead to pressure on environmental systems. Among so many laws that guide and foster the environment, we can highlight Law 9985/2000 which institutes the National System of Nature Conservation Units (SNUC). This law establishes the criteria and norms for the creation, implementation, and management of conservation units. From this, we infer that the conservation units (CUs) and their buffer zone are an essential instrument for the protection of biodiversity, natural processes, and environments involved.

Remote sensing, spaceborne data in particular, has been widely used for the dynamic monitoring of land use changes, biodiversity evolution, management of water resources, and the changes on the earth surface in general [8]. Various algorithms for processing remotely sensed data remove the barriers of the human visual system, facilitating effective interpretation of information contained in satellite imagery [9]. In this study, we used satellite data for detecting the land cover changes occurred in a state park located in southern Brazil and its surrounding areas within a 10 km buffer for the period between 1990 and 2016. This study period will allow us to compare the land cover changes occurred before and after the establishment of the park for natural preservation beside a hydroelectric dam in the region. The study area provides an excellent opportunity to understand how the ecosystem loss associated with dam construction can be compensated after the functioning or deactivation of the power plant. We mapped the regeneration of the natural species inside the park and how the establishment of the conservation area reduced the contamination of exotic species into the park while the surrounding areas of the park were cultivated with *Pinus* sp. for industrial raw materials.

2. Characterization of the Study Area

The study site is the Rio Canoas State Park and its surrounding areas, located in the municipality of Campos Novos, in the state Brazilian of Santa Catarina (Figure 1). The Rio Canoas State Park (RCSP) has an area of 1133 hectares and is situated within the Brazilian Atlantic Forest Limit. Its surrounding area, within a buffer zone of 10 km around the park, was also analyzed and encompasses a small portion of the municipalities of Abdon Batista, Anita Garibaldi, and Celso Ramos. The area of the RCSP was acquired by the company Campos Novos Energia S.A–ENERCAN and donated to the state of Santa Catarina to serve as compensation for the environmental loss during the construction of the Campos Novos Hydroelectric Power Plant and the dam.

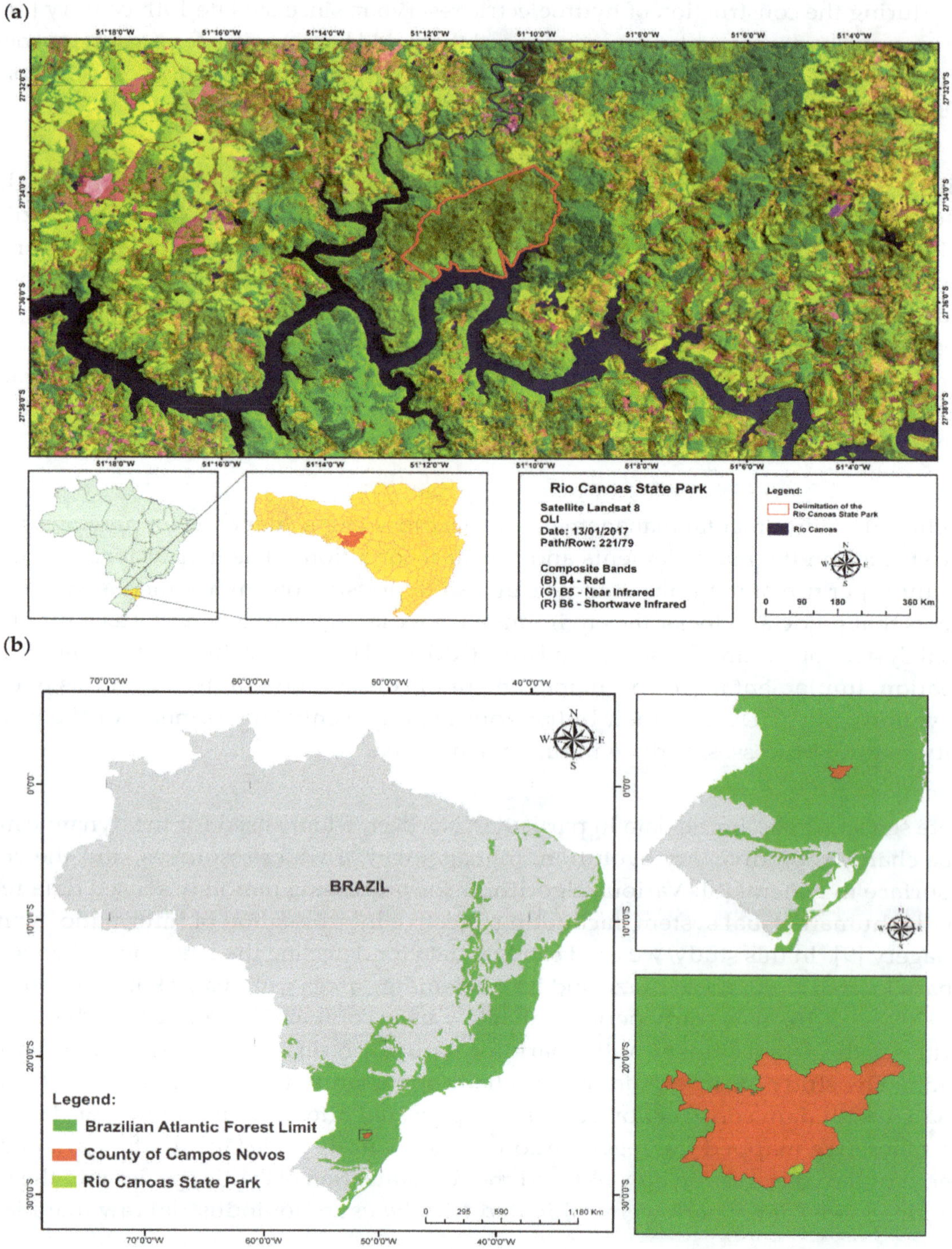

Figure 1. (**a**) Location of the study area in Brazil; (**b**) Location of Rio Canoas State Park (RCSP) within the Brazilian Atlantic Forest Limit.

The study area is characterized by flood basalt, presenting effusive acid rocks on the superior portion and a range of igneous and metamorphic rocks, and the sediments were recently found on the coast [10]. The RCSP is located at the Araucaria Plateau, in the center-west portion of the state. The geomorphology of the study area is, in general, situated on the Geomorphological Unit Rio Iguaçu/Rio Uruguai, characterized by the intense dissection of the plateau along with the main drainage, the Rio Canoas, with large slopes between valleys.

According to the data obtained from the City Hall of Campos Novos, the municipality is known as the breadbasket of Santa Catarina state, with an economy based on the agriculture and is also considered the largest producer of cereals in the state, with the prominent production of corn, soybean, wheat, and beans. The region also focuses on industries, such as cellulose and paper, metallurgy, furniture, and hydropower, which supplies 25% of the state´s consumption. It is worth noting that the raw materials (wood) for the paper and cellulose industries are provided by the pine forests in the surrounding areas of the park. The economy of nearby municipalities (Abdon Batista, Anita Garibaldi, and Celso Ramos) is also based on agriculture, especially corn, bean, soybean, and tobacco crops, and livestock. In Censo Ramos, the cultivation of sugarcane prevails.

According to the phytogeographic data provided by FATMA [11], the original vegetation of this region was represented, primarily, by a Mixed Ombrophilous Forest characterized by the expressive density and physiognomic uniformity of *Araucaria angustifolia* (Brazilian pine), with a sub-grove formed by an expressive number of three species belonging to the Lauraceae family.

3. Datasets

To perform this research, we used satellite imagery, rainfall data and complementary information on the economic and physical characterization of the study area as described below.

3.1. Satellite Data

The list of satellite data is constituted of:

- Three Landsat 8 OLI (Operational Land Imager) images with a 30 m spatial resolution (bands 4, 5, and 6) and band 8 (panchromatic) with spatial resolution of 15 m. Images where acquired on 16 April 2016, 23 September 2016, 9 October 2016 and are available at no cost from the United States Geological Survey (http://www.earthexplorer.usgs.gov).
- Four Landsat 5 TM images with 30 m resolution (bands 3, 4, and 5) and were acquired on 14 July 1990, 16 September 1990, 6 January 1997, 30 March 2004 and 8 October 2004, respectively.
- Global Land Survey (GLS) image of Landsat 5 TM for georeferencing the orbital images of the study area (Source: http://www.dgi.inpe.br/CDSR/).
- WorldView-2 satellite images acquired in 2010, which has a spatial resolution of 0.5 m, provided by FATMA (*Fundação do Meio Ambiente do Estado de Santa Catarina*) and ENERCAN.

3.2. Rainfall Data

The rainfall data near the Campos Novos Hydroelectric Power Plant were acquired from the Hydrological Information System—Hidroweb of the Waters National Agency (http://www.snirh.gov.br/hidroweb/). We used the average of rainfall from 1961 to 1990, historical series of 30 years with monthly information according to the National Meteorology Institute (*Instituto Nacional de Meteorologia*—INMET). Furthermore, we used the total monthly rainfall data obtained from the INMET for the years 1990, 2004 and 2016.

3.3. Complementary Data

In addition to the information mentioned above, we used hydrographic networks of the study area, highway networks, environmental conservation plan in and surrounding area of the dam, the Rio Canoas State Park management plan and information on the agricultural calendar of the main crops cultivated in the state of Santa Catarina, which is provided by the Socioeconomy and Agricultural Planning Center of the State of Santa Catarina (*Empresa de Pesquisa Agropecuária e Extensão Rural de Santa Catarina*–EPAGRI).

3.4. Image Processing System

We used SPRING (*Sistema de Processamento de Informações Georreferenciadas*), an open software package, for processing the images and structuring the databank. We also used ArcGis 10.2 for generating thematic maps and the Microsoft Office Excel 2007 for organizing the graphs and tables, and Garmin™ GPS (Global Positioning System) to collect the field data.

4. Methodology

The methodology adopted to develop this research encompasses various steps as described in the flowchart shown in Figure 2.

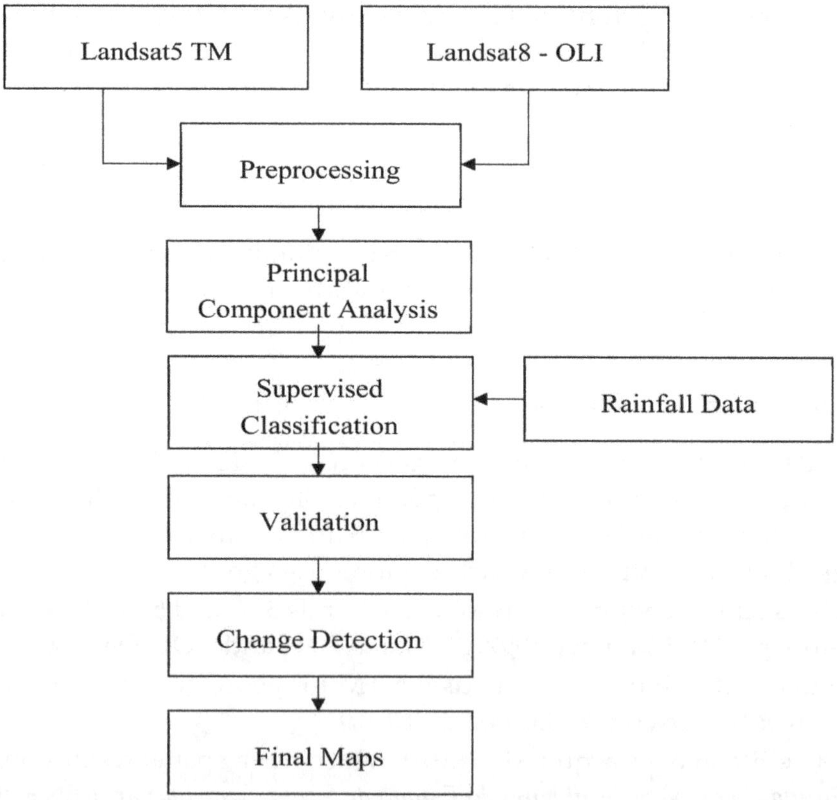

Figure 2. Flowchart of the procedures followed in this study (TM—Thematic Mapper, OLI—Operational Land Imager).

4.1. Image Co-Registration

Image co-registration has been conducted for Landsat *TM* and *OLI* data for eliminating the errors that may occur in the satellite data, due to image rotation, skew and scale. In this case, we proceeded to correct the images, since the study constituted of multitemporal dynamics, in which the images were compared and required to be perfectly coincident in space. Image co-registration is the adjustment of the coordinate system (pixel/lines) of one image to be equivalent to the other image of the same region [9,12]. In this work, we applied pixel resampling method using the nearest neighbor interpolation obtaining the GLS image from the National Institute of Space Research (INPE—*Instituto Nacional de Pesquisas Espaciais*) for choosing the control points. The image co-registration was conducted using SPRING software, in which the first phase was characterized by modifying the extension of the Landsat scenes (path/row: 221/79) for all the years studied with an average of 15 control points, accepting an error of less than one pixel, with a maximum value of 0.55.

4.2. Principal Component Analysis

After the co-registration of images, Principal Component Analysis (PCA) has been conducted with the objective of detecting the land use/land cover (LULC) changes that occurred within the 10 km buffer area surrounding the RCSP. This analysis will allow us to conclude, first, concerning the impact of the construction of the power plant and, second, whether the environmental compensation regarding the Park area was indeed incentivized and implemented properly. Another fact worth mentioning is the inclusion of the buffer zone to analyze is there were any changes in the area surrounding the park for this study to provide subsidy and aid in the future proposal of outlining the buffer zone, given that, until now, it is not present in the management plan or posterior projects. The study of the multitemporal dynamics in 1990, 2004, and 2016 were aided by the Principal Component Analysis to compare the changes before and after the formation of the Campos Novos Hydroelectric Power Plant.

The PCA is a digital image processing technique that uses statistical parameters is considered as efficient in detecting changes in the landscape. In general, the calculation of the principal components of a set of data is conducted by obtaining the eigenvalues and eigenvectors using the correlation matrix or the variance-covariance matrix between the variables of the set [13]. PCA reduces the dimensionality of satellite data that leads to improved data visualization and manageability of data analysis [14,15] and is recognized as one of the best methods for mapping and monitoring interannual and interdecadal vegetation anomalies [16]. The application of PCA, however, depends on the objective of the researcher who can analyze each component according to the work schedule. According to Duarte et al. [17], to identify LULC changes, comparing the differences in the information contained in two or more satellite images is necessary. This identification (of LULC changes) is possible by the application of PCA, which also has the function of determining the extension of the correlation of the image bands and removes them, reducing the dimension of the data and excluding the redundant information that is of no interest (to the user).

Vegetation mapping using automatic methods, such as band rations or vegetation indices are relatively straight forward. However, these methods are not always suitable for differentiating exactly where the changes occurred, but good for what types of changes occurred (native and reforested areas in this case) and PCA has the advantage of mitigating this limitation to some extent by reducing the data dimensionality [14]. Furthermore, the actual nature of changes occurred has been identified by a field visit in this study. It has been observed in a previous study that the overall accuracy and kappa coefficient in vegetation mapping increase when PCA has used along with the main bands of satellite data [18].

The first Principal Component is comprised of the information that is common to all original bands (PC1), the second (PC2) contains the most significant spectral feature of the set. The higher the order of the PCs are, the less significant the spectral features will be. The last principal component has only the information that remained from the set or the noise. We analyzed the correlations of the bands related to the red wavelength (TM3) of the Landsat 5 (1990 and 2004) along with near-infrared (TM4) of the year 2004, generating a new set of images denominated principal components, in a total of three PCs. The use of only three bands is justified by presenting the number of iterations and scenes distinct from when processed by the PCA, resulting in the information of change detection during the period, as confirmed by Lopes [19]. Furthermore, it has been affirmed that the inclusion of the first three PCs corresponds to more than 99.5% of the total variance of satellite data [20].

The same procedure was conducted with the same band combination for comparing the period between 2004 and 2016 after the construction of the RCSP in 2004. The false-color composite image used is comprised of the 2nd (G) and 3rd (R) principal component and the band related to the shortwave infrared (B5) of the original image referent to the most recent year. After applying the PCA, we proceeded to the supervised classification per region based on the Bhattacharyya distance (B). The Bhattacharyya distance can be used as a class separability measure for feature selection [21]. For two normally distributed classes, the Bhattacharyya distance (b) between two classes is defined as:

$$b = \frac{1}{8}(\mu_2 - \mu_1)^{\mathrm{T}}\left[\frac{\Sigma_1 + \Sigma_2}{2}\right]^{-1}(\mu_2 - \mu_1) + \frac{1}{2}\ln\frac{|(\Sigma_1 + \Sigma_2)/2|}{|\Sigma_1|_{1/2}|\Sigma_1|^{1/2}}, \tag{1}$$

where, μi and Σi are the mean vector and covariance matrix of class i, respectively. This algorithm is inbuilt in the SPRING software for image processing.

4.3. Supervised Classification

To proceed with the supervised classification, it is necessary that the user has previous knowledge of the study area. This classification requires the field observation of specific locations shown in the image, from which one can obtain ground-truth data [9,22]. The supervised classification is based on the statistical functions that analyze and compare the characteristics of the spectral reflectance of the pixels associated with a standard class defined by the user. Generally, we calculate the average values, and standard deviations of the defined classes and these values serve as criteria to group the pixels that fulfill the limits close to a specific class [8]. According to Novo [23], the classification process can be distinguished regarding the unit to be grouped. In this work, we used the region-growth algorithm from which we extract homogeneous regions according to the limits tested and established and from the group of contiguous pixels grouped.

The limit of similarity and area used were equal for all images (30). Despite the images being different and from different years, these limits were the most adequate for the identification of LULC. It is worth mentioning that the similarity limit demonstrates the smallest difference accepted between the average value of two pixels (or a set of pixels) and is considered the maximum distance between the spectral centers of two regions. The area limit represents the minimum size of the segment the user wishes to analyze.

In the supervised classification stage, we identified the areas with changes and no changes occurred comparing data acquired between 1990 and 2004, 2004 and 2016, and from 1997 to 2016. Using the false-color composite images from principal components, we observed the areas in which change occurred and those that remained with the same characteristics over the years using the multitemporal Landsat data. We determined the existence of cultivation and reforestation areas with the economic survey and fieldwork studies. However, when analyzing the images, we verified the presence of exposed soil in some regions, with well-formed texture and form. Based on the experience of the user knowledge of the area, allied to the analysis of images from different dates, we perceived that most regions were undergoing an exchange in monocrops or a shallow cutting of exotic species for replanting.

Furthermore, when analyzed the areas in which change occurred or not, we observed an intense increase of the bed of Rio Canoas, due to the creation of the Campos Novos Hydroelectric Power Plant managed by ENERCAN. The power plant is in operation since 2006 and provided approximately $\frac{1}{4}$ of the total energy consumption for the state of Santa Catarina. Therefore, we conducted a supervised classification of the areas in which change occurred on the riverbed before and after the creation of the power plant using a historical raifall data series encompassing 30 years in the Campos Novos station and the monthly total rainfall data for 2016 and 2004 to remove the influence of rainfall on the increase of the riverbed and analyze these differences.

4.4. Fieldwork

The fieldwork was conducted in two stages: From 26 November 2015 to 30 November 2015 and from 3 February 2017 to 4 February 2017 within the RCSP and its surrounding area aiming at a future elaboration of a proposal for buffer zones. The sampling points were selected using the Random Points tool of ArcGis® 10.2 and posteriorly guided according to the need for validating the multitemporal dynamics to identify the land cover at the locations presenting or not changes, as demonstrated on the map elaborated using the PCA. A total of 107 points were analyzed. The tools used in this work are Garmin® Etrex30 GPS, a digital camera (Sony Cyber Shot DSC H300), a clipboard for notes, and maps elaborated for the study area. To identify the areas with the need for change detection in the RCSP, we elaborated a spreadsheet with the features acquired from Landsat OLI images and compared the information with the photographs acquired at the location.

5. Results and Discussion

5.1. Principal Component Analysis

The Principal Component Analysis reduced the redundancy of the information between the spectral bands, which presented very similar behavior. The number of principal components is equal to the number of bands in which each component is associated with a variance of the digital levels, with the first component presenting the highest variance, successively decreasing the values [17,24]. Thus, the application of this technique, manipulated with multiple band association tests, can demonstrate the areas of use dynamic and land cover from 1990 to 2004 and 2004 to 2016.

Similar to the observations made by Ding et al. [20], the first three PCs of each year corresponds to more than 99.5% of the total covariance. For example, for the Landsat TM image in 2004, PC1 presented 68.6% of the total covariance of the set and PC2 and PC3 presented 20.5% and 10.7%, respectively (total 99.9%). Based on this information, we perceived that the second and third components are not correlated. This analysis associated with RGB combination tests between the bands allowed the identification of the changes that occurred in 1990 and 2004.

After the parameter analysis, we proceeded to study the combination of the principal components and the bands on the RGB composition. The composition that demonstrated the areas in which changes occurred or not was the second component regarding the green channel (G), the third component regarding the red channel (R), and band 5 (intermediate infrared–L5–2004) regarding the blue channel (B), as demonstrated in Figure 3a. This composition of bands best met the objectives after numerous tests for detecting changes in the reforestation areas from 1990 to 2004 and from 2004 to 2016.

(a)

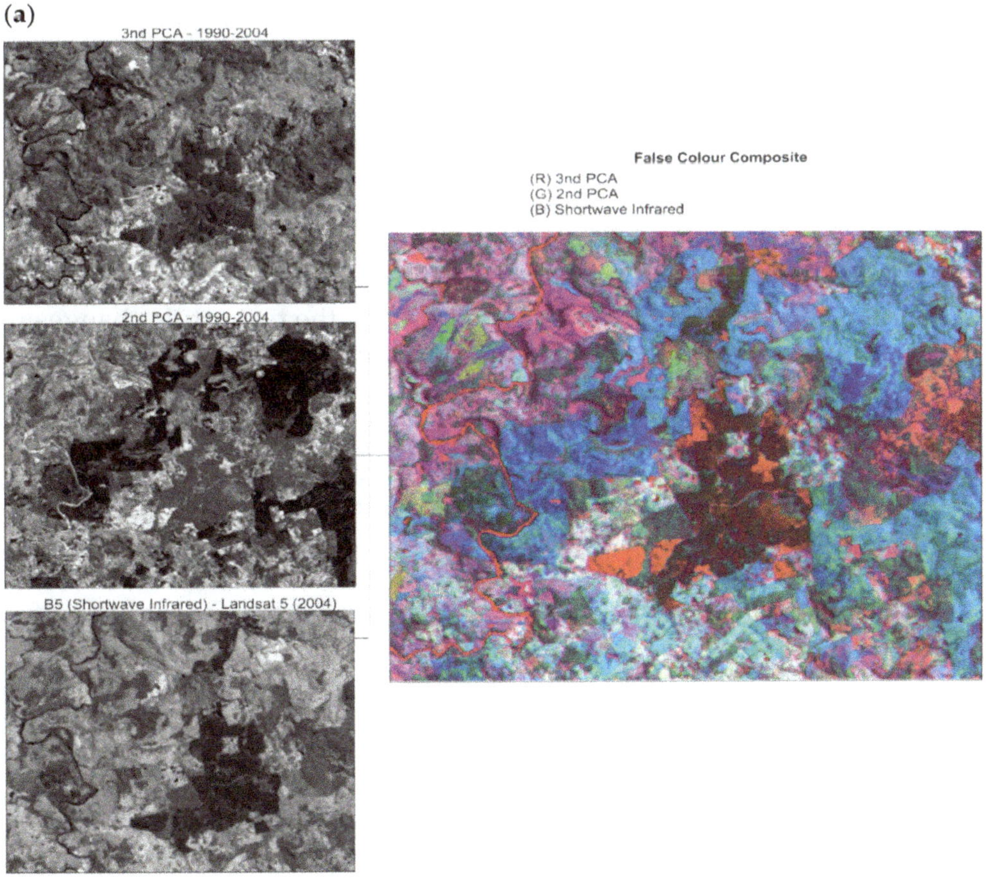

Figure 3. *Cont.*

(b)

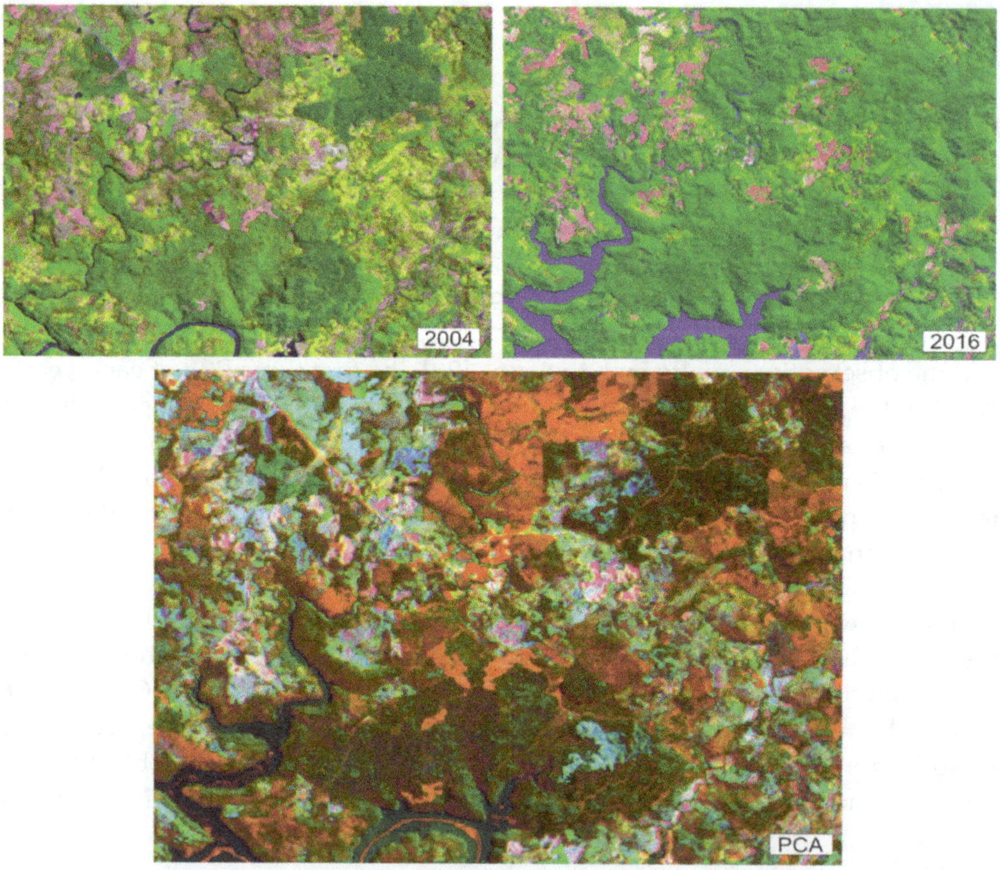

Figure 3. (**a**) False-colored composites using 2nd and 3rd Principal Components and Band 5 of the Landsat TM (2004); (**b**) Change detection in the reforested areas (2004 to 2016).

From Figure 3a, we can analyze that, the areas in dark green as LULC that had already existed in 1990 and continued in 2004. The areas colored in light and dark orange demonstrated the changes that occurred in the last 14 years. The areas in dark blue are the regions that existed in 1990 and showed changes in 2004. This combination indicated very effectively the respective changes/no-changes, as shown in Figure 3b, which established the comparison between the images acquired in 2004 and 2016.

It is worth mentioning that the result of the PCA highlighted the information measured and the changes that occurred between years (of image acquisition). However, if in an image of a specific date the soil was exposed, and, in another, the areas were cultivated, the color composition image will consequently present an area of land use and cover change. At this moment, the analyst must add the ground-truth data from the study area obtained during the fieldwork, images from other dates, which corroborated with the identification of areas with temporary crops and reforestation areas distinct from those used for the classification.

5.2. Supervised Classification

The supervised classification technique was used to elaborate on the LULC change detection map of the RCSP and its surrounding area. In this work, we applied the supervised classification based on the Bhattacharyya distance measurement. This classification divided the homogeneous regions of the images according to the area and similarity limits indicated. As Figure 4a shows, we used an area and similarity limit equal to 30 for both images originated from PCA.

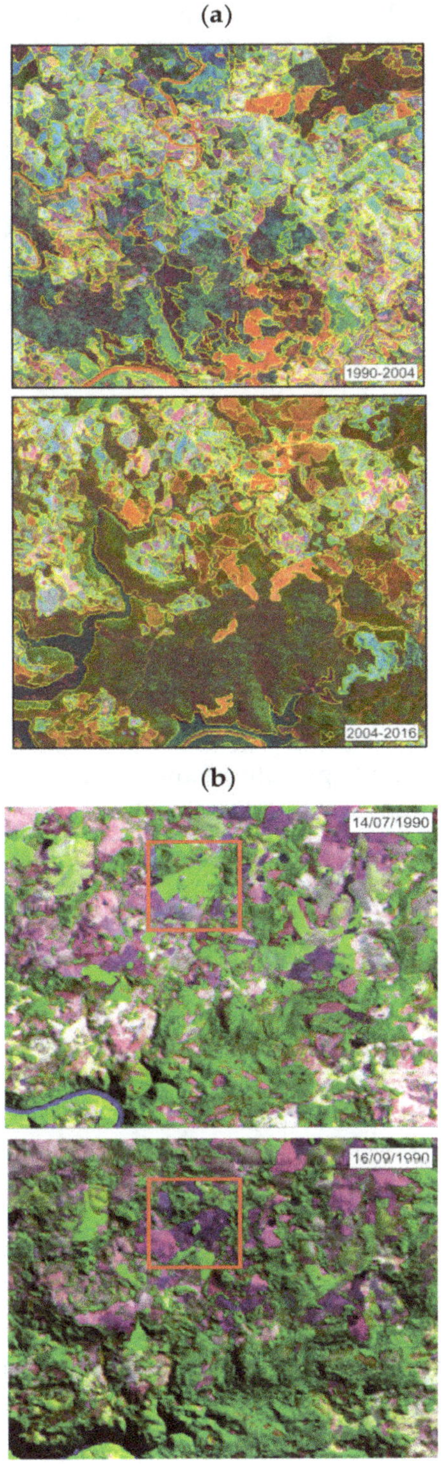

Figure 4. (**a**) Segmentation of the homogeneous areas of the images (limit of the area and similarity equal to 30) (**b**) Exemplification of crops in less than two months. Difference between the exposed soil and crop exchange.

The training samples were indicated after a previous study of the area and its LULC using the high-resolution image (WorldView-2) provided by ENERCAN, images from different dates to obtain a better measure, linear contrast and segmentation techniques, and fieldwork. Moreover, to aid in the classification of the areas that presented changes, but that, for their texture and well-defined form, were similar to temporary crops, we abandoned the use of the agricultural calendar from the Socioeconomic and Agricultural Planning Center (CEPA) from EPAGRI.

The data from the agricultural calendar are presented by micro-regions, encompassing the municipalities in the areas surrounding the Rio Canoas State Park. The study area presented higher representativeness of corn, soybean, tobacco, bean, and wheat crops. According to CEPA, in 2016, 75% of the corn crops and 95% of tobacco were planted in October, while the 85% of the soybean was planted between November and December, and 60% of the bean crop, in December. The winter crops, such as wheat, were planted in June, July, and August. This information associated with the orbital images from different dates was essential concerning the supervised classification because, depending on the date of the image, the cultivation areas could present exposed soil, indicating the times of production rest, areas improper for cultivation or the moment between crops.

Therefore, with the objective of more accurately classifying the agricultural areas and exposed soil, we used the images from different dates to befit the times of production for most crops in the region. We discriminated five categories established on the theme caption, change and no-change, monocrops, water bodies, and changes in water bodies. The monocrop class encompasses the areas with agriculture crops, reforestation, and pasture, since they often present a similar spectral behavior and distinction, not as an object of this work, but presenting or not changes in the land cover. The training and test samples were acquired in an average of 100 for the classification of the false-color composites of the principal components associated with the original band of the intermediate infrared. The PCA allowed the identification of changes and no-changes in the same image, also reducing the classification time of the area, and highlighting the distinct theme classes. After classifying the images, we conducted the post-classification procedures and matrix edition of the classes presented as changed areas with exposed soil, but that, according to many studies and analyses, we concluded that these areas were recently harvested or planted agriculture areas with no plant growth, as shown in Figure 4b.

After the classification of the image, validation has been performed visually and mathematically using the control points collected during the two field trips. More than 100 random points distributed in the image were analyzed using ground truth data from the study area with the aid of GPS. When considering all land cover classes defined, the overall accuracy index was 0.93 and the Kappa index was 0.88. This value of Kappa is associated with the quality of the classification is considered as good according to Landis and Koch [25].

5.2.1. Change Detection within the Rio Canoas State Park

The Rio Canoas State Park is a conservation unit of a mixed ombrophilous forest or araucaria forest with approximately 1200 hectares. Because of this, previous to 2004, the RCSP had not yet been created and, consequently, preservation was not required. The term Conservation Unit (CU) is defined by the MMA [26] as the territorial space and its environmental resources, including the jurisdictional waters, with relevant natural characteristics, legally instituted by the Public Power enterprise, with objectives of conservation and defined limits, under special administration regime, to which adequate protection guarantees are applied.

According to the Law nº 9985 of 18th July 2000, the conservation units integrating the National Nature Conservation Units System are divided into two groups with specific characteristics: The integral protection units and the sustainable use units. The objective of the integral protection units is to preserve nature and admits only the indirect use of its natural resources, except in certain legal cases. The general objective of the sustainable use units is to harmonize environmental conservation with the sustainable use of a portion of its natural resources.

Among the integral protection groups, the units considered are the Ecological Station, Biological Reserve, National Park, Natural Monument, and Wild Life Refuge. The National Park, the object of this work, has a specific objective of the preservation of natural ecosystems of high ecological relevance and scenic beauty, allowing the performance of scientific researches and development of environmental education activities, recreation in contact with nature, and ecological tourism. Based on this, we classified the images resultant from the PCA, comparing the periods between 1990 and 2004 and between 2004 and 2016, as presented in Figure 5a,b.

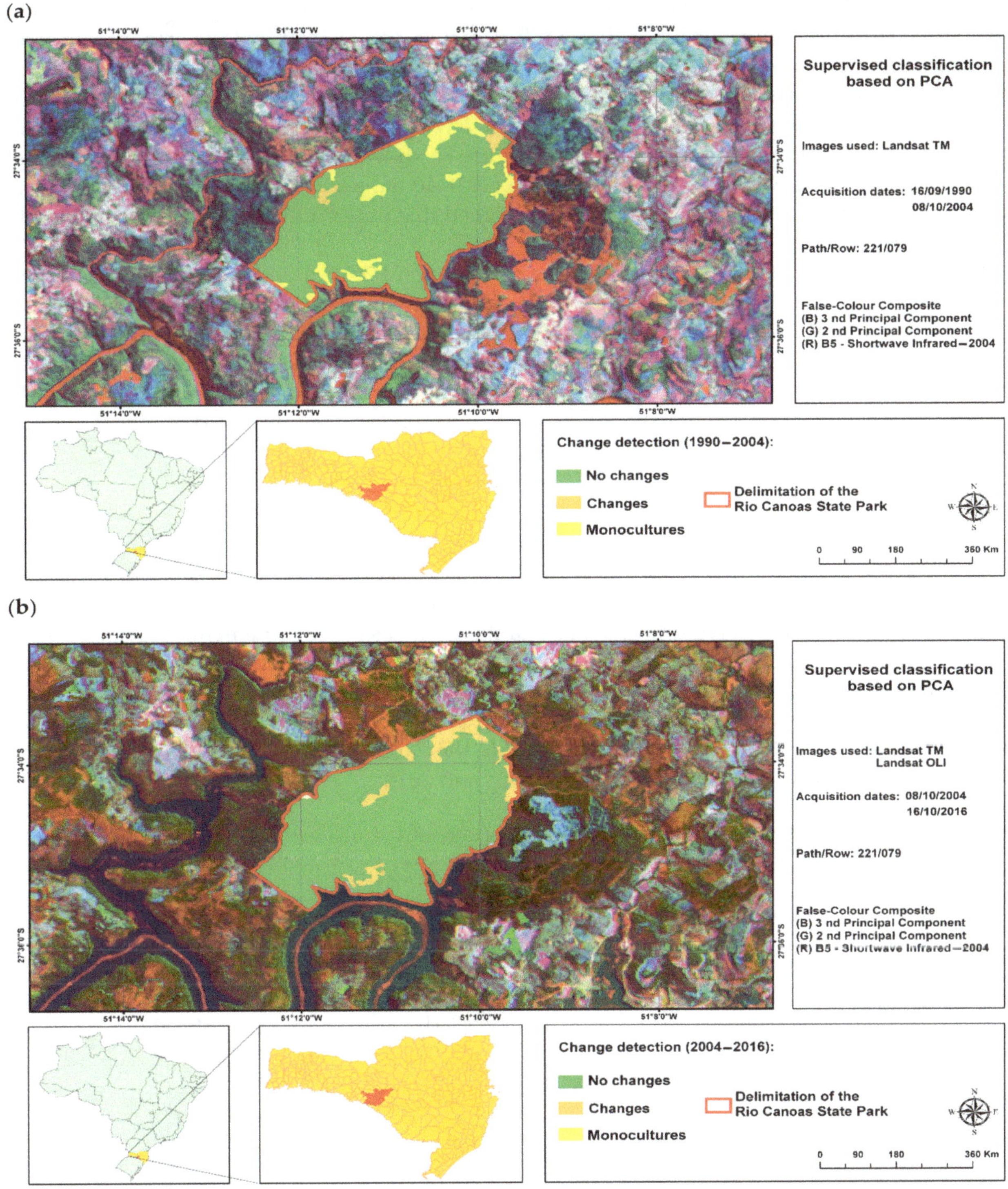

Figure 5. (**a**) Change detection map for Rio Canoas State Park (1990–2004); (**b**) Change detection map for Rio Canoas State Park (2004–2016).

Before the creation of the RCSP up until a few months after the Decree was approved, we verified that there were crops and plantations of exotic species within the park in areas significant to the conservation unit. However, with the consolidation of the park and over time, its interior was modified.

In Figure 4b, we observed the presence of monocrop class which encompasses areas of agricultural and silvicultural activities, such as soybean, corn, tobacco, and exotic species of the Pinus and Eucalyptus genre. In the change detection map referent to the comparison from 1990 to 2004, we verified that the areas with cultivations within the RCSP remained from 1990 until the date in which the image from 2004 (10 August 2004) was taken. These areas presented changes when comparing the images using the PCA. However, by studying other images, we verified that the area was undergoing crop exchange. Associated to this, the class measures from the SPRING software showed that 89.56% (1014.94 ha) were unchanged areas and 10.44% (118.32 ha) were areas with monocrops. Thus, we verified that, during the period from 1990 to 2004, there were no increments to the area of exposed soil, but changes of agricultural and forestry crops. Figure 5a shows an obvious distinction between the years of 2004 and 2016. We observed that 92.51% (1048.40 ha) are unchanged areas, 6.6% (75.45 ha) demonstrated changes, and 0.7% (7.92 ha) is the riverbed that entered the Park with the creation of the Campos Novos Hydroelectric Power Plant.

The multitemporal dynamic of the land cover of 2004 and 2016, aided with the fieldwork, images from other dates, and the experience of the managers of the conservation unit indicated that, over the years, the Park aimed for changing most monocrops in significant areas to regenerate other forestry species, which ratifies the objective of the conservation unit. As demonstrated in Figure 5b, we verified that two of the areas which the image-product classification of the PCA indicated as changed areas were real. These were areas previously occupied by Pinus species and removed to regenerate species native of the region.

5.2.2. Change Detection in the Surrounding Areas of RCSP

We analyzed the 10 km buffer area surrounding the RCSP to verify the vegetation coverage in the area after the formation of the Campos Novos Hydroelectric Power Plant. This verification allowed us to raise conclusion, first, on the impact the construction of the power plant and, second, if the environmental compensation regarding the PAs of the Park has been implemented correctly. Another fact concerns the inclusion of a Buffer Zone to analyze whether many changes occurred in the area surrounding the Park for this study to provide a subsidy and aid in the posterior proposal of a Buffer Zone, since it briefly mentioned in the management plan. According to the data presented in Table 1 and the same steps are taken to detect changes within the Park, the areas presenting the most changes in land cover were those of the images taken in 2004 and 2016.

Table 1. Classification table for change detection for the area surrounding the RCSP.

Changes from 1990 to 2004		
Classes	**Area (ha)**	**Area (%)**
Change	1436.56	3.10%
No-change	18,477.92	39.81
Monocrops No-change	25,738.98	55.46%
Rio Canoas	760.34	1.64%
Total	46,413.8 ha	100%
Changes from 2004 to 2016		
Classes	**Area (ha)**	**Area (%)**
Change	5663.78	12.20%
No-change	21,862.20	47.10%
Monocrops No-change	16,151.54	34.80%
Rio Canoas	796.04	1.72%
Increase of the Rio Canoas	1939.86	4.18%
Total	46,413.8 ha	100%
2016–Global Accuracy: 0.93		
2016–Kappa: 0.88		

During the period from 2004 to 2016, the land cover changes in 5663. 78 ha (12.20% of the area), mostly by reforestation of Pinus species (Figure 6b). The surrounding area of the park is characterized by cellulose and paper industries, such as *Iguaçu Celulose S.A*, and wood industries. The reforestation areas of exotic species surrounding the RCSP is old and a part of the region´s economy (Figure 6c,d).

However, this is a concerning subject for avoiding contamination with invasive species, due to the proximity to the park.

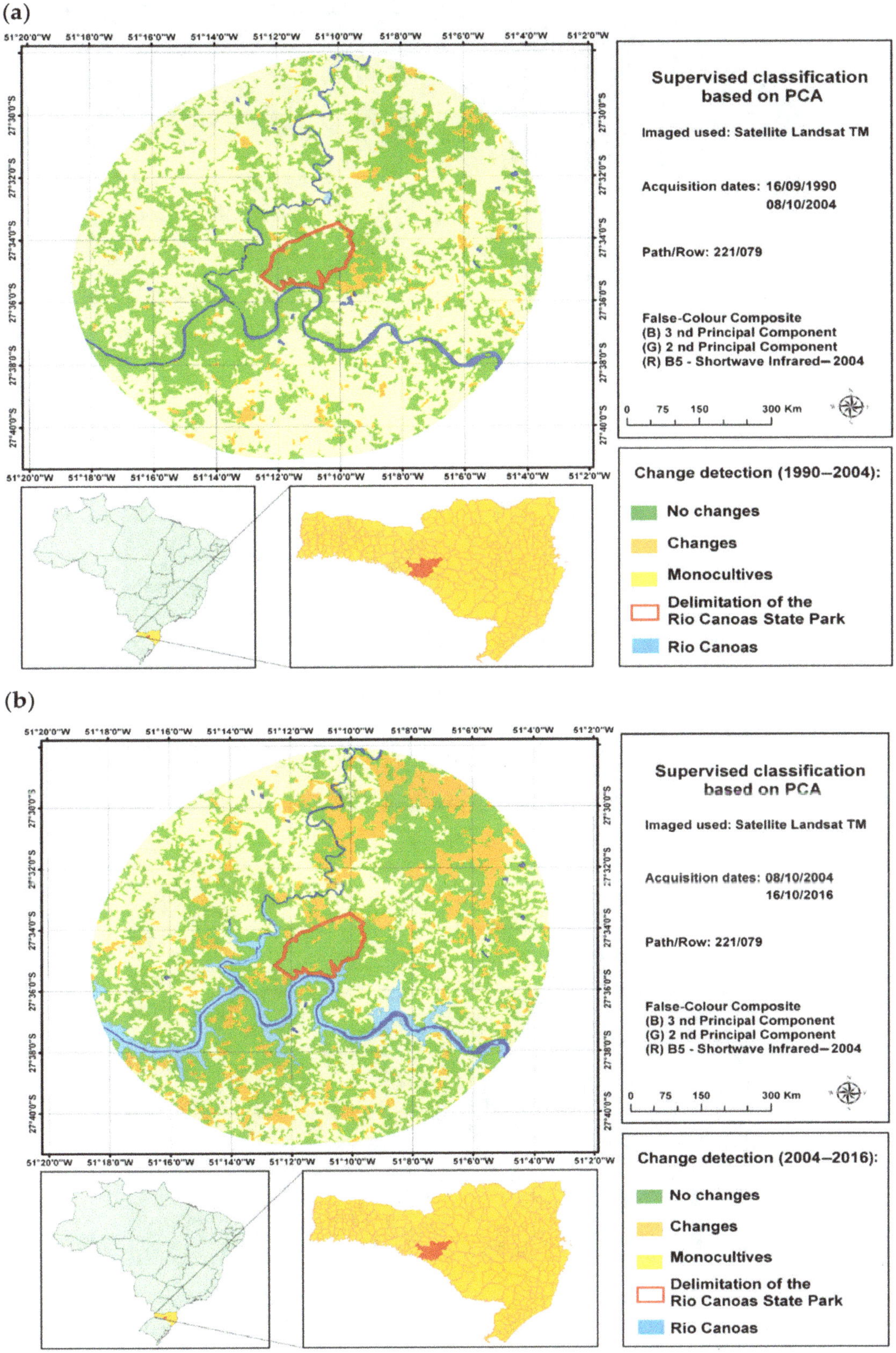

Figure 6. *Cont.*

(c)

(d)

(e)

Figure 6. (**a**) Change detection map for the period from 1990 to 2004; (**b**) Change detection map for the period from 2004 to 2016; (**c**) Worldview-2 image acquired in 2010 showing the areas of regenerated vegetation (Courtesy: ENERCAN); (**d**) Photograph of native species *Araucaria angustifolia* with the plantations of *Pinus eliotti*; (**e**) Photograph of various native species and *Pinus eliotti*.

In 1990 and 2004, the areas that remained unchanged were of approximately 95% of the study area and presented no changes in land cover during the 14 years (Figure 6a). This occurs mainly due to the continuous presence of agricultural monocrops in both summer and winter, with no great increments. Regarding the 12 years from 2004 to 2016, the changes were more recurrent. This occurs mostly due to the shallow cutting of the Pinus plantation areas, which differ from the crops, such as soybean, tobacco, and corn, and changes the entire landscape. The cutting of the exotic reforestation species take years and is performed in rotations that ranges from 5 to 8 years using shallow or partial cuts. Therefore, in these cases, there is change and, contrary to the soybean crops, for example, which can be replanted as soon as the wheat is harvested and germinates days later, the plantation of new seedlings takes time.

5.2.3. Change Detection in the Bed of the Rio Canoas

The Campos Novos Hydroelectric Power Plant was implanted on the Rio Canoas, approximately 20 km upstream of its confluence with Rio Pelotas at the border of the municipalities of Campos Novos and Celso Ramos, according to the data provided by ENERCAN [27]. According to ENERCAN, before finishing the construction at the margins of the Rio Canoas, the dam would be formed by the flooding of the peripheral areas to the Rio Canoas, characterizing a substantial extension of flooded land.

The PCA has demonstrated the transformations as pursuant to the study of the RCSP associated with the change detection. Because the data on the flooding of the marginal areas provided by the ENERCAN Company were provided before concluding the construction in 2004, we proceeded with the analyses of the area. Therefore, the data were classified in two images with specific dates and related to the rainfall data, derived from Landsat TM (acquired on 14 July 1990) and Landsat OLI (acquired on 16 April 2016).

To obtain a better accuracy of the image classification, we acquired the rainfall data from the Hydrological Information System—Hidroweb of the National Waters Agency (Figure 7). We used the average rainfall from 1961 to 1990, historical series of 30 years, with monthly information according to the National Meteorology Institute (INMET). To classify the images, they were selected according to the dates in which there was monthly accumulated rainfall nearly the same as the average for the 30 years analyzed to exclude the influences of rain in the demarcation of the riverbed before and after flooding the area.

As presented in Figure 7, the areas in light pink identify the dates with rainfall accumulations that could influence the increase of the riverbed. We used the images according to the availability and the value nearest to the average of rain accumulated in 30 years. The image acquired on 6 January 1997 demonstrates that the amount of rain accumulated was within the average and, therefore, there were not many influences of rain. Regarding the image acquired on 16 April 2016, when considering rainfall data in October, we verified a significant increase of accumulated rainfall, but we disregarded the daily averages of accumulated rainfall for October of 2016. With this detailed information from INMET, we verified that, until the 16th of October, the rainfall accumulation was of 78 mm, also having little effect in the area. According to the generated map (Figure 8), an area of 6.48 km^2 has been indicated as an approximate measure of the riverbed on 6 January 1997 and of 30.15 km^2 on 16 April 2016 with the creation of the Campos Novos Hydroelectric Power Plant. The objective of the study was to analyze the size of the area surrounding the RCSP that was flooded, due to the high difference made explicit in the change detection using the PCA technique.

Figure 7. Rainfall accumulated at the time of satellite data acquisition compared to the average of 30 years.

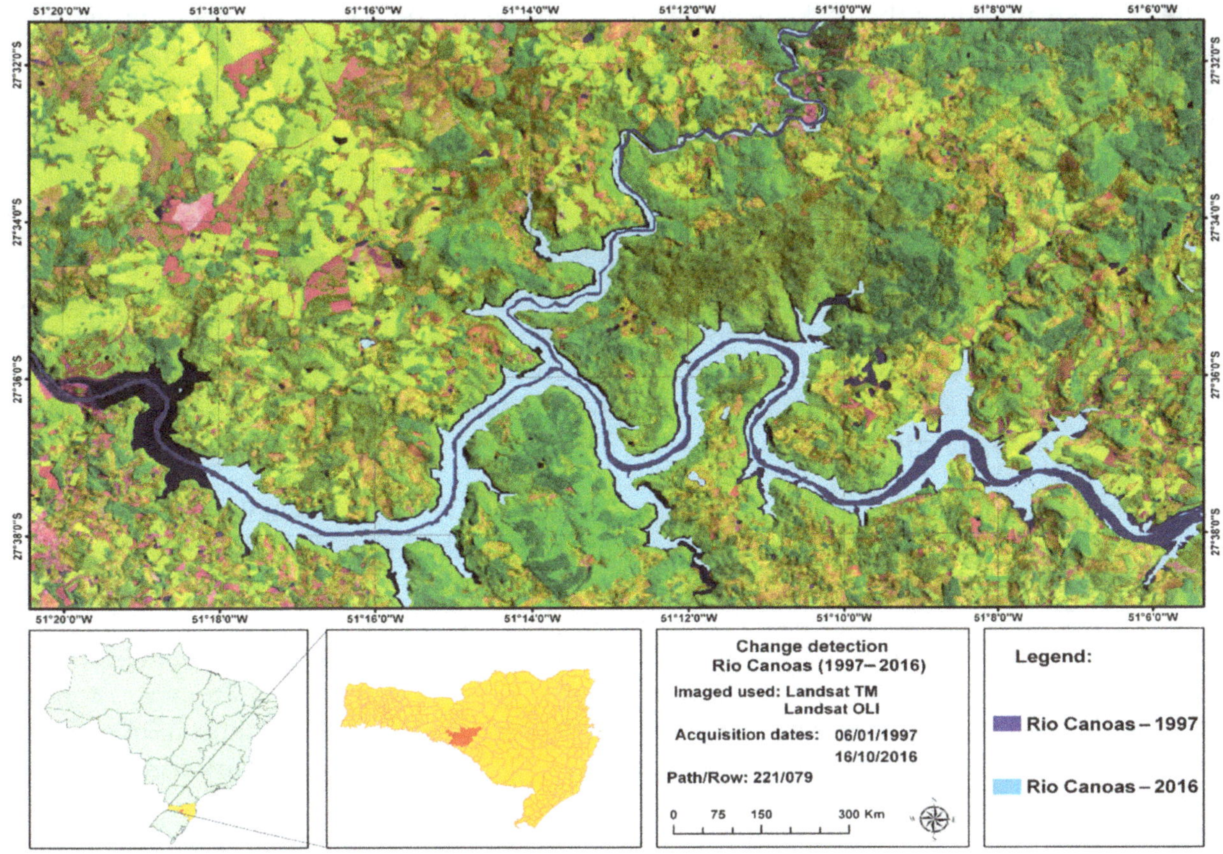

Figure 8. Map of the Rio Canoas in 1997 before the implementation of the Hydroelectric Plant of Rio Canoas, and in 2016.

6. Conclusions

The Rio Canoas State Park is the largest remaining portion of the Mixed Ombrophilous Forest (Araucaria Forest) surrounding the Campos Novos Hydroelectric Power Plant dam, in the southern region of Brazil. The forest ecosystem belongs to the Atlantic Forest Domain of South America, which is under the threat of extinction in the Santa Catarina state of Brazil.

Based on the work conducted, we conclude that the Principal Component Analysis associated with specific bands after conducting a field study is effective in detecting LULC changes when comparing two or more dates, particularly in understanding exactly where the changes occurred rather than what changes occurred.

The analysis conducted in the area surrounding the Rio Canoas State Park, which is established in 2004, for the period between 1990 and 2004 demonstrated that more than 95% of the park and the surrounding areas remained intact. The area, where the park is situated had no land cover change during this period and 3.1% of the surrounding areas showed changes, due to agriculture and silviculture practices. This change was minimal due to the continuous presence of agricultural monocrops in both summer and winter in the area surrounding the Park.

For the period between 2004 and 2016, it is observed that about 5663.78 ha (12.20% of the area) of the land cover surrounding the park changed, due to the reforestation of Pinus species. Within the park, we verified that 92.51% (1048.40 ha) are no-change areas, 6.6% (75.45 ha) presented changes (this includes regeneration of native species, land cover changes, due to the abandoning of agriculture within the park and invasive species), and the rest (0.89%) is the riverbed that entered the Park after the creation of the Campos Novos Hydroelectric Power Plant. This data indicates actions regarding the preservation of the vegetation cover within the Park to reduce the impacts of anthropic activities.

The areas surrounding the Park are characterized by cellulose, paper and wood industries. The reforestation areas with exotic species (e.g., *Pinus elliottii, P. taeda*) surrounding the Park are old and a part of the region´s economy. From an ecological point of view, it is important to avoid contamination of natural forest in the protected areas by such invasive species (*Pinus* sp. in particular), due to the proximity to the Park.

Furthermore, we demonstrated an increase in the Rio Canoas bed with the implementation of the Campos Novos Power Plant and concluded that there was an increase of 23.67 km^2 in 2016 when compared to 1997. This area advanced the initial limits of the Rio Canoas State Park, changed the landscape, caused the translocation of people, and flooding of the areas previously covered by ciliary forests. The results of this study can be used in the environmental management of similar kinds of parks and surrounding areas, allowing the monitoring of the environmental legislation and management plan, and guide plans of action.

In a nutshell, it was concluded that after the creation of the Rio Canoas State Park, some modifications were made to its LULC, such as the clearing of larger areas with Pinus plantations, in order to facilitate the regeneration of native species and to create projects to control the expansion of exotic species. The large-scale plantations of exotic species exist prior to the creation of the park and these areas contribute to the economy of the region. However, it is known that reforestation with these invasive species, such as *Pinus eliotti* and *P. taeda*, depending on how the plantations are conducted, may lead to the decrease in some of the native species and this may alter the natural diversity of the ecosystem. After the creation of the park, expansion of exotic species into the park has been reduced, due to the control measurements taken by the authorities. These factors project the importance of a time-series monitoring of the forest fragments of mixed ombrophilous forests belonging to the core zone of the Brazilian Atlantic Forest Biosphere Reserve.

Author Contributions: All authors contributed equally to the text; M.S.L. and D.L.S. conducted field studies and B.K.V. prepared the figures.

References

1. Palmeirim, A.F.; Peres, C.A. Rosas FCW Giant otter population responses to habitat expansion and degradation induced by a mega hydroelectric dam. *Biol. Conserv.* **2014**, *174*, 30–38. [CrossRef]
2. Lees, A.C.; Peres, C.A.; Fearnside, P.M.; Schneider, M.; Zuanon, J.A.S. Hydropower and the future of Amazonian biodiversity. *Biodivers. Conserv.* **2016**, *25*, 451–466. [CrossRef]
3. Rodriguez, J.F. Hydropower landscapes and tourism development in the Pyrenees. *J. Alp. Res.* **2012**. [CrossRef]
4. Rylands, A.B.; Brandon, K. Brazilian Protected Areas. *Conserv. Biol.* **2005**, *19*, 612–618. [CrossRef]
5. Sherren, K.; Beckley, T.M.; Parkins, J.R.; Stedman, R.C.; Kelity, K.; Morin, I. Learning (or living) to love the landscapes of hydroelectricity in Canada: Eliciting local perspectives on the Mactaquac Dam via headpond boat tours. *Energy Res. Soc. Sci.* **2016**, *14*, 102–110. [CrossRef]
6. Bernard, F.; de Groot, R.S.; Campos, J.J. Valuation of tropical forest services and mechanisms to finance their conservation and sustainable use: A case study of Tapantí National Park, Costa Rica. *For. Policy Econ.* **2009**, *11*, 174–183. [CrossRef]
7. Bernardi, D.; Disperati, A.A.; Santos, J.R.; Mendes, F.S. Monitoramento da dinâmica de paisagem através da análise por componentes principais (ACP) em imagens Landsat 5. *Simpósio Bras. Sens. Remoto* **2001**, *10*, 557–560.
8. Liu, W.T.H. *Aplicações de Sensoriamento Remoto*; Uniderp: Mato Grosso do Sul, Brazil, 2006; 881p.
9. Crósta, A.P. *Processamento Digital de Imagens de Sensoriamento Remoto*; UNICAMP: Campinas, Brazil, 1992; 170p.
10. Duzzioni, R.I. *Elaboração dos Planos de Manejo do Parque Nacional das Araucárias e da Estação Ecológica da Mata Preta*; Icmbio: Brasília, Brazil, 2005; 150p.

11. Fundação do meio ambiente de Santa Catarina (FATMA). Parque Estad ual das Araucárias–Plano de Manejo/Diagnósticos. Florianópolis-Santa Catarina. 2007. Available online: http://www.fatma.sc.gov.br/upload/ucs/araucarias/Plano_Manejo_PAEAR_Fase_II_final.pdf (accessed on 18 September 2018).
12. Vittek, M.; Brink, A.; Donnay, F.; Simonetti, D.; Desclée, B. Land Cover Change Monitoring Using Landsat MSS/TM Satellite Image Data over West Africa between 1975 and 1990. *Remote Sens.* **2014**, *6*, 658–676. [CrossRef]
13. Coutinho, M.D.L.; Brito, I.B. Análise de componentes principais com dados pluviométricos no estado do Ceará. In Proceedings of the Congresso Amazônia e o clima global, Belém-Pará, Brazil, 13 April 2010; Volume 1.
14. Munyati, C. Use of Principal Component Analysis (PCA) of remote sensing images in wetland change detection on the Kafue Flats, Zambia. *Geocarto Int.* **2004**, *19*, 11–22. [CrossRef]
15. Lattin, J.M.; Carrol, J.D.; Green, P.E. *Analyzing Multivariate Data*; Thomson Brooks/Cole: Singapore, 2004; p. 556.
16. Lasaponara, R. On the use of Principal Component Analysis (PCA) for evaluating interannual vegetation anomalies from SPOT/VEGETATION NDVI temporal series. *Ecol. Model.* **2006**, *194*, 429–434. [CrossRef]
17. Duarte, C.C.; Souza, S.F.; Galvíncio, J.D.; Melo, I.D.F. *Detecção de Mudanças na Cobertura Vegetal da Bacia Hidrográfica do Rio Tapacurá-PE Através da Análise por Componentes Principais*; Simpósio Brasileiro de Sensoriamento Remoto: São José dos Campos, Brazil, 2009; pp. 5765–5772.
18. Khedmatgozar, M.; EslamBonyad, A. Use of Principal Component Analysis in accuracy of classification maps (case study: North of Iran). *Res. J. For.* **2016**, *10*, 23–29. [CrossRef]
19. Lopes, M.S. Análise de vulnerabilidade natural à erosão como subsídio ao planejamento ambiental do oeste da bacia hidrográfica do Camaquã–RS. *Rev. Bras. Cartogr.* **2016**, *68*, 1689–1708.
20. Ding, J.L.; Wu, M.C.; Tiyip, T. Study on soil salinization information in arid region using remote sensing technique. *Agric. Sci. China* **2011**, *10*, 404–411. [CrossRef]
21. Choi, E.; Lee, C. Feature extraction based on the Bhattacharyya distance. *Pattern Recognit.* **2003**, *36*, 1703–1709. [CrossRef]
22. Guerschman, J.P.; Paruelo, J.M.; Bella, C.D.; Giallorenzi, M.C.; Pacin, F. Land cover classification in the Argentine Pampas using multi-temporal Landsat TM data. *Int. J. Remote Sens.* **2003**, *24*, 3381–3402. [CrossRef]
23. Novo, E.M.L. *Sensoriamento Remoto Princípios e Aplicações*; Edgard Blücher: São Paulo, Brazil, 2010; 388p.
24. Jensen, J.R. *Introductory Digital Image Processing*; Prentice Hall: Upper Saddle River, NJ, USA, 2005; 544p.
25. Landis, J.R.; Koch, G.G. The measurement of observer agreement for categorical data. *Biometrics* **1977**, *33*, 159–174. [CrossRef] [PubMed]
26. Ministério do Meio Ambiente (MMA). Plano Estratégico Nacional de Áreas Protegidas–PNAP. Brasília. 2006; 44p. Available online: http://www.mma.gov.br/estruturas/205/ arquivos/planonacionaareasprotegidas_205.pdf (accessed on 18 September 2018).
27. ENERCAN. *Campos Novos Energia S.A*; Plano de Manejo da Usina Hidrelétrica Campos Novos: Campos Novos, Santa Catarina, Brazil, 2004; 25p.

7

Extinction Risk of *Pseudotsuga menziesii* Populations in the Central Region of Mexico: An AHP Analysis

Javier López-Upton [1,†], J. René Valdez-Lazalde [1,†,*], Aracely Ventura-Ríos [1,†], J. Jesús Vargas-Hernández [1] and Vidal Guerra-de-la-Cruz [2]

1 Graduate School of Forest Sciences, Colegio de Postgraduados, Km. 36.5 Carr. México-Texcoco, Montecillo, Edo. de México 56230, Mexico; uptonj@colpos.mx (J.L.-U.); aracelivr@colpos.mx (A.V.-R.); jjesus.vargashernandez@gmail.com (J.J.V.-H.)

2 Experimental Site Tlaxcala, INIFAP, Km. 2.5 Carr. Tlaxcala-Santa Ana Chiautempan. Tlaxcala, Tlaxcala 90800, Mexico; guerra.vidal@inifap.gob.mx

* Author to whom correspondence should be addressed; valdez@colpos.mx;

† These authors contributed equally to this work.

Academic Editor: Diana F. Tomback

Abstract: Within the Analytic Hierarchy Process (AHP) framework, a hierarchical model was created considering anthropogenic, genetic and ecological criteria and sub-criteria that directly affect Douglas-fir (*Pseudotsuga menziesii* (Mirb.)) risk of extinction in central Mexico. The sub-criteria values were standardized, weighted, and ordered by importance in a pairwise comparison matrix; the model was mathematically integrated to quantify the degree of extinction risk for each of the 29 populations present in the study area. The results indicate diverse levels of risk for the populations, ranging from very low to very high. Estanzuela, Presa Jaramillo, Peñas Cargadas and Plan del Baile populations have very low risk, with values less than 0.25. On the other hand, Vicente Guerrero, Morán, Minatitlán, La Garita and Tonalapa populations have very high risk (>0.35) because they are heavily influenced by anthropogenic (close to roads and towns), ecological (presence of exotic species and little or no natural regeneration) and genetic (presence of mature to overmature trees and geographic isolation) factors. *In situ* conservation activities, prioritizing their implementation in populations at most risk is highly recommended; in addition, germplasm collection for use of assisted gene flow and migration approaches, including artificial reforestation, should be considered in these locations.

Keywords: Douglas-fir; risk analysis; conservation of populations; Analytic Hierarchy Process (AHP); multi-criteria evaluation; assisted gene flow

1. Introduction

Pseudotsuga menziesii (Mirb.) Franco barely survived in Mexico after the last ice age; the gradual increase in temperature forced this species to migrate from south to north and towards higher altitude in the mountains [1], which resulted in a fragmented and discontinuous distribution. In Mexico, this conifer is mainly distributed in the northern region, although in the central part of the country the species exists in small isolated stands [2]. At the southernmost limit, two isolated populations are located in the state of Oaxaca [3,4].

Added to this, the improper use and exploitation of natural resources affects forest species to the extent that some of them are endangered or threatened [5,6]. This is the case of *P. menziesii* located in the central region of Mexico, where the species grows in 29 small, fragmented populations suffering high anthropogenic pressure due to land-use changes, overgrazing, forest fires, inappropriate cone collecting, pest attack and illegal tree cutting and timber [7,8]. Also, Mexican populations of *P. menziesii* are exposed to high environmental stress due to their location at the southern end of its natural distribution [7,9]. The above-mentioned situation has caused a reduction in the size and density of populations, as well as low natural recruitment [8]; these circumstances are causing a reduction in genetic diversity and reproductive capacity associated with increased inbreeding [2,10].

P. menziesii is a valuable species for timber and wood production, but in Mexico this conifer is listed as protected by the Mexican government [11], so it is mainly used for Christmas-tree plantations. Therefore, the protection and proper management of each remaining population of the species is important for both natural recovery and establishment of commercial Christmas-tree plantations. It is imperative to implement *in situ* and *ex situ* conservation activities to ensure the preservation of the *P. menziesii* stands in its natural habitat, thereby allowing continuous evolution of the species and conservation of the remaining gene pool.

When designing conservation strategies for the species, we must understand and prioritize the current extinction risk experienced by the remaining populations. Risk of extinction is related to population viability and can be defined as the probability of the continued existence of populations over specified time periods [12], which can be estimated by qualitative and quantitative means. It refers to the likelihood of global or local extinction of target taxa. In this sense these authors suggest that no single measure is sufficient to assess population viability (risk) but different pools of data and type of analysis are necessary. Therefore, several methods to assess population risk might exist based on factors that potentially influence the viability of a species and upon its intrinsic characteristics.

The purpose of this study was to estimate, using a multi-criteria analysis technique, the Analytic Hierarchy Process (AHP), the degree of extinction risk faced by each natural *P. menziesii* population located in Central Mexico (states of Hidalgo, Tlaxcala and Puebla) influenced by anthropogenic, genetic, and environmental factors. Based on AHP, priorities to define a conservation strategy for the species in the region were determined.

2. Materials and Methods

2.1. Analytic Hierarchy Process (AHP)

Ventura-Ríos *et al.* [2] determined the spatial distribution of *P. menziesii* in central Mexico by characterizing 29 small, isolated populations. The present study examines the extinction risk of those populations located in the states of Hidalgo, Tlaxcala and Puebla (Figure 1). The extinction risk analysis was performed using the methodology known as Analytic Hierarchy Process (AHP), a structured approach to complex decision-making defined by Saaty [13] for decision-making in business environments based on the simultaneous analysis of multiple criteria. AHP is one of the most used tools in what is known as multi-criteria decision-making [14]. The tool has been successfully used to define strategies to mitigate the risks associated with decision-making in forestry [15,16] and environmental issues [17]. Theoretical details of the methodology can be found in Saaty [13] and Malczewsky [18]. Saaty and Niemira [14] presented a thorough summary of the technique. We describe the AHP applied to define extinction risk of *P. menziesii* populations located in central Mexico.

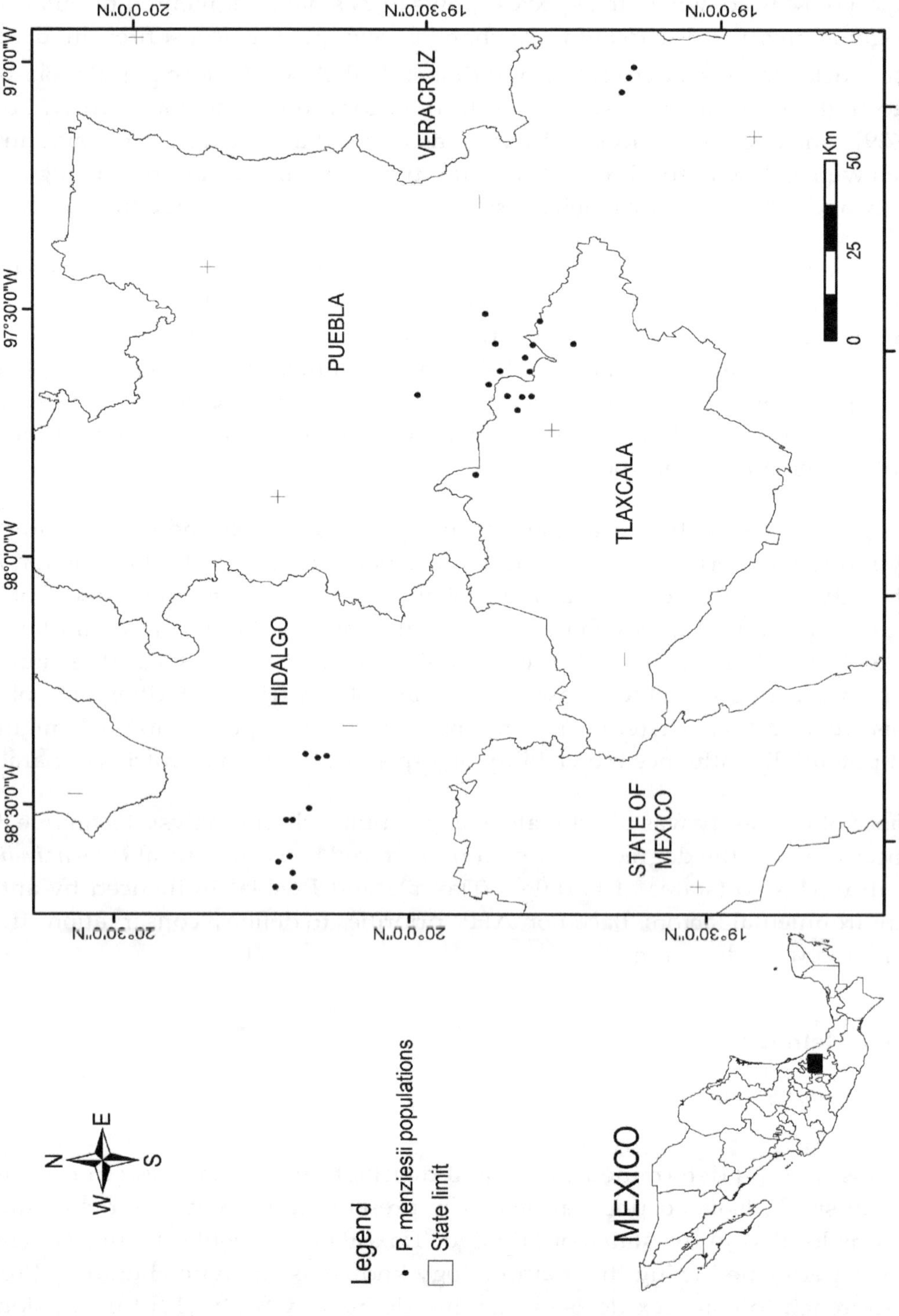

Figure 1. Geographical location of *Pseudotsuga menziesii* (Mirb.) Franco populations in central Mexico.

A model with three hierarchies was developed: definition of goal or objective, identification and definition of criteria for evaluating alternatives, and identification of decision alternatives (Figure 2). Three criteria or decision variables (anthropogenic, genetic and ecological) were considered to estimate the extinction risk of *P. menziesii* populations, each of them composed by several sub-criteria, 14 in total (Figure 2). The three criteria employed are not exhaustive, but these were considered as the most important issues to conduct an assessment of the current risk for these populations.

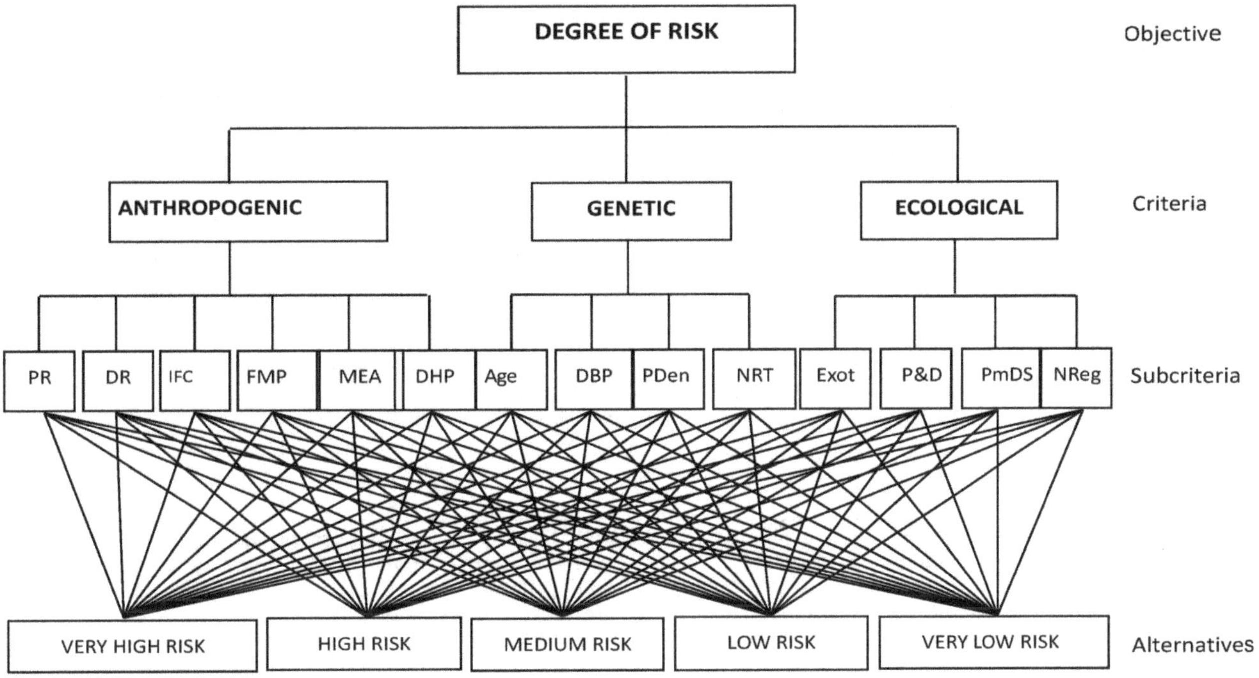

Figure 2. Decision problem hierarchization and sub-criteria considered. PR: Property regime; DR: Distance to roads; IFC: Interest in forest conservation; FMP: Forest management program; MEA: Main economic activity; DHP: Distance to human population; Age: Estimated age of Douglas-fir trees; DBP: Distance between populations of *Pseudotsuga*; PDen: Population density; NRT: Number of trees at reproductive age; Exot: Introduction of exotic species; P & D: Presence of pests and diseases; PmDS: *Pseudotsuga menziesii* is the dominant species; NReg: Natural regeneration.

2.2. Description and Application of Criteria and Sub-Criteria

In order to make a proper comparison using mathematical procedures, sub-criteria were standardized to a common scale or value range from 1 (low contribution) to 9 (high contribution) following the logic for the scale defined by Saaty [13], as shown in Table 1. The value assigned to each sub-criterion, considering the 1-to-9 scale (Table 1), was defined based on scientific literature and considering the results of the survey developed for the owners of *P. menziesii* populations under study (this is indicated following the description of the sub-criteria, according to the case). We used this combination of data sources in order to broaden the perspective of the analysis by including relevant social criteria in the Mexican context. Also, local and site-specific sampling provides more realistic information for the analysis.

Table 1. Standardized values for the sub-criteria used to determine the risk of 29 populations of *Pseudotsuga menziesii* in central Mexico.

DR	*Value*	MEA	*Value*	DHP	*Value*	DBP	*Value*	NRT	*Value*
>5	1	Conservation	1	>7.0	1	<2	1	>800	1
3.1–5.0	3	Ecotourism	3	4.1–7.0	3	2–5	3	401–800	3
1.6–3.0	5	Silviculture	5	2.1–4	5	5.1–10	5	201–400	5
0.5–1.5	7	Agriculture—livestock	9	0.5–2.0	7	10.1–30	7	51–200	7
<0.5	9			<0.5	9	>30	9	<50	9
PR		FMP		PDen		Age		NReg	
Communal	1	Yes	1	<10	7	Young	1	Good	1
Ejidal	3	Use rules	3	11–20	5	Mature	4	Low	5
Private	5	No	7	21–30	3	Overmature	4	None	9
				>30	1				
IFC		Exotic		P & D		PmDS			
Yes	1	Yes	7	Yes	7	Yes	1		
No	7	No	1	No	1	Moderate	4		
						No	7		

DR: Distance to roads (km); MEA: Main economic activity; DHP: Distance to human population (km); DBP: Distance between populations of *Pseudotsuga* (km); NRT: Number of trees at reproductive age; PR: Property regimen; FMP: Forest management program and/or forest-use rules; PDen: Population density (No. trees ha^{-1}); Age: Estimated age of Douglas-fir trees; NReg: Natural regeneration; IFC: Interest in forest conservation; Exotic: Introduction of exotic species; P & D: Presence of pest and diseases; PmDS: *Pseudotsuga menziesii* is the dominant species.

2.2.1. Anthropogenic Criteria

Considering that human activities positively or negatively impact the conservation of *P. menziesii populations*, six sub-criteria were included in the criteria (Table 1):

Distance to roads: Distance (km) between the stands (*P. menziesii* populations) and roads; the shorter the distance, the greater the risk of extinction [19].

Property regimen: The type of ownership (communal, private or "ejido") influences the conservation of populations. Based on conducted surveys, it is assumed that forest care is the best when the property regimen is communal, followed by the ejido. Forest in private ownership is the riskiest in the region. An ejido is a land concession given by the Mexican Government to a group of people after the Mexican Revolution (1918); is a collectively owned piece of land in which community individuals possess and farm a specific plot, but forestland is managed on a communal basis. Currently more than 70% of the forestland in Mexico is under this ownership regime.

Interest in forest conservation: If interest is expressed, a value of 1 was assigned; when such interest is lacking a value of 7 was designated. We used such extreme values because the lack of owners' interest in forest conservation makes a great difference in terms of success of a conservation program. This criterion was assessed by the survey taken by owners.

Forest management program (FMP) and/or forest-use rules: Populations that have a FMP approved by the Ministry of Environment and Natural Resources (Secretaría del Medio Ambiente y Recursos Naturales, SEMARNAT) or those under internal rules of management are most likely to be preserved. Data obtained by survey filled out by owners.

Main economic activity: Activities that cause the greatest impact to the forest are agriculture and livestock [20], followed by silviculture, ecotourism and conservation activities. Information based on field observations and survey taken by owners.

Distance to human population: Distance (km) between localities where *Pseudotsuga* grows and human population centers; the greater the distance, the less the risk of possible looting for wood and genetic material [19].

2.2.2. Genetic Criteria

The probability of continued existence is related to the gene pool of populations, because genetic variation helps species to adapt to environmental changes [21,22]. Reduced genetic variability is associated with increased vulnerability in the event of a sudden change in the environment. Four sub-criteria were considered (Table 1).

Estimated age of Douglas-fir trees: The younger the tree, the less the risk. An overmature population is at increased risk because of reduced vigor and decreased reproductive capacity [8]. Trees showing incipient reproductive events and vigorous growth were considered as young, and those that presented evidence of previous reproductive events were considered as mature. These data were obtained by sampling each population [2].

Distance between Pseudotsuga populations (genetic isolation): The longer the distance between populations, the greater the extinction risk. Isolated populations have limited gene flow [23] and genetic isolation may contribute to a reduced genetic variability.

Population density: The lower the density (Number of trees ha^{-1}), the greater the extinction risk. Decreased genetic exchange could generate low genetic variation and limited reproductive capacity, because the presence of few trees of the species leads to diminished pollen density. This criterion was assessed by field sampling.

Number of trees at reproductive age: The lower the number of individuals, the greater the extinction risk. A small population size increases the direct risk of extinction in the presence of a catastrophic event [24]. In addition, a small population has little gene flow and is subjected to random genetic drift, thereby increasing inbreeding, reducing genetic diversity and diminishing good development in the process of species evolution.

2.2.3. Ecological Criteria

Four sub-criteria linked to *P. menziesii* environment and organisms co-inhabiting with the species that could affect its performance and survival were included (Table 1).

Introduction of exotic species: It was considered that populations presenting evidence of *non-native* tree species are most at risk. Exotic species can displace native species by competition for resources [22]. Data evaluated through field surveys.

Presence of pests and diseases: Due to the high impact of cone and seed-feeding insects on Douglas-fir populations in Mexico, evidence that suggests the occurrence of some kind of pest or disease was graded with a higher level of extinction risk; absence of pests and diseases involves lower risk. Data are from field observation.

Pseudotsuga menziesii is the dominant species: The greater the dominance, the less the risk. Dominated species are most likely to disappear due to the strong competition by dominant species. *Pseudotsuga* dominance was determined by field sampling [2].

Natural regeneration: The greater the amount of natural regeneration, the less the risk. Thirty or more seedlings (50 cm in height) per 40 m^2 (average from 3 samples sites) were considered adequate recruitment, whereas less than thirty were considered as scarce. The criterion was assessed by sampling each population [2].

2.3. Construction of Pairwise Comparison Matrices (PCM)

The criteria (Table 2) and sub-criteria (Tables 3–5) were placed in a double entry matrix (called pairwise comparison matrix, PCM) employing the same order for both columns and rows. Subsequently, each pair of criteria and sub-criteria was compared using the 1-to-9 scale defined by Saaty [13]. The cells in the matrices were assessed by comparing the relative importance of the criteria (or sub-criteria) in each row against the criteria (or sub-criteria) listed in each column of the corresponding matrix; due to the principle of reciprocity defined as part of the procedure [13,18], it is only necessary to compare the upper half (above the diagonal) of the PCM. For example, when comparing the importance or contribution of sub-criteria distance to human population (DHP) and distance to roads (DR) for the extinction risk of *Pseudotsuga* populations, it was considered that DHP sub-criteria is moderately more important than DR sub-criteria, so a value of 5 was assigned (Table 3, row 6 and column 2 intersection); in the reciprocal case, the value is 1/5 (Table 3, row 2 and column 6 intersection). All the comparisons listed in Tables 2–5 were similarly performed.

Table 2. Pairwise comparison matrix and relative weights of the criteria used to estimate the degree of risk of the populations of *Pseudotsuga menziesii* in central Mexico †.

Criteria	Anthropogenic	Genetic	Ecological	*Weight*
Anthropogenic	1	3	3	0.60
Genetic	1/3	1	1	0.20
Ecological	1/3	1	1	0.20

† Consistency Ratio (CR) = 0.001; a quantitative measure of consistency associated with the defined pairwise comparisons of criteria.

Calculating the Criteria and Sub-Criteria Weight and Risk Estimation

Estimation of weights for each criterion and sub-criterion considered in the analysis (Tables 2–5) was performed using the algorithm implemented in the WEIGHT module of IDRISI; a software for the analysis and display of spatial data [25]. The weights result from computing a vector of priorities from each matrix; mathematically, the principal eigenvector is computed and normalized [13,14]. The algorithm computes a quantitative measure of consistency, a consistency ratio (CR), associated with the defined pairwise comparisons of criteria and sub-criteria. The CR evaluates the probability that the values in the PCM are not randomly assigned; its value must be less than 0.10, indicating that pairwise comparisons were consistent [13]. In this case, all CR values were lower than 0.10 (Tables 2–5).

Table 3. Pairwise comparison matrix and relative weights of anthropogenic sub-criteria used to estimate the degree of risk of the populations of *Pseudotsuga menziesii* in central Mexico †.

Anthropogenic Sub-Criteria	PR ‡	DR	IFC	FMP	MEA	DHP	*Weight*
PR	1	1/3	1/3	1/5	1/9	1/9	0.0301
DR	3	1	1	1/3	1/5	1/5	0.0689
IFC	3	1	1	1/3	1/5	1/5	0.0689
FMP	5	3	3	1	1/3	1/3	0.1532
MEA	9	5	5	3	1	1	0.3394
DHP	9	5	5	3	1	1	0.3394

† Consistency Ratio (CR) = 0.017; ‡ PR: Property regime; DR: Distance to roads; IFC: Interest in forest conservation; FMP: Forest management program; MEA: Main economic activity; DHP: Distance to human population.

Table 4. Pairwise comparison matrix and relative weights of genetic sub-criteria used to estimate the degree of risk of the populations of *Pseudotsuga menziesii* in central México †.

Genetic Sub-Criteria	Age ‡	DBP	PDen	NRT	*Weight*
Age	1	1/3	1/5	1/7	0.0595
DBP	3	1	1/3	1/5	0.1279
PDen	5	3	1	1/3	0.3639
NRT	7	5	3	1	0.4487

† Consistency Ratio (CR) = 0.031; ‡ Age: Estimated age of Douglas-fir trees; DBP: Distance between populations of *Pseudotsuga*; PDen: Population density; NRT: Number of trees at reproductive age.

Table 5. Pairwise comparison matrix and relative weights of ecological sub-criteria to estimate the level of risk of the *Pseudotsuga menziesii* populations' in central Mexico †.

Ecologic Sub-Criteria	Exot ‡	P&D	PmDS	NReg	*Weight*
Exot	1	1/3	1/7	1/7	0.0541
P & D	3	1	1/3	1/3	0.1431
PmDS	7	3	1	1	0.4014
NReg	7	3	1	1	0.4014

† Consistency Ratio (CR) = 0.003; ‡ Exot: Introduction of exotic species; P & D: Presence of pests and diseases; PmDS: *Pseudotsuga menziesii* is the dominant species; NReg: Natural regeneration.

To facilitate subsequent arithmetic operations and the interpretation of results by having final values between 0 and 1, standardized values for each sub-criteria (Table 1) were transformed (re-scaled) to values from 0 to 1 (0 = No risk, 1 = Maximum risk). The value of extinction risk for each population was calculated by adding the products of the transformed standardized values of the sub-criteria and their respective weighting values (linear weighted sum). The estimated risk values were ordered from lowest to highest and classified into five categories or degrees of risk: very low (values from 0 to 0.24), low (values of 0.25 to 0.27), medium (values from 0.28 to 0.30), high (values from 0.31 to 0.34) and very high (values greater than 0.35). The risk categories were defined by the authors based on the range of risk values calculated for populations (0.15 to 0.45, Table 6) and the need to clearly prioritize the activities for the conservation of the species. Notice that only one population (Estanzuela) had a risk value below 0.22 (Table 6), in other words, most of the risk values are between 0.22 and 0.45.

Table 6. Degree of risk of extinction for the *Pseudotsuga menziesii* populations in central Mexico.

Population	Value	Risk Level †	Population	Value	Risk Level
Estanzuela	0.15	VLow	Barranca Canoita	0.29	Med
Peñas Cargadas	0.22	VLow	Cuyamaloya	0.30	Med
Presa Jaramillo	0.23	VLow	Zapata	0.30	Med
Plan del Baile	0.24	VLow	La Caldera	0.31	High
Cañada El Atajo	0.26	Low	El Llanete	0.32	High
Cruz de Ocote	0.26	Low	Apizaquito	0.33	High
San José Capulines	0.26	Low	La Rosa	0.33	High
Capula	0.27	Low	Las Antenas	0.33	High
Cuatexmola	0.27	Low	Tlalmotolo	0.34	High
El Salto	0.27	Low	Tonalapa	0.36	Vhigh
Villareal	0.27	Low	La Garita	0.38	Vhigh
Axopilco	0.28	Med	Minatitlán	0.42	Vhigh
Buenavista	0.28	Med	Morán	0.44	Vhigh
San Juan	0.28	Med	Vicente Guerrero	0.45	Vhigh
Tlaxco	0.28	Med			

† VLow: very low risk; Low: low risk; Med: Medium risk; High: high risk; Vhigh: very high risk.

3. Results

Four of the twenty-nine evaluated populations (13.8%) had very low risk of extinction showing values less than 0.25 (Table 6). These populations, Estanzuela, Peñas Cargadas, Presa Jaramillo and Plan del Baile, are influenced to a lesser degree by anthropogenic factors, because the average distance to human population centers is 3.0 km, the average distance to roads is 3.0 km and the main economic activity for the first three populations is tourism and for the last one is forestry. Moreover, there is interest in preserving populations because the owners of the stands established rules for forest use. *Pseudotsuga menziesii* is the dominant species in these populations, with healthy trees (free of pests and diseases), ages range from young to mature, and presence of natural regeneration from poor to good.

The average distance to the nearest *P. menziesii* population is 4.0 km, the average number of trees at reproductive age is over 500, and the average population density is 23 trees ha^{-1}.

Seven *Pseudotsuga* populations had low risk with values varying from 0.26 to 0.27, including Cañada El Atajo, Cruz de Ocote, San José Capulines, Capula, Cuatexmola, El Salto, and Villareal (Table 6). These populations are influenced by anthropogenic factors because the average distance to human population centers is 2.4 km, the average distance to roads is 2.0 km and the main economic activities are agriculture and silviculture. However, the owners have interest in conserving the populations, since 50% of them have a Forest Management Program and the rest are managed under internal rules for forest use. The *P. menziesii* trees are young, but some populations are infested by pests and diseases. In these locations, Douglas-fir is the dominant species, but its natural regeneration is scarce, the average distance to the nearest *P. menziesii* population is 6.4 km and the average number of trees at reproductive age is 383, while the population density is 18 trees ha^{-1}.

Seven of the *Pseudotsuga* populations were at medium risk with values from 0.28 to 0.30, including Axopilco, Buenavista, San Juan, Tlaxco, Barranca Canoita, Cuyamaloya and Zapata (Table 6). These populations are negatively influenced by anthropogenic and ecological factors. They are at an average distance of 4.2 km away from human population centers and 3.3 km away from roads, but the main economic activity is agriculture. There is interest by owners to conserve the populations, since 50% of these have a Forest Management Program and the others have at least usage rules of the woodlands. Exotic species, pests and diseases are absent; *P. menziesii* trees are mature and dominant, but its natural regeneration is scarce to nonexistent. The average distance to the nearest *P. menziesii* population is 5.4 km, the average number of trees at reproductive age is 263 and the population density is 14 trees ha^{-1}, on average.

Six of the *Pseudotsuga* populations were at high risk, with values between 0.31 and 0.34, including La Caldera, El Llanete, Apizaquito, La Rosa, Las Antenas and Tlalmotolo (Table 6). These populations are strongly and negatively influenced by anthropogenic and genetic factors. They are at an average distance of 3.0 km away from human population centers and 1.8 km away from roads; also the main economic activity is agriculture. The populations are on private land; although some of these have a Forest Management Program and others have rules of use, two of the populations do not have any regulation. Three of these populations showed evidence of pests and diseases. Although *P. menziesii* trees are young and mature, they are not dominant and its natural regeneration is scarce. The average distance to the nearest *P. menziesii* population is 12.8 km, the average number of trees at reproductive age is 299 and the average population density is 11.8 trees ha^{-1}.

The five remaining populations had a very high risk, showing values above 0.35, including Tonalapa, La Garita, Minatitlán, Morán and Vicente Guerrero (Table 6). These populations are strongly and negatively influenced by anthropogenic and genetic factors, with an average distance of 1.2 km away from human population centers and 0.88 km away from roads; in addition, the main economic activity is agriculture. Four of these populations are privately owned and the other is part of an ejido, but none of them has a Forest Management Program. Although these stands varied from young to overmature and they are free from pests and diseases, natural regeneration is lacking and *P. menziesii* is not the dominant species. In this group are the smallest populations like Morán, La Garita and Vicente Guerrero with less than 20 reproductive trees; Minatitlán has 45 individuals, and Tonalapa shows 166 individuals of reproductive age. The average population density is 4.6 trees per ha, and the distance to nearest *P. menziesii* population is greater than 5 km. Furthermore, in these localities exotic pine species have been introduced through reforestation. A special case is the population of Morán, which is at very high risk. This relict of only four reproductive trees was originally described as *Pseudotsuga macrolepis* by Flous [26]. In addition, the owner is considering cutting down the trees because they are growing too close to his home and he is not interested in preserving them.

4. Discussion

The AHP, a mathematical process for measurement and decision-making, was used to evaluate the risk of extinction of fragmented *P. menziesii* populations in central Mexico. It allowed us to propose a hierarchy model that incorporates information from anthropogenic, genetic, and ecological factors or criteria, and offer a reference for future studies on the extinction risk of forest populations. The model was built considering elements that according to scientific literature and the authors experience on forest conservation practices are important to assess the risk of extinction of the species of interest in central Mexico. Neither the model, nor the criteria included in it are exhaustive or can be applied to other situations, but they might be used as a starting point to generate similar models for other tree species present in other latitudes.

AHP facilitated the task of calculating different weight values (importance) for each of the criteria identified as relevant for the risk assessment carried out, a critical issue in the modeling process. Both tasks, hierarchy model development and criteria weighting, are not trivial and must be carefully realized, AHP has widely demonstrated its value to perform such tasks [27,28].

Despite being a mature decision analysis methodology, the AHP is only one of many possible combinations of methods to standardize, weight, and rank available decision alternatives [18]. Other approaches to evaluate extinction risk exist, such as the so called population viability analysis [29]; however, it is also fair to mention that owing to its simplicity, ease of use, and theoretical foundation, the AHP has found wide acceptance among decision-makers [28]. It helps structure the decision problem in a manner that is simple to follow and analyze. Further, it has proved to be a methodology capable of producing results that agree with the decision-maker's expectations; that is the case of the study here reported.

Extinction of native forest tree species is caused by various processes, including reduction of and fragmentation of suitable environment, spread of diseases, genetic erosion and inbreeding [6]. The extinction of species also modifies and alters ecological processes involving it, affecting other species and causing changes in communities and ecosystems [20]. In the case of the *P. menziesii* populations in central Mexico, there is evidence of the effects of fragmentation and isolation by distance (9.4 km average distance between populations) as well as small size, from 4 to 1450 adult trees and 307 trees on average.

It is estimated that coniferous species require a population size of at least 180 reproductive trees to reduce the negative effects caused by low pollen levels and inbreeding, and to maintain seed production at sufficient levels [30]. In several of the stands included in the current study low natural regeneration of *P. menziesii* was observed [8]. In addition, seeds showed low germination and primary dormancy [7,31]. Moreover, a reduced level of genetic diversity with a high proportion of self-fertilization was found [10,32].

Habitat fragmentation in most of these populations is due to anthropogenic factors, but there is evidence that other factors such as environment and genetics play important roles in increasing the risks of extinction. The analysis conducted shows the relative effect of each of these factors over the current risks for each population, which is an important step to outline strategies for species conservation in the region. Risk factors for *Pseudotsuga menziesii* in the study area have their primary origin in the enormous social pressure that persists over populations as a result of historical changes in land use to fulfill human needs in the region, thereby increasing the fragmentation process of habitat.

One way to increase the genetic variability to counteract the effects of inbreeding over reproductive capacity [33,34] is through the promotion of gene flow between neighboring populations growing in similar environments (assisted gene flow as defined by Aitken and Whitlock [35]), which in turn could reduce adaptation problems (outbreeding depression). Populations from similar environments can also exchange genes through assisted migration by creating plantations that use more likely adaptable trees [32]; an action like this one is certainly possible to be carried out in central Mexico, which would increase genetic diversity by transferring new alleles or increasing the frequency of rare alleles in the recipient populations [23,33], and would provide long-term genetic stability to each population and a greater ability for adaptation to climate change or other risk factors. Performing movement of genetic material between related populations would counteract the effects of geographic isolation and fragmentation.

Conservation of remnant populations of *Pseudotsuga menziesii* in the central region of Mexico should be carried out in the natural environment where the species develops (*in situ*), because this would preserve the natural evolution processes of the species and its interactions with other organisms and ecological processes [22]. Such a conservation strategy is feasible in central Mexico due to the growing interest in managing the populations of the species as a genetic reserve—due to its current legal protection status. In this sense, the Mexican National Forestry Commission (CONAFOR), a federal agency, is currently promoting local and regional programs in order to enhance forest conservation to preserve the country's genetic resources.

The five populations with very high extinction risk and the six populations at high risk should have priority for conservation, due to the real possibility of losing their gene pool in case of extreme events. The populations at medium and low risk would also require protective actions such as grazing prevention, pest control and other actions to reduce current risks. *Ex situ* conservation should be considered as an alternative or complementary resource for populations at high and very high risk.

Other tree species present in central Mexico that are potentially relevant in this context and require studies of extinction risks include: *Pinus chiapensis*, *Fagus sylvatica*, and *Taxus globosa*, all of which have spatial distribution patterns and life history similar to that shown by *Pseudotsuga menziesii*.

5. Conclusions

The AHP-based analysis made it possible to show that 11 of the 29 studied populations were at high risk of extinction, seven were at medium risk and the remaining were at low risk; this suggest different strategies for conservation and protection of *Pseudotsuga menziesii* in the region. Tonalapa, La Garita and Minatitlán populations from the state of Puebla, as well as Morán and Vicente Guerrero from the state of Hidalgo, are at a very high risk of extinction because they are heavily impacted by anthropogenic and genetic factors. These populations are relatively small (less than 200 trees), they are very close to population centers (less than 1.5 km away) and to access roads (less than 1.0 km away) and these populations are located in areas where agriculture is the main economic activity. By contrast, the populations of Estanzuela, Presa Jaramillo, Peñas Cargadas and Plan del Baile are better preserved and they have a very low risk of extinction due to the lower human impacts because populations are at greater distance from human population centers and roads (more than 3.0 km away), in areas where ecotourism is the main economic activity, and some of them are within an ecological reserve.

Acknowledgments: This research was supported by the "Fondo Sectorial CONAFOR-CONACYT (Comisión Nacional Forestal—Consejo Nacional de Ciencia y Tecnología), México", grant 2002-C01-6416. We are grateful for the assistance of two anonymous reviewers that did much to improve the initial manuscript.

Author Contributions: Javier López-Upton and J. René Valdez-Lazalde conceived and designed the study and analyzed data; Aracely Ventura-Ríos generated the data; and all authors contributed in data analysis and discussion. The manuscript was written by J. René Valdez-Lazalde supported by Javier López-Upton and J. Jesús Vargas-Hernández.

References

1. Li, P.; Adams, W.T. Range-wide patterns of allozyme variation in Douglas-fir (*Pseudotsuga menziesii*). *Can. J. For. Res.* **1989**, *19*, 149–161. [CrossRef]
2. Ventura-Ríos, A.; López-Upton, J.; Vargas-Hernández, J.J.; Guerra de la Cruz, V. Caracterización de *Pseudotsuga menziesii* (Mirb.) Franco en el centro de México; Implicaciones para su conservación. *Rev. Fitotec. Mex.* **2010**, *33*, 107–116.
3. Rzedowski, J. *Vegetación de México*; Limusa: México, DF, México, 1978.
4. Del Castillo, R.F.; Pérez de la Rosa, J.A.; Vargas, G.; Rivera, R. Coníferas. In *Biodiversidad de Oaxaca*; García-Mendoza, A.J., Ordóñez, M.J., Briones-Salas, M., Eds.; Instituto de Biología UNAM-Fondo Oaxaqueño para la Conservación de la Naturaleza-World Wildlife Foundation: Mexico, Mexico, 2004; pp. 141–158.

5. CONABIO. *La Diversidad Biológica de México: Estudio de País*; Comisión Nacional para el Conocimiento y Uso de la Biodiversidad: México, DF, México, 1998.
6. Hakoyama, H.; Iwasa, Y.; Nakanishi, J. Comparing risk factors for population extinction. *J. Theor. Biol.* **2000**, *204*, 327–336. [CrossRef] [PubMed]
7. Mápula-Larreta, M.; López-Upton, J.; Vargas-Hernández, J.J.; Hernández-Livera, A. Reproductive indicators in natural populations of Douglas-fir in Mexico. *Biodivers. Conserv.* **2007**, *16*, 727–742. [CrossRef]
8. Velasco-García, M.V.; López-Upton, J.; Angeles-Pérez, G.; Vargas-Hernández, J.J.; Guerra de la Cruz, V. Dispersión de semillas de *Pseudotsuga menziesii* en poblaciones del centro de México. *Agrociencia* **2007**, *41*, 121–131.
9. Hermann, R.K.; Lavender, D.P. *Pseudotsuga menziesii* (Mirb.) Franco. In *Silvics of North America*; Burns, R., Honkala, B.H., Eds.; USDA Forest Service: Washington, DC, USA, 1990; Volume 1, pp. 527–540.
10. Cruz-Nicolás, J.; Vargas-Hernández, J.J.; Ramírez-Vallejo, P.; López-Upton, J. Patrón de cruzamiento en poblaciones naturales de *Pseudotsuga menziesii* (Mirb.) Franco, en México. *Agrociencia* **2008**, *42*, 367–378.
11. SEMARNAT 2010. NOM-059-SEMARNAT-2010, Protección Ambiental-Especies Nativas de México de Flora y Fauna Silvestres-Categorías de Riesgo y Especificaciones Para su Inclusión, Exclusión o Cambio-Lista de Especies en Riesgo. Diario Oficial de la Federación. 30 de Diciembre de 2010. Available online: http://www.profepa.gob.mx/innovaportal/file/435/1/NOM_059_SEMARNAT_2010.pdf (accessed on 10 January 2014).
12. Marcot, B.G.; Murphy, D.D. On population viability analysis and management. In *Biodiversity in Managed Landscapes*; Szaro, R.C., Johnson, D.W., Eds.; Oxford University Press: New York, NY, USA, 1996; pp. 58–76.
13. Saaty, T.L. *The Analytic Hierarchy Process*; McGraw-Hill: New York, NY, USA, 1980.
14. Saaty, T.L.; Niemira, M.P. A framework for making better decisions. *Res. Rev.* **2006**, *13*, 44–48.
15. Bustillos-Herrera, J.A.; Valdez-Lazalde, J.R.; Aldrete, A.; González-Guillén, M.J. Aptitud de terrenos para plantaciones de Eucalipto *Eucaliptus grandis* Hill ex Maiden): Definición mediante el proceso de análisis jerarquizado y SIG. *Agrociencia* **2007**, *41*, 787–796.
16. Olivas-Gallegos, U.E.; Valdez-Lazalde, J.R.; Aldrete, A.; Gonzalez-Guillén, M.J.; Vera-Castillo, G. Áreas con aptitud para establecer plantaciones de maguey cenizo: Definición mediante análisis multicriterio y SIG. *Rev. Fitotec. Mex.* **2007**, *30*, 411–419.
17. Wang, G.; Qin, L.; Li, G.; Chen, L. Landfill site selection using spatial information technologies and AHP: A case study in Beijing, China. *J. Environ. Manag.* **2009**, *90*, 2414–2421. [CrossRef]
18. Malczewski, J. *GIS and Multicriteria Decision Analysis*; John Wiley & Sons: Toronto, ON, Canada, 1999.
19. Vergara, G.; Gayoso, J. Efecto de factores físico-sociales sobre la degradación del bosque nativo. *Bosque* **2004**, *25*, 43–52. [CrossRef]
20. Lambin, E.F. *Modeling Deforestation Processes: A Review*; TREES Series: Research Report No. 1. EUR 15744 EN; European Commission: Luxemburg, Luxemburg, 1994.
21. Soulé, M.E. Conservation: Tactics for a constant crisis. *Science* **1991**, *253*, 744–750. [CrossRef]
22. Primack, R. *A Primer of Conservation Biology*; Sinauer-Sunderland: Sunderland, MA, USA, 1995.
23. Ledig, F.T.; Mápula-Larreta, M.; Bermejo-Velázquez, B.; Reyes-Hernández, J.V.; Flores-López, C.; Capó-Arteaga, M.A. Locations of endangered spruce populations in Mexico and the demography of *Picea chihuahuana*. *Madroño* **2000**, *47*, 71–88.
24. Robledo-Anuncio, J.J.; Alía, R.; Gil, L. Increased selfing and correlated paternity in a small population of a predominantly outcrossing conifer, *Pinus sylvestris*. *Mol. Ecol.* **2004**, *13*, 2567–2577. [CrossRef] [PubMed]
25. Eastman, J.R. *IDRISI Selva Manual*; Clark University: Worcester, MA, USA, 2012.
26. Flous, F. *Diagnoses D'especes et Variétés Nouvelles de Pseudotsuga Americains*; Travaux du laboratorie Forestier de Toulouse: Toulouse, France, 1934; Volume 2, p. 18.
27. Bhushan, N.; Rai, K. *Strategic Decision Making: Applying the Analytic Hierarchy Process*; Springer: London, UK, 2004.
28. Brunelli, M. *Introduction to the Analytic Hierarchy Process*; SpringerBriefs in Operations Research: New York, NY, USA, 2015.
29. Keedwell, R.J. *Use of Population Viability Analysis in Conservation Management in New Zealand*; Science for Conservation 243; New Zealand Department of Conservation: Wellington, New Zealand, 2004.
30. O'Connell, L.M.; Mosseler, A.; Rajora, O.P. Impacts of forest fragmentation on the reproductive success of white spruce (*Picea glauca*). *Can. J. Bot.* **2006**, *84*, 956–965. [CrossRef]

31. Juárez-Agís, A.; López-Upton, J.; Vargas-Hernández, J.J.; Sáenz-Romero, C. Variación geográfica en la germinación y crecimiento inicial de plántulas de *Pseudotsuga menziesii* de México. *Agrociencia* **2006**, *40*, 783–792.
32. Cruz-Nicolás, J.; Vargas-Hernández, J.J.; Ramírez-Vallejo, P.; López-Upton, J. Genetic diversity and differentiation of *Pseudotsuga menziesii* (Mirb.) Franco populations in Mexico. *Rev. Fitotec. Mex.* **2011**, *34*, 233–240.
33. Mosseler, A. Minimum viable population size and the conservation of forest genetics resources. In *Tree Improvement: Applied Research and Technology Transfer*; Puri, S., Ed.; Science Publishers: Enfield, NH, USA, 1998; pp. 191–205.
34. Mosseler, A.; Major, J.E.; Simpson, J.D.; Daigle, B.; Lange, K.; Park, Y.S.; Johnsen, K.H.; Rajora, O.P. Indicators of population viability in red spruce, *Picea rubens*. I. Reproductive traits and fecundity. *Can. J. Bot.* **2000**, *78*, 928–940.
35. Aitken, S.N.; Whitlock, M.C. Assisted gene flow to facilitate local adaptation to climate change. *Annu. Rev. Ecol. Evol. Syst.* **2013**, *44*, 367–388. [CrossRef]

8

Evaluation of the Operational Environment Factors of Nature Conservation Policy Implementation: Cases of Selected EU and Non-EU Countries

Špela Pezdevšek Malovrh, Alessandro Paletto, Stjepan Posavec, Zuzana Dobšinská , Ilija Đorđević, Bruno Marić, Mersudin Avdibegović, Emil Kitchoukov, Aleksandar Stijović, Pande Trajkov and Tomislav Laktić

[1] Department of Forestry and Renewable Forest Resources, Biotechnical Faculty, University of Ljubljana, 1000 Ljubljana, Slovenia

[2] Consiglio per la ricerca in agricoltura e l'analisi dell'economia agraria, Research Centre for Forestry and Wood (CREA), 38123 Trento, Italy; alessandro.paletto@crea.gov.it

[3] Department of Forest Inventory and Management, Faculty of Forestry, University of Zagreb, 10000 Zagreb, Croatia; sposavec@sumfak.hr

[4] Department of Forest Economics and Management, Technical University in Zvolen, Faculty of Forestry, 96001 Zvolen, Slovakia; zuzana.dobsinska@tuzvo.sk

[5] Department of spatial planning, GIS and Forest Policy, Institute of Forestry, 11030 Belgrade, Serbia; ilija.djordjevic@forest.org.rs

[6] Faculty of Forestry, University of Sarajevo, 71000 Sarajevo, Bosnia and Herzegovina; b.maric@sfsa.unsa.ba (B.M.); m.avdibegovic@sfsa.unsa.ba (M.A.)

[7] Department of Management and Alternative Tourism, Faculty of business management, University of Forestry-Sofia, 1756 Sofia, Bulgaria; ekitchoukov@yahoo.com

[8] Institute of Forestry of Montenegro, 81000 Podgorica, Montenegro; stijovicaleksandar@gmail.com

[9] Department of Forest Management, Faculty of Forestry, Ss. Cyril and Methodius University in Skopje, 1000 Skopje, North Macedonia; ptrajkov@sf.ukim.edu.mk

[10] Ministry of the Environment and Spatial Planning, Water and Investments Directorate, Cohesion Policy Division, 1000 Ljubljana, Slovenia; tomislav.laktic@gov.si

* Correspondence: spela.pezdevsek.malovrh@bf.uni-lj.si.

Abstract: The complex policy decision-making situation around nature conservation requires examination of the operational environment. This study develops and tests a three-phase analytical framework for the evaluation of operational environment factors influencing nature conservation policy implementation. The four important operational environment factors (legal, policy, economic, and social) have been identified, to build up a framework. The framework was tested in selected countries and includes experts' opinions. Experts (n = 44) from five EU countries (Bulgaria, Croatia, Italy, Slovakia, and Slovenia) and four non-EU countries (Bosnia and Herzegovina, North Macedonia, Montenegro, and Serbia) defined and evaluated the factors and sub-factors that affect the operational environment related to nature conservation policy implementation. The results show policy changes arising from the new governance requirements introduced by changed political regime and Europeanization are key driving factors for changes in the nature conservation operational environment. For nature conservation, these wide-reaching changes have led to new political and legal frameworks, new institutional set-ups and multilevel governance frameworks, new establishment of protected areas and Natura 2000 network, and the re-allocation of financial resources and inclusion of non-state actors in policy decision-making. However, there are also some challenges and unsolved problems that need further attention from policy decision-makers and institutions, especially related to the institutional gap, sustainable financing of nature conservation, transposition of the EU Directives into legal systems, designation of sites or improving their implementation, implementation of innovative funding schemes, and a transparent participatory process. This analytical framework can be applied to various problems related to any environmental issues or other policy implementation or management, and other sectors where public decision-making is combined with stakeholders' engagement.

Keywords: biodiversity conservation; nature conservation policy; operational environment; legal; policy; economic and social factors; evaluation framework; protected areas; Natura 2000 network

---

1. Introduction

Environmental and nature conservation awareness in today's sense began to grow in the second half of the 20th century, when the consequences of human activity in nature became more apparent [1]. Therefore, nature conservation has become a political issue at the global level [2,3].

In order to achieve nature conservation objectives of European wildlife and natural habitats, laws on nature protection have been adopted and PAs were established. Later on the Emerald Network at the national level has been set up in each contracting party of the Bern Convention, which involves also all the EU states [4]. In order to fulfil its obligations arising from the Bern Convention, the EU implemented the "Habitats" Directive [5] in 1992, which together with "Birds" Directive [6] set up the EU's network of PAs called Natura 2000. Natura 2000 is at the moment the largest network of PAs in the world and the core pillar of the EU's nature conservation policy [3,7,8]. Therefore, it is considered an important legal policy framework in achieving a favourable conservation status of nature in the EU Biodiversity Strategy by 2020 [9].

In a broader context, the Natura 2000 network is just one in a wide range of EU policies and legislation related to nature conservation to which the EU member countries must adhere. In the case of legally binding legislation specifically, it is required that the individual EU member countries translate them into national legislation (transposition) while leaving the details of implementation to the national authorities (enforcement) [8].

In the case of EU acceding countries, membership in the EU requires the adoption, implementation and enforcement of the 'acquis communautaire'—the body of the EU law and regulation [10]. The 'acquis' consists of different thematic chapters, where environmental law and regulation constitute one of them. Transposing the environmental chapter does not require a simple transformation of 'acquis communautaire', but also the development and adoption of institutions and structures by which legally binding legislation can be effectively implemented [10,11]. This process is commonly known as Europeanization of domestic environmental governance, which has led to changed policies, values, and norms while empowering new stakeholders in policy decision-making, leading to effective environmental governance [10,12]. On the other hand, in these countries, traditional command-and-control policy decision-making by state actors still dominates quite often as a consequence of post-socialistic governance type [11,13].

The implementation of EU nature conservation policy—de facto, not necessarily de jure—did not harmonize with traditional domestic nature conservation policies in the most of the EU member countries [14]. It introduced several new (in)formal rules such as the required participation of a broad range of stakeholders, and cooperation between government actors from various sectors and across several jurisdictional levels [13,15,16]. Moreover, the implementation and success of nature conservation policies depends on a range of various factors from the operational environment "... political, economic and socio-psychological factors related to the design of the instrument as well as to the historical, local and sectoral contexts have an influence on the success of any conservation programme" [8].

The analysis of operational environments of the nature conservation policies and adjusting to changes in the environments are crucial parts of effective policy implementation [17–19]. Several factors can be expected to influence the behaviour of nature conservation policy implementation actors and outputs. These factors correspond to the different dimensions of the conservation—e.g., social, cultural, economic, policy, environmental, but also institutional and individual-level ones [20]. According to Blicharska, et al. [21] to achieve a good functionality of the nature conservation, there is a need for knowledge not only on the ecological conservation and management issues, but also on key social, economic, political, and managerial realities potentially influencing policy implementation. Many previous studies have shown that national implementation of nature conservation policies differs across the EU [7–9,22–25]. Differences in implementation pathways are related to differences in policy culture in each country and consequently policy integration [14,26–29], financing of conservation measures and compensation payments [29–33], management practices [34,35], institutional factors [3,13,23,29,36–38], social factors [3,15,39–45], and environmental factors [46,47]. In a recent review of studies related to nature conservation, Popescu, et al. [48], and Blicharska, Orlikowska, Roberge and Grodzinska-Jurczak [21], concluded that research related to ecological factors prevail, while social, economic, and policy ones are underrepresented.

Studies focusing on national and sub-national operational environment factors related to nature conservation policies implementation are available in central-eastern European countries and south-eastern European countries, but most studies fail to systematically evaluate the whole operational environment factors related to nature conservation policies implementation [7,12,14,39,40,49–53]. To fill the above-mentioned gaps, this research aims to develop a framework to analyse the operational environment factors that affected nature conservation policies implementation in selected EU and non-EU member countries. Selected EU members' countries were Italy (IT), Slovakia (SK), Slovenia (SI), Bulgaria (BG), and Croatia (CRO), and non-EU members' countries were Serbia (SRB), Montenegro (MNE), North Macedonia (NM), and Bosnia and Herzegovina (BiH). These countries were selected as they were involved in COST Targeted Network TN1401 "Capacity Building in Forest Policy and Governance in Western Balkan Region (CAPABAL)". In addition, these countries represent good cases as in these countries political, legal, economic and social operational environment changed considerably in a relatively short period, making a substantial impact on nature conservation. These differences influence nature conservation policy implementation in selected countries and make a cross-country comparison interesting. The results of this study are useful for understanding the influences that operational environmental factors have on nature conservation policy implementation in selected countries and are likely to be relevant for policy decision-makers, institutions responsible for management and planning of PAs, and other stakeholders to increase the quality of future nature conservation policy implementation processes.

2. Materials and Methods

To identify operational environment factors, while explicitly addressing different political and socio-economic context in case countries, this study developed a three-phase analytical framework for analyzing the operational environment factors related to nature conservation policies implementation. The literature review in the introduction showed that institutional theories and governance theory are frequently used to explain nature conservation policy implementation. Our analytical framework draws on these findings, and the factors are therefore based on neo-institutional theory [54]. They are complemented by governance elements (participation) because the implementation of nature conservation policies is highly dependent on stakeholder engagement. As a tool, framework identifies the elements and general relationships among them that one needs to consider for institutional analysis. They provide a general set of variables that can be used to analyze all types of institutional arrangements [55]. The institutional environment is dynamic and constantly changing, so neo-institutionalism tries to find a balance between actors and structure. It seeks to answer the question of whether historical, social, and political outcomes are the result of the intentions, motivations, and behavior of actors; or whether these are shaped by political institutions, power hierarchies, and cultural conventions [54]. The proposed framework is based on the basic elements of these theories namely rules (legal and political), institutions (political, economicl) and governance elements (participation), and it is applicable for different regions and conditions both across Europe and worldwide as it uses general factors to analyze the institutional setting. Simultaneously, it allows a better understanding of the country/region specific situation by using sub-factors. It was developed in a series of steps involving nature conservation experts and applied in selected countries as described in more detail below.

2.1. *Phase 1—Identification of Operational Environment Factors and Sub-Factors*

To be able to analyze the main operational environment factors that affected nature conservation policies implementation in selected countries four groups of operational environment factors were identified: (1) Legal—including sub-criteria such as international conventions and agreements, national legal framework, PA legal status, Emerald network, Transposition of the EU "Birds" and "Habitats" Directives, Natura 2000 management plans; (2) Policy—including sub-criteria such as policy framework and institutional set-up; (3) Economic—including sub-criteria such as financing mechanisms of PAs and Natura 2000 and compensatory measures for Natura 2000; and (4) Social—including sub-criteria such as stakeholders' involvement in establishment and management of PAs and stakeholders' involvement in the implementation of Natura 2000 network. As legal and policy factors often overlap they were merged into one category in our study, which is commonly done also in other studies [56]. As these general factors do not enable detailed analysis of the operational environment of each country a preliminary list of sub-factors under each factor was prepared. The general factors and sub-factors were selected based on the theoretical background and literature review [9,21,48], where it was found that mainly legal issues, governance settings, policy integration, conservation priority setting, management, and participation evaluation have influence on nature conservation policy implementation.

The identified list of factors and sub-factors was discussed afterwards among researchers, university professors, and nature conservation experts. After that, it was modified in order to clarify the defined factors and sub-factors. Based on these factors and sub-factors, a template was developed as a semi-structured, problem-centred interview. The template included basic information about the aim of the study and a brief explanation of factors to ensure a common understanding among the experts.

2.2. *Phase 2—Expert Selection and Data Collection*

The prepared template was sent by e-mail to nature conservation experts in each country to get the information about the influence of operational environmental factors on nature conservation policy implementation. When selecting the experts, a priority was given to participants who were experienced with respect to nature conservation in their country or have an institutional influence on nature conservation policy formulation or implementation (Table 1). The experts (n = 44) have filled the template based on their expertise and the information gaps have been filled with information of other experts who had a good overview of nature conservation policies implementation in the country or other literature sources. The data were collected between March and May 2019.

Table 1. Experts involved in the data collection.

Country	Name of Institution/Organization/Association	Total No. of Experts
IT	Council for agricultural research and economics (CREA) (1), Forestry officials of the Regions and Autonomous Provinces (2), Forest consultants involved in the drafting of Natura 2000 site management plans (2)	5
SK	Faculty of Forestry (2), National Forestry Centre (1), State Nature Conservancy (6), Ministry of Agriculture and Rural Development (1)	10
SI	Faculty of Forestry (1), Ministry of Environment and Spatial Planning (3)	4

Table 1. *Cont.*

Country	Name of Institution/Organization/Association	Total No. of Experts
BG	University of Forestry (1), Ministry of Environment and Water (1), Executive Forest Agency (2)	4
CRO	Faculty of Forestry University of Zagreb (2), State forest company Hrvatske šume Ltd. (1), Ministry of Agriculture (1)	4
SRB	Public enterprise "Srbijašume" (1), Public enterprise "Vojvodinašume" (1), Institute of Forestry (1), Ministry of Agriculture and Environmental Protection (1)	4
MNE	Institute of Forestry of Montenegro-Natura 2000 expert for forest habitats (1), Ministry of Sustainable Development and Tourism (1), Ministry of Agriculture and Rural Development (1), Public Enterprise National Parks of Montenegro (1)	4
NM	Public Forest Enterprise (1), Faculty of Forestry (1), National Association of Private Forest Owners (1), NGO Connecting Natural Values and People - CNVP Macedonia (1), Ministry of Agriculture, Forestry and Water Management (1)	5
BiH	Public Forest Enterprise (1), Faculty of Forestry (1), National Park Una (1), Ministry of Agriculture, Water Management and Forestry of the Federation of Bosnia and Herzegovina (1)	4

2.3. *Phase 3—Qualitative Content Analysis*

A qualitative content analysis was applied on collected information about operational environment factors that influence nature conservation policies implementation. The qualitative content analysis includes contextual information, latent content, as well as formal aspects of the analysis information [57]. For analysis, we first identified and coded the parts of the documents that include general codes (factors) and then codes related to the sub-factors. Coded elements were extracted into a standardized Excel table, which enables a simplified overview of relevant operational environment factors and sub-factors and their interpretation and cross-country comparison in terms of influence on nature conservation policy implementation.

3. Results

3.1. *Legal and Policy Factors*

3.1.1. International Conventions and Agreements

In the analyzed countries, the main drivers for nature conservation came from the international environment as all countries are signatories of the most important international conventions and agreements related to nature conservation such as Ramsar Convention, Convention on International Trade in Endangered Species of Wild Flora and Fauna, Boon Convention, Bern Convention, Convention on biological diversity, United Nation Framework Convention on climate change and European Landscape Convention. All these conventions and agreements have been ratified by specific laws or decrees (Italy, Slovenia, Croatia, Bulgaria, Serbia, North Macedonia, and Bosnia and Herzegovina) and/or were implemented in Nature Conservation Act (Slovakia, Bulgaria, and Montenegro).

3.1.2. National Legal Framework of Nature Conservation

All analyzed countries have a legal framework in place to ensure and support nature conservation (Table A1). Strategies that influence nature conservation exist in all countries. They are mainly related to sustainable development (Italy, Slovakia, Slovenia, Croatia, Serbia, Montenegro, and North Macedonia), environmental protection (Slovakia, Bulgaria, Croatia, Republic of Srpska, Federation of Bosnia and Herzegovina), environmental and climate changes (Bulgaria and North Macedonia), nature protection (Croatia, Bulgaria, North Macedonia, Republic of Srpska), biodiversity (Italy, Slovakia, Slovenia, Bulgaria, Serbia, Montenegro, North Macedonia, Republic of Srpska, Federation of Bosnia and Herzegovina), and forestry (Italy, Slovakia, Croatia, Bulgaria, Serbia, Montenegro, North Macedonia, Republic of Srpska). Moreover, from the aspect of nature conservation, national programs related to this area are also very significant in some analyzed countries. They are mainly related to environmental protection—National environmental protection programs (Slovenia, Serbia, Republic of Srpska, Federation of Bosnia and Herzegovina); and forestry—National forest programs (Italy, Slovenia, Bulgaria, Montenegro, and Federation of Bosnia and Herzegovina).

The main laws that regulate nature conservation in the countries analyzed are nature protection, nature conservation law, environmental protection law, or biological diversity law. Apart from specific nature conservation-related laws all countries also have a number of ordinances, decrees, rulebooks, that further regulate specific nature conservation issues. Additionally, all analyzed countries have also adopted specific legislation governing the proclamation of PAs, mainly national parks (see Section 3.1.4).

All analyzed countries also have other sectoral laws that are relevant for nature conservation—such as forest law, water law, game management and hunting law—as nature conservation has an indispensable position in the context of other cross-sectoral policies, especially related to forest sector and therefore cannot be seen as an isolated policy sector [58].

3.1.3. Institutional Set-Up

Strong institutional set-up is one of the key prerequisites for effective nature conservation policy implementation. Table 2, summarize the institutional set-up in analyzed countries. Evidently, not all of the institutions are presented in all countries, which potentially create a gap for the successful nature conservation policy implementation.

In all analyzed countries, one central authority plays the most important role in nature conservation. In most of the analyzed countries, the Ministry of Environment with jurisdiction over environmental protection is responsible for nature conservation policies, including PAs and Natura 2000 sites. There are exceptions in Montenegro, Bosnia and Herzegovina, and the Republic of Srpska where other ministries are responsible for environmental protection and nature conservation policies (e.g., Ministry of Foreign Trade and Economic Relations in Bosnia and Herzegovina and Ministry of Spatial Planning; Civil Engineering and Ecology of the Republic of Srpska in Republic of Srpska; and Ministry of Sustainable Development and Tourism in Montenegro). Only in Slovakia and Serbia there is an independent Ministry of Environment or Ministry of Environmental Protection, while in other analyzed countries the Ministry of Environment has broader competences, e.g., protection of land and sea, spatial and physical planning, energy, water, tourism.

The presence of Institutes for nature conservation varies from country to country. They exist in Italy, Slovakia, Slovenia, Bulgaria, Croatia, Serbia, and Bosnia and Herzegovina. In countries like Croatia, Montenegro, and North Macedonia, they have experienced changes in relation to formally independent institutes as in most countries' government adopted a decree merging institutes to some other institution. All countries, except Bosnia and Herzegovina and North Macedonia, have Environmental agencies in place at the national level and also regional level as in case of Italy, with similar duties being the collection, integration and processing of environmental data and submission of reports to the European Environmental Agency as European Environmental Agency member countries. In the case of Bosnia and Herzegovina and North Macedonia they are a part of European Environmental Agency partnership network as cooperating countries.

One of the institutional mechanisms that could substantially contribute to the funding of nature conservation, particularly its tasks, is an environmental fund. This funds usually receives finances from different sources (e.g., environmental or eco-taxes, national or regional budgets, regulation fees). An environmental fund exists in all analyzed countries, except Italy, North Macedonia, and Montenegro.

In all analyzed countries, Ministry responsible for nature conservation plays an important role in monitoring of different types of PA and are responsible for the approval of management plans and programs for PA. Although governance by government is a predominant form of governance in analyzed countries, all analyzed countries have established public institutions (PA authorities) for management of PA at the national, regional or local levels. Unlike in other countries, Montenegro and Slovakia have one central public institution at the national level for management of all PAs. In Italy, Slovenia, Croatia, Bulgaria, North Macedonia, Serbia, and Bosnia and Herzegovina, separate public institutions are set up with the responsibility to manage a specific PA. Moreover, in some analyzed countries, the government can delegate PAs management responsibility to other actors, both public and private ones. The transfer of management by delegation is possible "de jure" in Slovakia, Slovenia, Bulgaria, Serbia, Montenegro, and North Macedonia. Delegation can be done either on a contract basis (Slovakia) or by a legal act, at the designation of PA (Slovenia, Bulgaria, Croatia, Serbia, North Macedonia, and Montenegro). Delegation to different types of actors (e.g., NGO, local association, local municipalities or counties, public enterprises, private companies, and churches or monasteries) was found in some analyzed countries (Slovenia, Serbia, North Macedonia, Montenegro, and Croatia).

3.1.4. Legal Status of Protected Areas

The idea of setting aside areas to safeguard for nature conservation was initiated in most of the analyzed countries in the middle of the 20th century, by designating the first national park (Table 2).

Table 2. Legal status of protected areas and their coverage.

Country	Year of the Establishment of First NP	No. of NP	Existence of Other Statutory Designated Categories of PA	% of the Country's Territory Covered by PAs
IT	1922	24	Yes	9.5
SK	1949	9	Yes	23.4
SI	1981	1	Yes	14.0
BG	1992	3	Yes	5.3
CRO	1949	8	Yes	8.5
SRB	1960	5	Yes	7.5
MNE	1952	5	Yes	12.5
NM	1948	3	Yes	8.9
BiH	1962	4	Yes	2.7

Legal protection of national park is in majority of the analyzed countries provided by law on national parks (Italy, Slovenia, Serbia, North Macedonia, Montenegro, and Bosnia and Herzegovina) or Protected areas law (Bulgaria). In countries like Italy, Bulgaria, Serbia, and Republic of Srpska, laws regarding national parks/protected areas exist on a country/entity level. In Italy, Slovenia, Montenegro, North Macedonia, and the Federation of Bosnia and Herzegovina, specific laws on the establishment of individual national park exist. In Slovakia and North Macedonia national park establishment in terms of activities granted and prohibited is regulated by nature conservation/protection law, while the national park itself is created by the government ordinance/law. In Croatia, PAs are established by the government or parliament ordinance.

Other statutory designated categories of PA also exist in these countries, mostly harmonized with the IUCN categorization. According to the national data, PAs cover between 2.7% in Bosnia and Herzegovina and 23.4% in Slovakia of the country's territory.

3.1.5. Emerald Network

The Emerald network of the Council of Europe is declaratory complementary to the Natura 2000 network outside EU.

The creation of the Emerald Network was started with a help of different pilot projects in all analyzed countries, except Italy. The main purpose of those projects was to initiate the process, create the expert teams, and set up a database of pilot project sites and proposed Areas of Special Conservation Interests (Table 3).

Table 3. Status of the Emerald network.

Country	Emerald Network Status	Approved by Standing Committee of Bern Convention	Approved by the National Authority
IT	Emerald network is not implemented.	-	-
SK, SI, BG, CRO [1]	Participated as pilot countries in the first round of Emerald projects; Emerald network is implemented (Areas of Special Conservation Interests)	Yes	No
SRB, MNE, NM, BiH	Proposed Areas of Special Conservation Interests, officially nominated	No	No

[1] Areas of Special Conservation Interests were approved by the national authority.

In Slovakia, Slovenia, Croatia, and Bulgaria, the pilot project of Emerald network creation started in 1999, but stopped, due to the countries' orientation to the EU process and implementation of the Natura 2000 network. All relevant data and results gained within the Emerald pilot project were used for the implementation of Natura 2000 that later became the formal contribution to the Emerald network.

In case of Croatia, Serbia, Montenegro, North Macedonia, and Bosnia and Herzegovina, EU CARDS Regional Project "Development of Emerald Network in South-Eastern Europe" was implemented with the financial contribution of the European Environmental Agency to the Council of Europe [59]. Further activities in these countries were performed under the EU IPA projects [60]. As a result of these projects, countries have nominated their candidate sites. All future activities in these countries related to the Emerald network are dependent on the further national allocation of money.

3.1.6. Transposition of the EU Birds and Habitats Directives

The countries involved in this study are at different stages of transposition of the "Birds" and "Habitats" Directives into national legislation (Table 4), considering the country's status regarding the EU accession. EU Member States countries have transposed the Directives and implemented Natura 2000 network. This network has been implemented gradually, starting in 1997 in Italy, followed by other countries joining the EU afterwards (Slovenia and Slovakia in 2004, Bulgaria in 2007 and Croatia in 2013).

The EU candidate and potential candidate countries have, in compliance with the acquis obligations, to implement and enforce environmental policies in their legal framework, but enforcement and implementation of the nature conservation policy are still at the early stage because there are still significant gaps in transposition and site designation. For example, these countries have designated Emerald network sites (future Natura 2000 sites), but more effort should be put into the selection of Special Protection Areas (SPAs) and Sites of Community Importance (SCI) under the Directives. In countries like Montenegro, North Macedonia, and Bosnia and Herzegovina, potential SPAs and SCI sides are proposed, while in SRB only potential SPAs are proposed. These proposed sites will in the future act as potential sites for inclusion in the Natura 2000 network.

In most of the analyzed countries, the "Habitats" Directive is mainly transposed through the Law on nature protection/conservation (Slovakia, Slovenia, Serbia, Montenegro, North Macedonia, and Bosnia and Herzegovina) and its decrees (as in case of Serbia, Italy, Slovenia, Croatia, and the Federation of Bosnia and Herzegovina). The "Birds" Directive is also transposed by the Law on nature protection/conservation (Italy, Slovenia, Croatia, Serbia, and North Macedonia) and Law on Game and

Hunting (Italy, Serbia, and Bosnia and Herzegovina) and Forest Law (Croatia). In Bulgaria, Natura 2000 sites are not considered as PAs, they are called protected sites and therefore designated under the biological diversity law, not the protected areas law. Therefore, also the Directives are transposed through the biological diversity law.

According to EEA [61], the Natura 2000 network now covers 18.0% of EU's land territory in EU-28. Currently, the Natura 2000 network covers 37.2% of land territory in Slovenia, followed by Croatia (36.7% of the land territory), Bulgaria (34.4% of the land territory), and Slovakia (30.0% of the land territory). These countries are also ranked as countries with the highest share of Natura 2000 area within EU-28. In Italy, the Natura 2000 network covers about 19.0% of country's land territory. In analyzed 'new' member states in Eastern and Southeastern Europe, the share of national land territory under Natura 2000 is well above the EU average. In the rest of the study area, larger Natura 2000 sites were designated as a consequence of overlapping with (but sometimes also exceeding) existing nature PA, but in the old member states, like Italy, smaller sites were established [8].

Table 4. Status of the Natura 2000 network [61].

Country	No. SPA	No. SCI	% of the Country's Territory under Natura 2000	% of the Forests inside Natura 2000
IT	631	2.335	19.0	35.0
SK	41	642	30.0	48.8
SI	31	324	37.2	71.0
BG	119	233	34.4	56.5
CRO	38	741	36.7	36.0
	No. pSPA	**No. pSCI**		
SRB [1]	43	-	-	-
MNE [2]	X	X	16.0	9.0
NM [1]	3	6	-	-
BiH [1]	1	72	18.7	-

[1] Data about Natura 2000 coverage of country territory and % of forest inside Natura 2000 are not available. [2] Some SPA and SCI sides are proposed but the process is not finished.

Forests are of crucial importance for Natura 2000. According to the European Commission [62] it has been estimated that the Natura 2000 network includes approximately 375,000 km^2 of forests, which is about 50% of the total Natura 2000 network and about 21% of total forest resource in EU. In Slovakia, Slovenia, and Bulgaria, forest is included in more than half of all proposed Natura 2000 sites (48.8%, 71.0%, and 56.5% respectively) [63,64]. The high percentage of forests in Natura 2000 reflects not only the wide distribution of forests in these countries but also their overall importance for biodiversity. In Croatia, forests cover approximately 36.0% of Natura 2000 network and in Italy 35.0% of the Natura 2000 network [65].

3.1.7. Management Plans for Natura 2000 Network

After the designation of the Natura 2000 sites, policy enforcement and practical management becomes the primary task for national authorities [8]. According to the "Birds" and "Habitats" Directives, the preparation of management plans for the Natura 2000 sites is not obligatory, but the "Habitats Directive" recommends their use as a means to secure the beneficial conservation status of the sites. In spite of the soft regulation, the preparation of site-level management plans is promoted in all selected EU countries as the main tool to identify conservation measures at the site level (Table 5). Some of the analyzed EU countries consider the development of a management plan for sites as a legal obligation (Slovenia, Slovakia), or/and make a possibility to include specific management measures in other sectoral plans related to management and use of natural resources (forests, water) or include them in contractual obligations, as recommended by Article 6 of the "Habitats Directive" (Slovenia, Bulgaria, Croatia). In Italy, according to the Decree no. 224 (2002), there is a legal obligation to define the conservation measures for the Natura 2000 sites, while management plan is an additional instrument aimed to protect threatened species and habitats.

Natura 2000 management plans are prepared under the responsibility of the ministry responsible for nature conservation in almost all analyzed EU members' countries, except Croatia, assisted by facilitators (experts), with the provision of EU financial sources and the national budget. In Croatia, directives are implemented in forest management plans under the responsibility of the Ministry of Agriculture. These management plans are developed at the national (Slovakia, Slovenia, Bulgaria) or regional/local level (Italy).

Table 5. Synthetic overview of legal requirements in Natura 2000 management planning.

Country	Legal Obligation for Management Plans	Management Plans are Developed
IT	No	Yes (R)
SK, SI	Yes	Yes (N)
BG	No	Yes but not for all sites (N)
CRO	No	No

N: national level, R: regional level.

3.2. *Economic Factors*

There are numerous economic factors that must be considered in the analyses of nature conservation policies. The most noticeable is funding as this was identified in many studies as "the biggest issue", and that it is the key factor of further development of nature conservation [8,32,66].

3.2.1. Financing Mechanisms of Protected Areas

Well-developed legal frameworks are in place in the analyzed countries (see Section 3.1), which underline the need for and importance of funding of PAs. All analyzed countries have ratified international conventions, which call in some way for contracting parties to allocate funds for nature conservation. Therefore, in analyzed countries laws (mainly nature protection/conservation law) and decrees related to PAs prescribe different financing mechanisms.

Financial resources for PAs may be generated by different sources (Table 3); most commonly these are external sources and market-based fees for goods and services.

In all analyzed countries, financing of PAs is ensured from the state budget funds, through the ministries responsible for nature conservation. To a much lesser extent, financing comes from municipal budget (Slovenia, Croatia, Serbia, Montenegro, North Macedonia, and Bosnia and Herzegovina) or province/regional budget (Italy, Croatia, Serbia, and Bosnia and Herzegovina) in the case where the PAs are established by the municipality or province/regions. In addition, Slovenian law on nature conservation permits a possibility of public/private partnership on the management of PAs. Moreover, in countries like Slovakia, Serbia, Montenegro, North Macedonia, and Bosnia and Herzegovina, international assistance and funding or private voluntary donations (Slovakia, North Macedonia, and Bosnia and Herzegovina) of PA is provided. In Slovakia, Croatia, and Serbia, the Environmental Protection Fund is established to provide financial support for environmental protection and sustainable development to applicants in the form of grants or loans in support of projects and activities aimed at achieving environmental policy goals at national, regional, or local levels.

PA management bodies are offered the possibility to supplement their budget financing through income-generating activities—market-based fees for goods and services (e.g., entrance fees in PAs, entrance fees in visitor or information centres, informational materials, tourism activities, etc.), such revenues represent in most cases a fairly small share from their total annual budget (Italy, Slovakia, Slovenia, Croatia, Serbia, North Macedonia, and Bosnia and Herzegovina), except when income is generated from management of lands and resources inside the PAs (Italy, Slovakia, Slovenia, Bulgaria, Serbia, Montenegro, North Macedonia, and Bosnia and Herzegovina) or concessions of management or resource use rights (Bulgaria, Croatia, Serbia). Another financing opportunity of PAs is represented by the existence of grants that can be accessed through projects, which becomes a quite common and important source of funds in all analyzed countries.

In countries like Slovakia, Bulgaria, Croatia, North Macedonia, Montenegro, and Bosnia and Herzegovina, a general characteristic of financing nature conservation and PAs is their large dependency on external funding sources, which finance various nature conservation projects. The European Fund for Southeast Europe has also played an important role in financing different activities in Slovakia. In addition, external funding also comes from EU funds (Italy, Slovakia, Slovenia, and Bulgaria), or EU-accession funds (Croatia, North Macedonia, and Bosnia and Herzegovina). Moreover, in Italy, Croatia, Montenegro, and North Macedonia, innovative instruments—such as payments for ecosystem services (hereafter PES)—have been developed and represent a financing source of PAs.

3.2.2. Financing Mechanisms of the Natura 2000 Network

The EU's integrated approach to financing Natura 2000 has resulted in a complex funding structure [30]. In accordance with the Treaty on the Functioning of the EU [67], the responsibility for implementing and financing environmental policy, including the Natura 2000 network, lies with the Member States. Article 8 of the "Habitats" Directive states that in a case in which the respective Member States face exceptionally high costs, these costs can be co-financed by any relevant EU co-financing instrument and foresees the need to develop a prioritized action framework [5]. According to the aforementioned article, implementing institutions have two options for funding the necessary measures in the Natura 2000 network: to use either their existing state nature conservation budgets or the EU co-financed instruments.

National public funding is the most frequently mentioned source of financing Natura 2000 objectives and measures in all analyzed EU member countries (Table 6), because funding of nature conservation measures is prescribed by nature protection/conservation law. In Slovakia and Croatia, national environmental protection funds have been mentioned as an important additional source of financing Natura 2000 objectives and measures.

Table 6. Overview of funding instruments for Natura 2000 objectives and measures for 2014–2020 in analyzed EU member countries.

Funding Instruments	Country				
	IT	SK	SI	BG	CRO
National public funds	X	X	X	X	X
National environmental protection fund	-	X	-	-	X
European Agricultural Fund for Rural Development	X	X	X	X	X
European Agricultural Guarantee Fund	X	-	-	-	-
European Regional Development Fund	X	X	X	X	X
European Social Fund	X	-	-	X	-
European Maritime and Fisheries Fund	X	X	X	X	X
Cohesion Fund	X	X	-	X	-
Financial Instrument for the Environment (LIFE/LIFE+)	X	X	X	X	X
Framework Programme for Research and Innovation (FP7, Horizon 2020)	X	X	X	X	X
Public/Private Partnership financing schemes	X	-	X	-	-

Moreover, the objectives and measures of Natura 2000 in analyzed EU member countries are supported through the European structural and investment funds, with the majority of their investment being handed out through national governmental institutions (Table 6).

In analyzed countries objectives and measures of Natura 2000 were co-funded through several EU instruments. These available EU-level funding instruments can cover only a small amount of the estimated costs of the implementation of Natura 2000 and its measures. Therefore, in countries like

Slovakia and Croatia National environmental protection funds promote advances in environmental protection through the award of credits or other financings.

Additionally, in Slovenia the legislation on nature conservation allows public/private partnership on the management of Natura 2000 sites. Such is the case in the management of Sečoveljske soline Landscape Park. In Italy, 76 Natura 2000 sites located in the WWF Oasis are managed by WWF in accordance with the national and regional/provincial guidelines.

In the EU candidate and potential candidate countries, initial work on the establishment of Natura 2000 network started in the frame of the different project, mostly financed by EU IPA program (Serbia, North Macedonia, Montenegro, and Bosnia and Herzegovina), foreign governments (Bosnia and Herzegovina), bilateral aid agencies (North Macedonia and Bosnia and Herzegovina), and NGOs (WWF in Montenegro).

The transposition of the "Habitats" Directive requires compensatory measures necessary to ensure that the overall coherence of Natura 2000 is protected (Article 6). Therefore, different compensation measures are assured in current EU funding programs (2014–2020), such as European Structural and Investment Funds, the Rural Development Fund, the Cohesion Policy Funds, and the European Maritime and Fisheries Fund.

As regards the compensatory measures included in the Rural Development Programme 2014–2020, that are relevant for Natura 2000 three measures have been identified in analyzed EU member countries. The most widely used measures for the conservation or recovery of natural habitats and species are the agri-environment and climate payment measures (M10) which exists in all of the analyzed countries. In Italy and Slovakia, payments for Natura 2000 areas in combination with the Water Framework Directive (M12), which is aimed at activities on agricultural land (M12.1) and forest land (M12.2), also exists. In particular, the measure M12.2 (Payments Natura 2000 for forest areas) has been implemented in five regions (Basilicata, Liguria, Marche, Piemonte, Umbria) in Italy. Forest-environmental and climate commitment payment measures (M15) are included in Bulgaria, Slovakia, and certain regions of Italy. In addition, in Slovenia a budgetary forest fund was established in 2016 based on the changes in the organizational structure of the state forest management body and the adoption of the Management of State Forest Act (Article 33). The forest fund intended to cover compensation measures in private forests within Natura 2000 areas, in accordance with the Natura 2000 management program and program of investments in forests, based on the national forest program prepared by the public forestry service in accordance with the forest act.

3.3. *Social Factors*

Stakeholder participation, cooperation, and their ability to influence the policy decision-making process are, as identified in numerous studies (e.g., [8,40–42,52]), the key factors for policy implementation success.

3.3.1. Stakeholders' Involvement in the Establishment and Management of Protected Areas

Views about PAs and stakeholders' participation (mainly local population and public) in decision making concerning the establishment and management of PAs have changed considerably over the last decades in all analyzed countries.

As former socialist states with communist experiences, each of the analyzed countries, (with the exception of Italy), has its own peculiarities in the development of participatory decision-making related to nature conservation matters [68]. In these countries, a self-management system was based on decentralized powers of local authorities, but the true decision-making power was never given to the local people in the areas of nature conservation and PAs. After becoming independent in the early 1990s, countries started to implement international agreements related to nature conservation and EU nature conservation policies. Consequently, the concept of participation changed, and PAs started being treated differently. Moreover, in these countries, the legal framework has changed, which requires increased stakeholder involvement in PAs management. Therefore, in these countries, a shift from the 'traditional paradigm' in which PAs were established by the central government, to the 'new paradigm', where the cooperation among different stakeholders (i.e., local stakeholders, NGOs, private companies, regional, and local authorities) in the governance of PAs has taken place. As a result, local stakeholders are no longer passive recipients of the top-down approach; rather, they are active partners. In Italy, the participatory process is related to the administrative decentralization established by Decree no. 616 (1977). The legal competences in agriculture, forestry, and land management have been transferred to the regions and autonomous provinces. Regarding nature conservation, Law no. 394 of 1991 states the statute of each public authority delegated to the management of the parks establishes the criteria of public participation in the decision-making process. Generally, the local public administrations—e.g., provinces and municipalities—are involved in the environmental governance, while the involvement of other stakeholder changes from case to case (from information to collaboration).

In the case of Italy, Croatia, Montenegro, and North Macedonia, establishment and management of PAs has been mostly characterized as a top-down approach, although PAs' establishment and management followed participatory procedures in these countries. In Slovakia, Slovenia, Bulgaria, Serbia, and Bosnia and Herzegovina, also a top-down approach in establishment and management of PAs prevails, but in some cases a bottom-up approach has been applied. In countries like Serbia, North Macedonia, Montenegro, and Bosnia and Herzegovina, the participatory approach is a relatively new concept for PAs as most of the PAs were established with very low participation of other stakeholders; especially on the local level. This practice has been changing mostly in the case of management of PAs in these countries, but the value of involving different stakeholders has yet to be fully recognized. Certain changes are evident in these countries as well as Slovakia and Slovenia because the legal framework has changed, which necessitates increased stakeholder involvement in the management of PAs (Table 7).

Local stakeholder involvement is a known important factor in the establishment and management of PAs, and their participation in the decision-making process is ensured in almost every analyzed country (Table 7).

Table 7. Overview of the participatory process in the case of protected areas.

Country	Approach	Participatory Process Organized	Local Stakeholders Involved
IT	Top-down	Variable from case to case	Variable from case to case
SK, SI, BG, SRB [1], BiH	Different from case to case (from top-down to bottom-up)	Variable from case to case	Yes
CRO, MNE, NM	Top-down	Yes	Yes

[1] Local stakeholders were involved in the case of national park; but in the case of other PAs, local stakeholders' involvement depends on activities of PA managers.

3.3.2. Stakeholders' Involvement in the Implementation of the Natura 2000 Network

The habitats directive as the legal basis for the Natura 2000 network does not pay attention to the issue of public participation. This leaves countries with considerable leeway as to how they deal with the new possibility of participatory approaches to the implementation of the Natura 2000 network [41,69]. In all analyzed countries, the participatory process has been applied (Table 8); however, performance is still low in terms of engaging key stakeholders and sharing decision-making power with them.

In all analyzed EU countries public actors have ultimate decision-making power. Therefore, the process of implementing the Natura 2000 network was a controversial top-down process in a majority of EU analyzed countries. An exception is Italy, where most of the implementation responsibilities and tasks were, according to the Decree no. 357 of 1997, delegated to the regions and autonomous provinces. Consequently, the approach differed from one region to another.

Table 8. Overview of the participatory process in the case of the Natura 2000 network.

Country	Approach	Degrees of Involvement [1]
IT	From top-down to bottom-up	Consultation Information
SK, SI, BG, CRO	Top-down	Co-decision Collaboration Information

[1] A four-level system was adopted by Herwig (2008) [69] which distinguishes between four degrees of involvement: information, consultation, collaboration, co-decision.

The analyzed countries varied in their implementation styles regarding the degree of stakeholders' involvement. For example, in countries like Slovakia, Slovenia, Bulgaria, and Croatia, almost all degrees of involvement, except consultation, were adopted during Natura 2000 network implementation. In Italy, consultation and information were the primary degrees of stakeholder involvement adopted during Natura 2000 network implementation.

4. Discussion

The cross-country analysis of national nature conservation systems and policy implementation reveals several similarities and only a few differences among the analyzed countries. What makes these systems similar is first that they are a part of strong EU nature conservation efforts and second, that the EU acquis has provided opportunities and financial (material) benefits to environmental state authorities and non-state actors through funding and capacity building projects supported by the European Commission and other old EU Member States. However, analyzed countries differ in socio-economic and political contexts backgrounds. The differences exist mostly in the institutional set-up and the method with which the countries transposed international obligations (non-EU countries) together with EU legislation (EU members' countries) into their respective national legislation, primarily regarding management plans for PAs.

4.1. Socio-Economic and Political Context as the Main Driver of Nature Conservation Changes

All analyzed countries, except Italy, share a similar history in terms of regime changes. They have transitioned from socialism to democratization with a more recent period of Europeanization. Changes after the political and economic transition in the 1990s shifted public focus to both nature conservation and the use of natural resources in these countries. Moreover, the changes made nature conservation a higher priority within the state, which resulted in the expansion of PAs. The transition to democracy led to new political and legal frameworks while also bringing change to institutional settings. Accession to the EU became an additional layer in the transition process that brought new policy elements, such as the Natura 2000 network.

The legal issues of nature conservation are dealt with in a comprehensive political and legislative process that spans from the international and European level down to the national legal framework. Despite the fact that a legal framework of nature conservation was in place in all analyzed countries even before the negotiation process with the EU started, the gap between national nature conservation systems and the EU's requirement was quite wide. It required the adoption of a whole set of new legal rules. The Environmental Protection or Nature Conservation Act established new environmental principles and the legal basis for nature protection. Based on the new legal framework in each of the analyzed countries, all important nature conservation-related international conventions and agreements have successfully been transposed and several types of PAs have been established in harmony with the IUCN. Therefore, PAs are one of the most important nature and forest conservation measures in all of the analyzed countries. Moreover, countries have successfully transposed the "Habitats" and "Birds" Directives into national legal order that fulfills the requirements under the acquis but the implementation process needs to be improved. Also, the comprehensive evaluation of the Directives (known as the "Fitness Check") undertaken by the European Commission highlighted that the full potential of Directives can only be accomplished by substantially improving their implementation [70].

After the changes in the political system and following the EU requirements, each of the analyzed countries began to adapt their respective institutional context of nature conservation gradually. According to our study, which is in line with observations of Kluvánková-Oravská et al. (2009) [13] and Falkner et al. (2008) [28], an institutional misfit was inevitable because there was a need to change crucial domestic institutions. The primary approach of eliminating this misfit has been through the incremental-transformation type since some institutions already existed, albeit most were rearranged, merged, and given more resources. In Italy, nature conservation policy has changed from a centralized first phase—where the main actor was the Ministry of Environment—to a second decentralized phase where the main actors are regions and autonomous provinces. In almost all analyzed countries, a well-rounded institutional set-up exists, which include a central authority for the environment, a ministry related to environment or nature conservation/protection, institutes for nature conservation, and management authorities for PAs (public institutions) at the national, regional, or local levels. Additionally, in some countries these institutional set-ups are complemented by a specialized environmental agency and an environmental fund.

In countries like Croatia, Montenegro, North Macedonia, and Bosnia and Herzegovina not all of the institutions are represented, which potentially creates a gap for the successful implementation of nature conservation policy. These findings are in line with observations of other studies [14,60] where it was found that strengthening institutional structures and administrative capacities are necessary to ensure continued and long-term efficiency.

4.2. Economic Factors as the Major Obstacle of Nature Conservation Development

By establishing a network of PAs, analyzed countries governments become responsible for funding the conservation and maintenance of these areas. Many of these pledges to fund PAs are enforced by national policies. In all analyzed countries, financing of PAs is ensured from the state budget funds and it is expected that these funds will remain at the core of long-term funding. Although the PA management institutions have the possibility to supplement their budget through income generated activities, such revenues represent a fairly small share of their total annual budget in analyzed countries. For that reason, current sources of PA funding in analyzed countries are not sufficient to maintain and expand PA networks; therefore, financial constraints act as a major obstacle for effective nature conservation development in PAs both now and in the future. In order to overcome this, it is important that countries develop and expand the innovative PA financing mechanisms that have emerged in some of the analyzed countries. Noted innovative financing mechanisms with high potential in the analyzed countries include grants or loan schemes, public/private partnership and PES. In other studies (e.g., [71–73]) PES specifically has shown high potential as a mechanism to generate funding for PAs and nature conservation in general and therefore could be applied also in our analyzed countries; especially as those kinds of innovative mechanism already exist. Such mechanisms definitely offer the greatest chance of substantially increasing PA funding in the future. Additionally, they can help stimulate broader improvements in PA management and sustainability. Moreover, in the analyzed countries other opportunities to improve the financial sustainability of PAs also exist; particularly, in some analyzed countries a delegation of PAs management to other actors is possible de jure. In many cases, these actors seek to mobilize different financial resources mainly through commercial and market-based activities. While there are many opportunities in analyzed countries to improve PAs financial sustainability, there are also some challenges that need to be overcome. Notable among these challenges is the common dependence of PAs on international funding assistance and funding from projects. This opportunity is important not only for EU countries, but also countries which are in the process of EU integration as they have opportunities to be involved in the EU project. However, this possibility is often held back by the limited capacity of the PA staff and the possibility of the state institutions to ensure co-financing of projects. The same limitations were reported in other studies (see e.g., [66,74]). Based on that it can be said that diversification of funding is a solution to ensuring the long-term financial sustainability of PAs.

The responsibility of implementing and financing the Natura 2000 lays with the EU member states. Therefore, national public funding is the most frequently mentioned source of financing Natura 2000 objectives and measures in all analyzed EU member countries as well as the scientific literature [9,23,30,32,75]. Additionally, the EU offers funding opportunities for Natura 2000 through a so-called 'integrated approach'. Currently, the financing of Natura 2000 is delivered through several EU instruments in all analyzed EU countries. The availability of funding to cover the costs of Natura 2000 implementation is frequently insufficient in analyzed countries. This lack of funding is viewed as a major obstacle for the effective implementation of Natura 2000 not only in analyzed countries but also in other EU countries (see e.g., [20,23,24,30]). Since the use of EU funds seems to be hardly guided by the aim of compensating for the disadvantages of the Natura 2000 designation for forest owners, their effectiveness and efficiency is questionable. A better understanding of the reasons behind this use of EU funds requires an in-depth analysis.

In addition, in some analyzed countries additional innovative funding instruments have been developed. For example, in Slovakia and Croatia National environmental protection funds help finance management costs, and public/private partnerships in Slovenia and Italy offset the price of maintaining PAs. These innovative financing instruments are also suggested by the European Commission, although such instruments are covering a rather small portion of the total financing requirements. This claim for innovative financing mechanisms is also supported by the scientific literature [16,75]. In analyzed EU candidate and potential candidate countries, with the EU Accession, huge financial resources have become available for different nature conservation-related projects. The incoming EU funds started to play an important role in financing nature conservation activities. In addition, foreign governments, bilateral aid agencies, and NGOs were noted as funding sources, but they play minor roles.

4.3. Stakeholders' Engagement and Participation—Main Failure of Effective Nature Conservation Implementation

On the European level, there are no legally binding obligations to organize a participatory process in establishing either PAs or Natura 2000 network. This in turn leaves the countries unlimited freedom to introduce elements of participation on their own. In analyzed countries, the participatory process was dependent on the national political context and the historical development of participatory approaches in public policy, which was also reported in other studies [15,37].

The analysis of participatory processes shows that they are in an initial phase in the case of PAs. To a certain extent it can be said that, a shift from the 'traditional paradigm' in which PAs were established by the central government, to the 'new paradigm', where the cooperation among different actors (i.e., local stakeholders, NGOs, private companies, regional, and local authorities) in the governance of PAs has happened in analyzed countries. This shows a positive trend from a command-and-control approach in establishing the PAs towards a participatory approach, not only in analyzed countries, but also in other EU countries [14,76–78]. This paradigm shift is the result of a growing social demand for a direct involvement of citizens in environmental governance. Concerning the management planning of PAs, a wide variety of actors from different levels are most of the times, at least de jure, involved in the process. However, not all of the requisite conditions for successful participation were created (i.e., equally engagement of stakeholders, communication with some of key stakeholders) in analyzed countries, as quite often still a top-down approach was applied in final decision-making. Other studies [51,79–81] also show the same conditions for successful participation (i.e., stakeholders' engagement, planning, and implementing the communication) were not achieved. Therefore, both managers of PAs and actors need to develop their capacities for successful stakeholder participation.

The situation is slightly better in case of Natura 2000 network implementation. In the implementation of the Natura 2000 network, analyzed EU Member countries differ in their implementation style with regard to stakeholder involvement through both the level of involvement and the approach that was adopted. However, the overall impression is that, the level of participation is strictly linked to the stakeholders' power to influence the final decision in the participatory process. According to the results, the ultimate decision-making power was held by public actors (ministries related to nature conservation) and they were the ones deciding who may participate in what form by aiming to fulfil legal requirements rather than empowering the stakeholders. In all analyzed countries, the initial process was a controversial top-down conservation science-based approach. Only later and mainly in response to the need to manage the substantial conflicts, the implementation approach shift to a more socially inclusive and participatory bottom-up approach came about [8]. Sotirov et al. (2015) [14] called such effect "symbolic transformation", where informal institution and practical behavior did not change in line with formal domestic policy and institution. Many studies have shown similar problems regarding the influence and the power of stakeholders that emerged as a consequence of the adopted top-down approach [13,29,37,38,40,41].

The degree of stakeholder participation varied among analyzed countries and even from one Natura 2000 site to another (Italy). The results point out that public actors have been involved through the co-decision during the participatory process with the supervision of the ministry responsible for nature conservation. Non-state actors have been involved through consultation or collaboration. The general public and local stakeholders have been involved mainly through information. Other transition countries (the Czech Republic, Poland, Romania, and Hungary) have shown similar problems with regards to stakeholders' participation [13,27,38,39,42].

4.4. *Methodological Viewpoints*

When interpreting the results of this study, it should be noted that the analytical framework for evaluation of operational environment factors related to nature conservation policy implementation included a relatively small number of experts mainly from state-level institutions and included only a few local experts. The results may thus not be considered as statistically representative for the case countries. However, the number of experts (4–10 experts in each country) was big enough to show that the developed framework is functional and provides aspired results. In comparison to other studies—e.g., [82–85]—the total number of experts was satisfactory. In addition, the background and expertise profiles of experts varied between countries, which may have affected the results. However, the experts were selected based on their experiences in the field of nature conservation and consulted to objectively evaluate the operational environment factors. Moreover, the present study does not reveal the opinion of the local stakeholders or the general public. A strength point of the study concerns the large number of different countries involved in the analysis (n = 9), addressing different political and socio-economic context. Besides, the analyzed countries differentiate regarding EU membership status. These differences among countries allowed a wider comparison of the nature conservation policy at national level in a context of increasing Europeanization.

The developed analytical framework for evaluation of operational environment factors influencing nature conservation policy implementation showed to be functional as the methodological approach followed rather similar procedures in different countries. The use of the common framework and factors enabled comparisons between countries. Therefore, presented framework may be used and replicated in other countries. In the future, the developed framework needs to be tested in different socio-economic and political contexts by including additional case studies countries to allow for a more comprehensive applicability. The weakest point of the developed framework is that the framework allows including only the most important general factors and sub-factors in the analysis (based on a literature review, but including expert opinion), but did not allow countries to modify them; therefore, inclusion of country-specific sub-factors was limited. Diverging factors such as environmental factors, ownership structures and resource rights, resource management, as well as perceptions, attitudes and values of various stakeholders can be additional factors that potentially might influence nature conservation policy implementation. Therefore, these factors might be included in the future in an analytical framework, for a broader applicability and robustness of the proposed framework.

5. Conclusions

This study developed and tested a new analytical framework for the evaluation of four separate operational environmental factors (legal, policy, economic, and social) that influence the implementation of nature conservation policy in several EU and non-EU countries.

Our analysis has shown that the legal framework is well established in all countries taking into account international, national, sub-regional, and regional obligations in nature conservation. Therefore, in all analyzed countries, legislative framework related to nature conservation is very complex as numerous new strategies and laws related to nature conservation have been adopted and integrated into different sectors. In all analyzed countries, a well-rounded institutional framework exists in analyzed countries—but still not all of the required institutions—are represented, which potentially creates a gap for the successful implementation of nature conservation policy.

Economic factors were seen as the major obstacle for nature conservation development. In all analyzed countries, diversification of funding of PAs exists and may be seen as a prerequisite for ensuring the long-term financial sustainability of PAs, even if financing of PAs is nowadays ensured mainly from state budget. This is evidence that there are many opportunities in the analyzed countries to improve PAs' financial sustainability (from state budget funding to market-based funding), but there are also some challenges that need to be overcome. Establishment of PAs is often dependent on international funding assistance and funding from projects, and PAs staff has a limited capacity for implementing these projects. With regard to Natura 2000, even though the EU offers funding opportunities for Natura 2000 through a so-called 'integrated approach', the availability of funding to cover the costs of Natura 2000 implementation and further management is frequently insufficient in analyzed countries. This lack of funding is viewed as a major obstacle for the effective implementation of the Natura 2000. Thus, in some analyzed countries, additional innovative funding instruments have been developed as for example national environmental protection funds and public/private partnership.

Stakeholders' engagement and participation has shown as the main failure of effective nature conservation implementation since not all the requisite conditions for successful participation were created in analyzed countries. Commonly, a top-down approach was applied in final decision-making and not all stakeholder groups were engaged. It can be concluded that there is a possibility for improvement in terms of continuity and transparency of participatory processes, especially by involving the general public and non-state actors from the beginning and throughout the whole policy decision-making process.

Author Contributions: Š.P.M., T.L., and A.P. designed the research; A.P. collected data for IT; Z.D. collected data for SK; Š.P.M. and T.L collected data for SI; E.K. collected data for BG; S.P. collected data for CRO; I.Đ. collected data for SRB; A.S. collected data for MNE; P.T. collected data for NM; and B.M. and M.A. collected data for BiH; Š.P.M., T.L., and A.P. wrote the manuscript; Z.D., I.Đ., and B.M. reviewed the first version of the manuscript and made editing.

Acknowledgments: First authors wish to thank to all the interviewees who took part in this research and make it possible. Additionally, authors would like to thank Andrew Franks of the University of Tennessee for his proof-reading. Finally, the authors wish to thank to the editors and anonymous reviewers for their contributions to the improvement of our manuscript.

Appendix A

Table A1. National legal framework relevant for nature conservation.

Regulation Area	National Legislation	Country									
		IT	SK	SI	BG	CRO	SRB	MNE	NM	BiH RS 2	BiH FBiH
Sustainable development	National sustainable development strategy	X	X	X	-	X	X	X	X	-	-
Environmental protection	Environmental protection strategy	-	X	-	-	X	-	-	-	X	X
	Environmental and climate change strategy	-	-	-	X	-	-	-	X	-	-
	National environmental protection program	-	-	X	-	-	X	-	-	X	X
	Environmental protection law	X	X	X	X	X	X	X	X	X	X
Nature protection	National strategy for nature protection	-	-	-	X	X	-	-	X	X	-
	Nature protection/conservation law	X	X	X	-	X	X	X	X	X	X
Biodiversity conservation	Biodiversity conservation strategy	X	X	X	X	-	X	X	X	X	X
	Biodiversity law/biological diversity law	-	-	-	X	-	X	X	-	-	-
Protected areas	Protected areas act or decree	X	-	X	X	X	X	X	X	-	X
	Law on national parks	X	-	X	X	-	X	X	-	X	X
Forestry	Forestry development strategy	-	X	-	X	X	X	X	X	X	-
	National forest program	X	-	X	X	-	-	X	-	-	X
	Forest law	X	X	X	X	X	X	X	X	X	-
Water resources	Law on water	X	X	X	X	X	X	X	X	X	X
Game management	Law on game and hunting	X	X	X	X	X	X	X	X	X	X

[12] The complex BiH political set-up has led to an absence of the nature conservation legislation at the national level. Hence, nature conservation is regulated through the laws adopted and implemented at the entity level (Republic of Srpska—BiH RS, the Federation of BiH —FBiH and Brčko District as separate district).

Table 2. Main institutions responsible for nature conservation.

Institutions	Country								
	IT	SK	SI	BG	CRO	SRB	MNE	NM	BiH
Ministry	X	X	X	X	X	X	X	X	X
Institutes for nature conservation	X	X	X	X	-	X	-	-	X
Environmental funds	-	X	X	X	X	X	-	-	X
Environmental agency	X	X	X	X	X	X	X	-	-
Public institution (protected area authority)	X	X	X	X	X	X	X	X	X

Table 3. Overview of financing mechanisms according to the source of funds in protected areas.

Source of Funds	Country								
	IT	SK	SI	BG	CRO	SRB	MNE	NM	BiH
External sources									
State budget	X	X	X	X	X	X	X	X	X
Province/regional budget	X	-	-	-	X	X	-	-	X
Municipal budget	-	-	X	-	X	X	X	X	X
Environmental protection funds	-	X	-	X	X	X	-	-	X
International assistance/funding (agencies, foreign governments)	-	X	-	-	-	X	X	X	X
Private voluntary donations/private foundations	-	X	-	-	-	-	-	X	X
Public/private partnership	-	-	X	-	-	-	-	-	-
Market-based fees for goods and services									
Tourism charges (entrance fees, parking)	X	X [1]	X	-	X	X	-	X	X
Management of land and resources inside PAs	X	X	X	X	-	X	X	X	X
PES	X	-	-	-	X	-	X	X	-
Revenues generated in performing activities (providing lodging, food and beverage, guiding tourists)	X	X	X	-	X	X	-	X	X
Concession of management or resource use rights	-	-	-	X	X	X	-	-	-
Funds provided for implementation of projects	X	X	X	X	X	X	-	X	X

[1] Fee is collected only for parking. According to the legislation entrance fees into national park and PA cannot be charged.

References

1. Evans, D. Building the European Union's Natura 2000 network. *Nat. Conserv.* **2012**, *1*, 11–26. [CrossRef]
2. Šobot, A.; Lukšič, A. The Impact of Europeanisation on the Nature Protection System of Croatia: Example of the Establishment of Multi-Level Governance System of Protected Areas NATURA 2000. *Soc. Ekol. Časopis Ekološku Misao I Sociol. Istraživanja Okoline* **2016**, *25*, 235–270. [CrossRef]
3. Kati, V.; Hovardas, T.; Dieterich, M.; Ibisch, P.L.; Mihok, B.; Selva, N. The challenge of implementing the European network of protected areas Natura 2000. *Conserv. Biol.* **2015**, *29*, 260–270. [CrossRef] [PubMed]
4. Federation, E. Bern Convention & Emerald Network. Available online: https://www.europarc.org/european-policy/bern-convention/ (accessed on 25 April 2019).
5. EEC. *European Council Directive 92/43/EEC on the Conservation of Natural Habitats and of Wild Fauna and Flora*; EEC: Brussels, Belgium, 1992.
6. EC. Directive 2009/147/EC of the European Parliament and of the Council on the conservation of wild birds. In *2009/147*; EC: Brussels, Belgium, 2009.
7. Lovrić, M.; Lovrić, N.; Schraml, U.; Winkel, G. Implementing Natura 2000 in Croatian forests: An interplay of science, values and interests. *J. Nat. Conserv.* **2018**, *43*, 46–66. [CrossRef]
8. EFI. *Natura 2000 and Forests-Assessing the State of Implementation and Effectiveness*; EFI: Joensuu, Finland, 2017.
9. Geitzenauer, M.; Hogl, K.; Weiss, G. The implementation of Natura 2000 in Austria—A European policy in a federal system. *Land Use Policy* **2016**, *52*, 120–135. [CrossRef]
10. Carmin, J.; Vandeveer, S.D. Enlarging EU Environments: Central and Eastern Europe from Transition to Accession. *Environ. Politics* **2004**, *13*, 3–24. [CrossRef]
11. Börzel, A.T. *Coping with Accession to the European Union: New Modes of Environmental Governance in Southern, Central and Eastern Europe*; Palgrave Macmillan: London, UK, 2009.
12. Dimitrova, A.; Buzogány, A. Post-Accession Policy-Making in Bulgaria and Romania: Can Non-state Actors Use EU Rules to Promote Better Governance? *JCMS J. Common. Mark. Stud.* **2014**, *52*, 139–156. [CrossRef]
13. Kluvánková-Oravská, T.; Chobotová, V.; Banaszak, I.; Slavikova, L.; Trifunovova, S. From Government to Governance for Biodiversity: The Perspective of Central and Eastern European Transition Countries. *Envion Policy Gov.* **2009**, *19*, 186–196. [CrossRef]
14. Sotirov, M.; Lovrić, M.; Winkel, G. Symbolic transformation of environmental governance: Implementation of EU biodiversity policy in Bulgaria and Croatia between Europeanization and domestic politics. *Environ. Plan. C Gov. Policy* **2015**, *33*, 986–1004. [CrossRef]
15. Rauschmayer, F.; Van den Hove, S.; Koetz, T. Participation in EU Biodiversity Governance: How Far beyond Rhetoric? *Environ. Plan. C Politics Space* **2009**, *27*, 42–58. [CrossRef]
16. Paavola, J. Protected Areas Governance and Justice. Theory and the European Union's Habitats Directive. *Environ. Sci.* **2004**, *1*, 59–77. [CrossRef]
17. Kotler, P. *Marketing Management*, 13th ed.; Pearson Education Limited: Edinburg, UK, 2009.
18. Ansoff, H.I.; Kipley, D.; Lewis, A.O.; Helm-Stevens, R.; Ansoff, R. *Implanting Strategic Management*, 3rd ed.; Springer: Cham, Switzerland, 2018.
19. Dess, G.G.; Miller, A. *Strategic Management*; McGraw-Hill International Editions: New York, NY, USA, 1993.
20. Fernandes, J.P.; Guiomar, N.; Gil, A. Identifying key factors, actors and relevant scales in landscape and conservation planning, management and decision making: Promoting effective citizen involvement. *J. Nat. Conserv.* **2019**, *47*, 12–27. [CrossRef]
21. Blicharska, M.; Orlikowska, E.H.; Roberge, J.M.; Grodzinska-Jurczak, M. Contribution of social science to large scale biodiversity conservation: A review of research about the Natura 2000 network. *Biol. Conserv.* **2016**, *199*, 110–122. [CrossRef]
22. Grodzinska-Jurczak, M.; Cent, J. Expansion of nature conservation areas: Problems with Natura 2000 implementation in Poland? *Environ. Manag.* **2011**, *47*, 11–27. [CrossRef]
23. Ferranti, F.; Beunen, R.; Speranza, M. Natura 2000 Network: A Comparison of the Italian and Dutch Implementation Experiences. *J. Environ. Policy Plan.* **2010**, *12*, 293–314. [CrossRef]
24. Winkel, G.; Blondet, M.; Borrass, L.; Frei, T.; Geitzenauer, M.; Gruppe, A.; Jump, A.; de Koning, J.; Sotirov, M.; Weiss, G.; et al. The implementation of Natura 2000 in forests: A trans- and interdisciplinary assessment of challenges and choices. *Environ. Sci. Policy* **2015**, *52*, 23–32. [CrossRef]

25. Julien, B.; Lammertz, M.; Barbier, J.M.; Jen, S.; Ballesteros, M.; de Bovi, C. Voicing interests and concerns: NATURA 2000: An ecological network in conflict with people. *For. Policy Econ.* **2000**, *1*, 357–366. [CrossRef]
26. McCauley, D. Sustainable development and the 'governance challenge': The French experience with Natura 2000. *Eur. Environ.* **2008**, *18*, 152–167. [CrossRef]
27. Mocsari, J. The Implementation of the Habitats Directive in Hungary. Missing Details Behind the Big Picture. Available online: http://www.oeue.net/papers.asp (accessed on 25 May 2019).
28. Falkner, G.; Treib, O. Three Worlds of Compliance or Four? The EU-15 Compared to New Member States. *J. Common. Mark. Stud.* **2008**, *46*, 293–313. [CrossRef]
29. Apostolopoulou, E.; Drakou, E.G.; Pediaditi, K. Participation in the management of Greek Natura 2000 sites: Evidence from a cross-level analysis. *J. Environ. Manag.* **2012**, *113*, 308–318. [CrossRef]
30. Geitzenauer, M.; Blondet, M.; de Koning, J.; Ferranti, F.; Sotirov, M.; Weiss, G.; Winkel, G. The challenge of financing the implementation of Natura 2000–Empirical evidence from six European Union Member States. *For. Policy Econ.* **2017**, *82*, 3–13. [CrossRef]
31. Hily, E.; Garcia, S.; Stenger, A.; Tu, G. Assessing the cost-effectiveness of a biodiversity conservation policy: A bio-econometric analysis of Natura 2000 contracts in forest. *Ecol. Econ.* **2015**, *119*, 197–208. [CrossRef]
32. Sarvašová, Z.; Ali, T.; Đorđević, I.; Lukmine, D.; Quiroga, S.; Suárez, C.; Hrib, M.; Rondeux, J.; Mantzanas, K.T.; Franz, K. Natura 2000 payments for private forest owners in Rural Development Programmes 2007–2013—A comparative view. *For. Policy Econ.* **2019**, *99*, 123–135. [CrossRef]
33. Sarvašová, Z.; Quiroga, S.; Suárez, C.; Ali, T.; Lukmine, D.; Đorđević, I.; Hrib, M. Understanding the drivers for Natura 2000 payments in forests: A Heckman selection analysis. *J. Nat. Conserv.* **2018**, *46*, 28–37. [CrossRef]
34. De Meo, I.; Brescancin, F.; Graziani, A.; Paletto, A. Management of Natura 2000 sites in Italy: An exploratory study on stakeholders' opinions. *J. For. Sci.* **2016**, *62*, 511–520. [CrossRef]
35. Beunen, R.; de Vries, J.R. The governance of Natura 2000 sites: The importance of initial choices in the organisation of planning processes. *J. Environ. Plan. Manag.* **2011**, *54*, 1041–1059. [CrossRef]
36. Borrass, L.; Sotirov, M.; Winkel, G. Policy change and Europeanization: Implementing the European Union's Habitats Directive in Germany and the United Kingdom. *Environ. Politics* **2015**, *24*, 788–809. [CrossRef]
37. Cent, J.; Grodzińska-Jurczak, M.; Pietrzyk-Kaszyńska, A. Emerging multilevel environmental governance–A case of public participation in Poland. *J. Nat. Conserv.* **2014**, *22*, 93–102. [CrossRef]
38. Cent, J.; Mertens, C.; NiedziaŁKowski, K. Roles and impacts of non-governmental organizations in Natura 2000 implementation in Hungary and Poland. *Environ. Conserv.* **2013**, *40*, 119–128. [CrossRef]
39. Stringer, L.C.; Paavola, J. Participation in environmental conservation and protected area management in Romania: A review of three case studies. *Environ. Conserv.* **2013**, *40*, 138–146. [CrossRef]
40. Laktić, T.; Pezdevšek Malovrh, Š. Stakeholder Participation in Natura 2000 Management Program: Case Study of Slovenia. *Forests* **2018**, *9*, 599. [CrossRef]
41. Blondet, M.; de Koning, J.; Borrass, L.; Ferranti, F.; Geitzenauer, M.; Weiss, G.; Turnhout, E.; Winkel, G. Participation in the implementation of Natura 2000: A comparative study of six EU member states. *Land Use Policy* **2017**, *66*, 346–355. [CrossRef]
42. Kovács, E.; Kelemen, E.; Kiss, G.; Kalóczkai, Á.; Fabók, V.; Mihók, B.; Megyesi, B.; Pataki, G.; Bodorkós, B.; Balázs, B.; et al. Evaluation of participatory planning: Lessons from Hungarian Natura 2000 management planning processes. *J. Environ. Manag.* **2017**, *204*, 540–550. [CrossRef] [PubMed]
43. Dimitrakopoulos, P.G.; Jones, N.; Iosifides, T.; Florokapi, I.; Lasda, O.; Paliouras, F.; Evangelinos, K.I. Local attitudes on protected areas: Evidence from three Natura 2000 wetland sites in Greece. *J. Environ. Manag.* **2010**, *91*, 1847–1854. [CrossRef] [PubMed]
44. Pietrzyk-Kaszyńska, A.; Cent, J.; Grodzińska-Jurczak, M.; Szymańska, M. Factors influencing perception of protected areas—The case of Natura 2000 in Polish Carpathian communities. *J. Nat. Conserv.* **2012**, *20*, 284–292. [CrossRef]
45. Jones, N.; Filos, E.; Fates, E.; Dimitrakopoulos, P.G. Exploring perceptions on participatory management of NATURA 2000 forest sites in Greece. *For. Policy Econ.* **2015**, *56*, 1–8. [CrossRef]
46. Howes, M.; Wortley, L.; Potts, R.; Dedekorkut-Howes, A.; Serrao-Neumann, S.; Davidson, J.; Smith, T.; Nunn, P. Environmental Sustainability: A Case of Policy Implementation Failure? *Sustainability* **2017**, *9*, 165. [CrossRef]

47. Siebert, R.; Toogood, M.; Knierim, A. Factors Affecting European Farmers' Participation in Biodiversity Policies. *Sociol. Rural.* **2006**, *46*, 318–340. [CrossRef]
48. Popescu, D.V.; Rozylowicz, L.; Niculae, M.I.; Cucu, L.A.; Hartel, T. Species, Habitats, Society: An Evaluation of Research Supporting EU's Natura 2000 Network. *PLoS ONE* **2014**, *9*, e113648. [CrossRef]
49. Lovrić, M.; Lovrić, N.; Schraml, U. Modeling policy networks: The case of Natura 2000 in Croatian forestry. *For. Policy Econ.* **2019**, *103*, 90–102. [CrossRef]
50. Gallo, M.; Pezdevšek Malovrh, Š.; Laktić, T.; De Meo, I.; Paletto, A. Collaboration and conflicts between stakeholders in drafting the Natura 2000 Management Programme (2015–2020) in Slovenia. *J. Nat. Conserv.* **2018**, *42*, 36–44. [CrossRef]
51. Rodela, R.; Udovč, A. Participation in nature protection: Does it benefit the local community? A Triglav National Park case study. *Int. J. Biodivers. Sci. Manag.* **2008**, *4*, 209–218. [CrossRef]
52. Brescancin, F.; Dobšinská, Z.; De Meo, I.; Šálka, J.; Paletto, A. Analysis of stakeholders' involvement in the implementation of the Natura 2000 network in Slovakia. *For. Policy Econ.* **2017**, *78*, 107–115. [CrossRef]
53. Nonić, D.; Avdibegović, M.; Nedeljković, J.; Radosavljević, A.; Ranković, N. *Održivo upravljanje u šumarstvu i zaštiti prirode. Glasnik Šumarskog Fakulteta, Specijalno Izdanje Povodom Naučnog Skupa "Šume Srbije i Održivi Razvojoj"*; University of Belgrade, Faculty of Forestry: Belgrade, Serbia, 2014; pp. 113–140.
54. Arts, B. Forests policy analysis and theory use: Overview and trends. *For. Policy Econ.* **2012**, *16*, 7–13. [CrossRef]
55. Ostrom, E. Background on the Institutional Analysis and Development Framework. *Policy Stud. J.* **2011**, *39*, 7–27. [CrossRef]
56. Fozer, D.; Sziraky, F.Z.; Racz, L.; Nagy, T.; Tarjani, A.J.; Toth, A.J.; Haaz, E.; Benko, T.; Mizsey, P. Life cycle, PESTLE and Multi-Criteria Decision Analysis of CCS process alternatives. *J. Clean. Prod.* **2017**, *147*, 75–85. [CrossRef]
57. Krippendorff, K. *Content Analysis: An Introduction to Its Methodology*, 2nd ed.; SAGE Publications: Thousand Oaks, CA, USA, 2004; p. 422.
58. Sarvašova, Z.; Šalka, J.; Dobšinska, Z. Mechanism of cross-sectoral coordination between nature protection and forestry in the Natura 2000 formulation process in Slovakia. *J. Environ. Manag.* **2013**, *127*, S65–S72. [CrossRef]
59. Directorate of Culture and of Cultural and Natural Heritage. *Development of the Emerald Site Network in the West-Balkana under the CARDS Program*; Council of Europe: Strasbourg, France, 2007; p. 38.
60. Vasiljević, M.; Pokrajac, S.; Erg, B. *State of Nature Conservation Sytems in South-Eastern Europe*; IUCN: Gland, Switzerland; Belgrade, Serbia, 2018; p. 58.
61. EEA. *The Natura 2000 Barometer*; EEA: Kobenhavn, Denmark, 2019; Available online: https://www.eea.europa.eu/data-and-maps/dashboards/natura-2000-barometer#tab-based-on-data (accessed on 7 May 2019).
62. European Commission. *Natura 2000 and Forests, Part I–II*; European Union: Brussel, Belgium, 2015; p. 114.
63. Petkovšek, M. Slovenian Natura 2000 network in numbers. *Varst. Narave* **2017**, *30*, 99–126.
64. Kapusta, P. Forests and Protected Areas. Available online: https://www.enviroportal.sk/indicator/detail?id=1103 (accessed on 1 October 2019).
65. Mariano, A.; Gasparini, P.; De Natale, F.; Romano, R.; Ammassari, P.; Liberati, D.; Ballin, M.; Vitullo, M. *Italy-Global Forest Resources Assessment 2015–Country Report*; FAO: Rome, Italy, 2015; p. 105.
66. Emerton, L.; Bishop, J.; Thomas, L. *Sustainable Financing of Protected Areas: A Global Review of Challenges and Options*; IUCN: Gland, Switzerland; Cambridge, UK, 2006; p. 109.
67. Communities, C.O.T.E. *Treaty on European Union*; Office for Official Publication of the European Communities: Luxembourg, 1992; p. 260.
68. Elliott, C.; Udovč, A. Nature conservation and spatial planning in Slovenia: Continuity in transition. *Land Use Policy* **2005**, *22*, 265–276. [CrossRef]
69. Herwig, U. Public Participation in the Establishment and Management of the Natura 2000 Network—Legal Framework and Administrative Practices in Selected Member States. *J. Eur. Environ. Plan. Law* **2008**, *5*, 35–68.
70. Ministry of Environment and Spatial Planning. *Support for the Organization of Bilateral Dialogues with Slovenia in the Context of Action 5 of the Action Plan for Nature, People and the Economy*; Ministry of Environment and Spatial Planning: Ljubljana, Slovenia, 2018; p. 66.

71. Perrot-Maître, D. *The Vittel Payments for Ecosystem Services: A Perfect" PES Case?* International Institute for Environment and Development: London, UK, 2006; p. 24.
72. Marino, D.; Pellegrino, D. Can Payments for Ecosystem Services Improve the Management of Natura 2000 Sites? A Contribution to Explore Their Role in Italy. *Sustainability* **2018**, *10*, 665. [CrossRef]
73. Schomers, S.; Matzdorf, B. Payments for ecosystem services: A review and comparison of developing and industrialized countries. *Ecosyst. Serv.* **2013**, *6*, 16–30. [CrossRef]
74. Jones-Walters, L.; Čivić, K. European protected areas: Past, present and future. *J. Nat. Conserv.* **2013**, *21*, 122–124. [CrossRef]
75. Klassert, C.; Möckel, S. Improving the Policy Mix: The Scope for Market-Based Instruments in EU Biodiversity Policy. *Environ. Policy Gov.* **2013**, *23*, 311–322. [CrossRef]
76. Niedziakowski, K.; Paavola, J.; Jedrzejewska, B. Participation and Protected Areas Governance: The Impact of Changing Influence of Local Authorities on the Conservation of the Bialowieza Primeval Forest, Poland. *Ecol. Soc.* **2012**, 17. [CrossRef]
77. Keulartz, J. European Nature Conservation and Restoration Policy—Problems and Perspectives. *Restor. Ecol.* **2009**, *17*, 446–450. [CrossRef]
78. Ferranti, F.; Turnhout, E.; Beunen, R.; Behagel, J.H. Shifting nature conservation approaches in Natura 2000 and the implications for the roles of stakeholders. *J. Environ. Plan. Manag.* **2014**, *57*, 1642–1657. [CrossRef]
79. Nastran, M. Stakeholder analysis in a protected natural park: Case study from Slovenia. *J. Environ. Plan. Manag.* **2014**, *57*, 1359–1380. [CrossRef]
80. Nastran, M. Why does nobody ask us? Impacts on local perception of a protected area in designation, Slovenia. *Land Use Policy* **2015**, *46*, 38–49. [CrossRef]
81. Nastran, M.; Pirnat, J. Stakeholder participation in planning of the protected natural areas: Slovenia. *Sociol. I Prost.* **2012**, *50*, 141–164. [CrossRef]
82. Dwivedi, P.; Alavalapati, J.R.R. Stakeholders' perceptions on forest biomass-based bioenergy development in the southern US. *Energy Policy* **2009**, *37*, 1999–2007. [CrossRef]
83. Pezdevšek Malovrh, Š.; Kurttila, M.; Hujala, T.; Kärkkäinen, L.; Leban, V.; Lindstad, B.H.; Peters, D.M.; Rhodius, R.; Solberg, B.; Wirth, K.; et al. Decision support framework for evaluating the operational environment of forest bioenergy production and use: Case of four European countries. *J. Environ. Manag.* **2016**, *180*, 68–81. [CrossRef] [PubMed]
84. Grošelj, P.; Zadnik Stirn, L. The environmental management problem of Pohorje, Slovenia: A new group approach within ANP–SWOT framework. *J. Environ. Manag.* **2015**, *161*, 106–112. [CrossRef] [PubMed]
85. Huber, P.; Hujala, T.; Kurttila, M.; Wolfslehner, B.; Vacik, H. Application of multi criteria analysis methods for a participatory assessment of non-wood forest products in two European case studies. *For. Policy Econ.* **2019**, *103*, 103–111. [CrossRef]

9

Evaluating WorldClim Version 1 (1961–1990) as the Baseline for Sustainable use of Forest and Environmental Resources in a Changing Climate

Maurizio Marchi [1,*], Iztok Sinjur [2], Michele Bozzano [3] and Marjana Westergren [2]

[1] CREA—Research Centre for Forestry and Wood, I-52100 Arezzo, Italy
[2] Slovenian Forestry Institute, Vecna pot 2, 1000 Ljubljana, Slovenia; iztok.sinjur@gozdis.si (I.S.); marjana.westergren@gozdis.si (M.W.)
[3] European Forest Institute, 53113 Bonn, Germany; michele.bozzano@efi.int
* Correspondence: maurizio.marchi@crea.gov.it.

Abstract: WorldClim version 1 is a high-resolution, global climate gridded dataset covering 1961–1990; a "normal" climate. It has been widely used for ecological studies thanks to its free availability and global coverage. This study aims to evaluate the quality of WorldClim data by quantifying any discrepancies by comparison with an independent dataset of measured temperature and precipitation records across Europe. BIO1 (mean annual temperature, MAT) and BIO12 (mean total annual precipitation, MAP) were used as proxies to evaluate the spatial accuracy of the WorldClim grids. While good representativeness was detected for MAT, the study demonstrated a bias with respect to MAP. The average difference between WorldClim predictions and climate observations was around +0.2 °C for MAT and −48.7 mm for MAP, with large variability. The regression analysis revealed a good correlation and adequate proportion of explained variance for MAT (adjusted $R^2 = 0.856$) but results for MAP were poor, with just 64% of the variance explained (adjusted $R^2 = 0.642$). Moreover no spatial structure was found across Europe, nor any statistical relationship with elevation, latitude, or longitude, the environmental predictors used to generate climate surfaces. A detectable spatial autocorrelation was only detectable for the two most thoroughly sampled countries (Germany and Sweden). Although further adjustments might be evaluated by means of geostatistical methods (i.e., kriging), the huge environmental variability of the European environment deeply stressed the WorldClim database. Overall, these results show the importance of an adequate spatial structure of meteorological stations as fundamental to improve the reliability of climate surfaces and derived products of the research (i.e., statistical models, future projections).

Keywords: spatial analysis; 1961–1990 normal period; spatial interpolation; geostatistics; ecological mathematics

---

1. Introduction

Easy access to standardized climate data with global coverage is paramount for the advancement of many ecological studies and to understand future ecosystem services provided by forest systems [1–3] and productive lands in agriculture [4,5]. One of the main aims for researchers dealing with environmental resources has become to forecast possible impacts of climate change on organisms and to evaluate possible mitigation [6–9]. In the past few decades, many conservation strategies have been suggested in order to maintain human well-being and ensure an adequate level of welfare [10] from (relatively) simple management strategies [4,11], including "assisted migration" [12–14], a controversial protocol that includes translocating more adapted or resilient genotypes for conservation or to improve the resilience of ecosystems. Such efforts are often driven by statistical models [14–17] and management simulators [18,19], with both genetic variation and phenotypic plasticity included in the statistical models as covariates [20–22]. However, despite modelling efforts, such studies always require absolutely reliable climate data to be used as both baseline (e.g., 30-year average climate data) and for future predictions. Furthermore, while the uncertainty around GCMs and future trajectories is well known [23–25], information on current ecological limits of forest tree species has also been questioned [26]. In this context, the interest of researchers in gridded climate datasets has grown strongly.

The interpolation method, the spatial resolution and the coverage are the three main features that researchers use to select the most suitable datasets for their research [27–30]. The first release of the WorldClim dataset [31] is probably the most famous gridded climate dataset, widely used for ecological studies and freely available from (www.worldclim.org). Thanks to its high resolution (30 arc-second in the WGS84 reference system and approximately 1 km at the equator), global coverage, and availability, it has been used and cited more than 5200 times since publication [31].

The dataset is suitable for basic and applied studies in ecology, including forestry and ecological modeling [32–34], as well as to construct related datasets such as bio-geographical zones or environmental stratifications [35]. One of the main products of this database is "version 1", representative of the 1961–1990 climate normal period for the whole globe, including Antarctica. This version 1 dataset was generated by interpolating weather station data with the ANUSPLIN software (version 4.3) using latitude, longitude, and elevation as independent variables. The software implements a thin-plate smoothing spline procedure, using every station as a data point. A second-order spline function was fitted by the Authors using the above three variables, which produced the lowest overall cross-validation errors [31].

Considering the ANUSPLIN program creates a continuous surface projection, the LAPGRD program was used to create a global grid of climate surfaces with 30 arc-seconds horizontal and vertical resolution commonly referred to as 1 km^2 resolution. Raster maps for monthly precipitation amount and mean, maximum, and minimum air temperature were then provided. Raw data came from weather stations retrieved from various databases including GHCN, WMO climatological normals, FAOCLIM 2.0, CIAT, and regional databases and, where possible, restricted to the period 1950–2000. Quality control measures were taken to remove duplicate records, giving precedence to the GHCN database. After the quality control check and cleaning, the database consisted of precipitation records from 47,554 locations and mean air temperature from 24,542 locations [31]. Then elevation bias in weather stations was related to latitude and presence of mountain ranges. However, local records from many European countries were not easily accessible and WorldClim climate surfaces for Europe were constructed using 1263 records for air temperature and 2116 for precipitation.

WorldClim version 1 has recently been acknowledged to be representative of the 1961–1990 climate normal period. This time-slice has been widely used as the pre-industrial climate in many papers about the potential impact of climate change on ecosystems [1,3,26,28,36,37] and other ecological fields. Nevertheless, given the detailed description provided by the Authors in their paper, the question remains whether the quality of the WorldClim climate surfaces as a proxy of the climate baseline is adequate in complex environments such as, for instance, the European environment.

The present study aims to assess and quantify the reliability of WorldClim climate raster maps for Europe. We compared WorldClim with observed average values for mean annual temperature and total annual precipitation for the period 1961–1990. Data were retrieved for the whole of Europe building an independent dataset with data from many meteorological services. Then statistical analysis was run in order to evaluate the reliability of this dataset across the study area.

2. Materials and Methods

2.1. Construction and Description of the Database Used for Comparison

To investigate WorldClim's reliability in predicting baseline climate conditions we compiled an independent climate dataset by collecting data from weather services across Europe which were already freely available or delivered upon request (Table 1). All data were specifically requested or downloaded as monthly averaged values over the 30-year normal period (1961–1990). Local monthly air temperature averages (MAT) and precipitation sums (MAP) were aggregated to calculate annual values. In total we retrieved data from 6659 meteostations across Europe, with 1759 records for temperature and 6526 records for precipitation (Figure 1). Most of the records were retrieved for Germany and Sweden with 4825 and 1391 meteostations, respectively, while for some countries, records were much fewer (e.g., Spain, France, Italy) or totally absent (e.g., Serbia, Poland, Romania).

Nevertheless, even if not equally distributed geographically, neither balanced concerning the ecological regions of Europe, we considered the distribution of the collected data as adequate for the purpose. Despite the lack of uniform coverage of both geography and ecological regions, we considered the data collected to be adequate for subsequent analysis.

Moreover we tested the random distribution of MAT and MAP with the randtest package of the R statistical language [38]. The database was carefully checked and cleaned to remove entries with missing data and to geo-reference each record. Very few points (112), corresponding to less than 1% of all the records, lay outside country borders or land masses due to coordinate uncertainties, which reflects the high-quality of the new database. Such records were removed completely from the database in order to avoid any influences on the calculations.

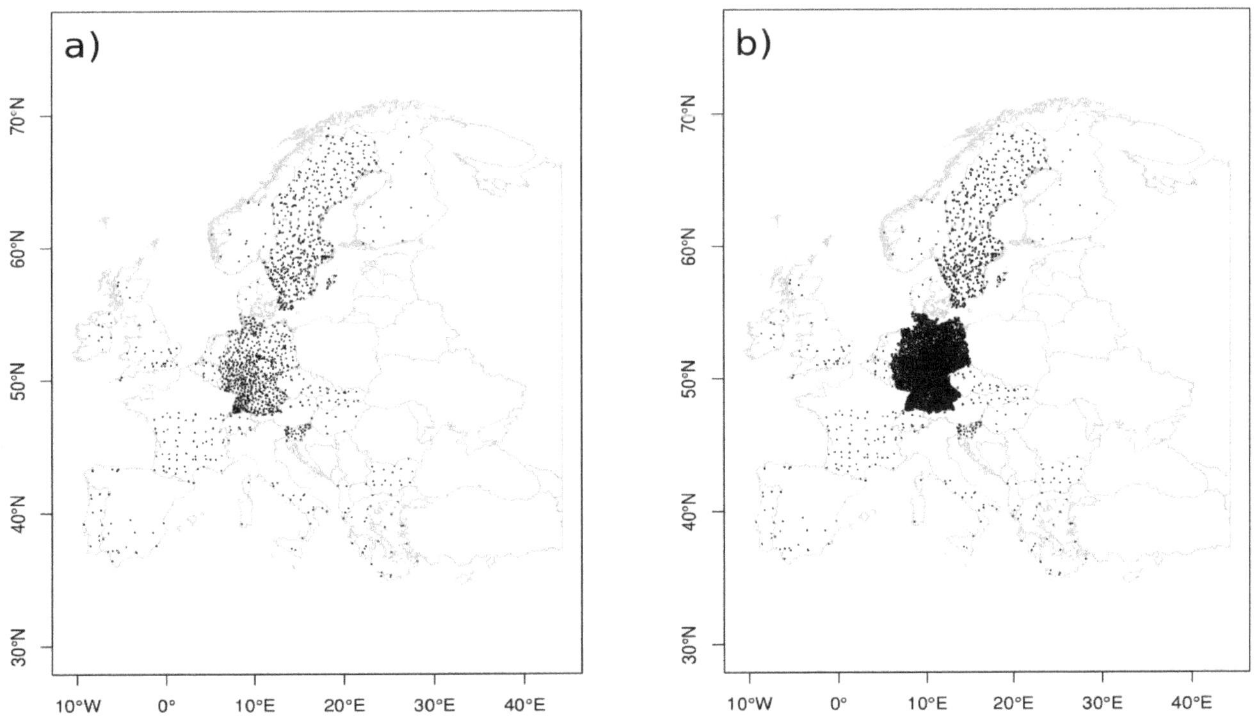

Figure 1. Spatial distribution of the compiled dataset for: (**a**) temperature records and (**b**) precipitation records. Each dot represents a meteorological station. The darker the area, the more data were retrieved.

Table 1. Structure of the compiled database.

Country	Total Meteostations	MAT Records	MAP Records	Data Source
Albania	3	3	3	
Austria	23	21	23	
Belgium	9	9	9	
Bulgaria	18	18	18	[39]
Croatia	1	1	1	
Czech	20	19	20	
Denmark	4	4	4	
Finland	18	17	18	
France	76	75	76	
Germany	4825	719	4733	[40]
Greece	26	25	26	[41]
Hungary	9	9	9	
Ireland	6	6	6	
Italy	30	30	30	
North Macedonia	1	1	1	[39]
Montenegro	1	1	1	
Netherlands	5	5	5	
Norway	18	18	18	
Portugal	18	18	18	
Slovakia	14	14	14	
Slovenia	42	42	42	[42]
Spain	51	51	51	[39]
Sweden	1391	604	1351	[43]
Switzerland	12	11	12	[39]
United Kingdom	38	38	37	
TOTAL	*6659*	*1759*	*6526*	
MEAN	*266*	*70*	*261*	
ST. DEV	*988.64*	*179.54*	*969.08*	

2.2. *Comparisons and Statistical Procedures*

The BIO1 (mean annual air temperature) and BIO12 (mean total annual precipitation) variables of WorldClim were used as proxies to evaluate the spatial accuracy of raster surfaces. The strata were first downloaded from the official WorldClim web portal. Then, using an overlay function, the corresponding values of the two climate variables were extracted for each meteorological station in our database. A linear regression analysis was then applied to analyze the relationships between the predicted WorldClim value and the observed value in our dataset. The adjusted R^2 was used to measure the amount of environmental variability expressed by WorldClim. Then the difference between the WorldClim value and the observed value (30-years normal value from our database) was calculated for each location of our database. To avoid confusion and mathematical balancing between positive and negative values, which might seriously affect the analysis, both the raw discrepancy (BIAS) and its absolute value (ABIAS) were calculated. To study possible trends across the data, we looked at the relationships between BIAS and the predictors used by the authors of WorldClim during the spatial interpolation process (i.e., latitude, longitude, elevation). Then, we retrieved the complete database of meteorological stations used by the WorldClim authors from www.arcgis.com/home/item.html?id=7644c6e78c1644b4bde2edfc44787520) and clipped to the European environment (Table 2).

We calculated the average distance of each meteorological station in our database from the geographically closest five stations in the WorldClim dataset. We expected a smaller difference where WorldClim stations were denser. Finally, the spatial autocorrelation of BIAS was evaluated using geostatistical analysis implemented in R using the gstat package [44] and modelling the semivariance of BIAS as a function of the spatial distance between records.

The whole structure of the data collection and analysis procedure is graphically reported on Figure 2.

Table 2. Number of meteorological stations per country used by Hijmans et al. [31] in Europe.

Country	Temperature	Precipitation	Country	Temperature	Precipitation
Albania	0	7	Latvia	3	9
Andorra	0	0	Liechtenstein	0	0
Armenia	2	2	Lithuania	16	19
Austria	3	25	Luxembourg	1	6
Belarus	8	22	North Macedonia	7	7
Belgium	3	18	Malta	1	3
Bosnia and Herz.	7	10	Moldova	2	3
Bulgaria	4	15	Monaco	0	0
Croatia	13	13	Montenegro	5	2
Czech Republic	7	16	Netherlands	7	10
Denmark	19	41	Norway	8	54
Estonia	3	12	Poland	18	63
Faeroe Islands	1	1	Portugal	16	18
Finland	19	32	Romania	11	28
France	82	107	Russia	44	124
Georgia	1	20	San Marino	0	0
Germany	89	116	Serbia	23	12
Gibraltar	0	1	Slovakia	3	10
Greece	26	48	Slovenia	6	2
Guernsey	0	0	Spain	60	117
Hungary	8	20	Sweden	16	60
Ireland	16	51	Switzerland	8	20
Isle of Man	0	1	Turkey	513	548
Italy	133	151	Ukraine	22	81
Jersey	0	3	UK	29	188

Summary statistics	Temperature records	Precipitation records
TOTAL	1263	2116
MEAN	25	42
SD	74.86	84.89

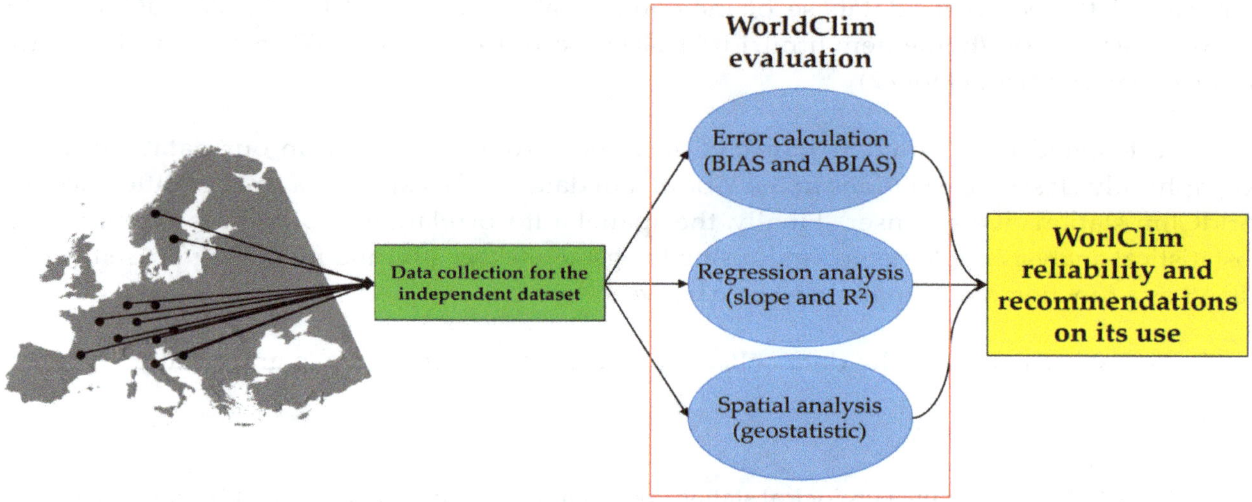

Figure 2. Flowchart of the data collection and statistical analysis we made to test the reliability of WorlClim version 1 data.

3. Results

The compiled database included 25 European countries, albeit with an unbalanced distribution. Overall, an average of 266 records per country (both MAT and MAP) was included in the database. However, the difference among countries was huge, with a standard deviation of ± 988.64 records per country. This large standard deviation was caused by the disproportionate number of records for Sweden and Germany. Temperature (MAT) values ranged from −5.8 °C to 21.2 °C, while precipitation (MAP) was between 104.8 mm and 3318 mm. The mean difference between the interpolated WorldClim values and the observed values was 0.22 °C for temperature and −48.7mm for precipitation (Table 3), with a high coefficient of variation (6.82 for MAT and 3.40 for MAP). BIAS ranged between −10.6 °C and 13.2 °C for MAT and between −1578.1 mm and 950.8 mm for MAP. Mean ABIAS was 0.76 for MAT and 98.56 for MAP.

Results of the regression analysis for MAT and MAP are shown in Figure 3. Residuals of linear models were randomly distributed for both of the analyzed variables and were highly significant ($p < 2.2 \times 10^{-16}$). Concerning MAT, the good correlation and adequate proportion of explained variance point to a low discrepancy between the two datasets; WorldClim explained 86% of the variance (adjusted $R^2 = 0.856$) with a residual random standard error of 1.50°C, intercept of −0.202 °C and slope almost equal to 1 (0.996). The regression line and the expected regression line for a perfect match between the two datasets almost overlapped. For MAP, 64% (adjusted $R^2 = 0.642$) of the variance of the precipitation dataset was explained by a linear regression model, with a residual standard error of 159.6 mm. The match between the two regression lines was considerably low (Figure 3, right) with the slope of the regression coefficient higher than 1. WorldClim was characterized by higher values than observed under 500 mm precipitation and lower values above this threshold. As overall, a general overestimation of MAP values was detected in dry areas (<500 mm) with an underestimation in the remaining zones.

Table 3. Difference between local data and WorldClim's surfaces.

Variable	AVR	SD	CV	MAX	MIN	ABSAVR
MAT [°C]	0.22	1.50	6.82	−10.62	13.21	0.76
MAP [mm]	−48.70	165.35	3.40	−1578.10	950.80	98.56

AVR = average value; SD = standard deviation; CV = coefficient of variation; MAX = maximum difference; MIN = minimum difference; ABSAVR = average of absolute values.

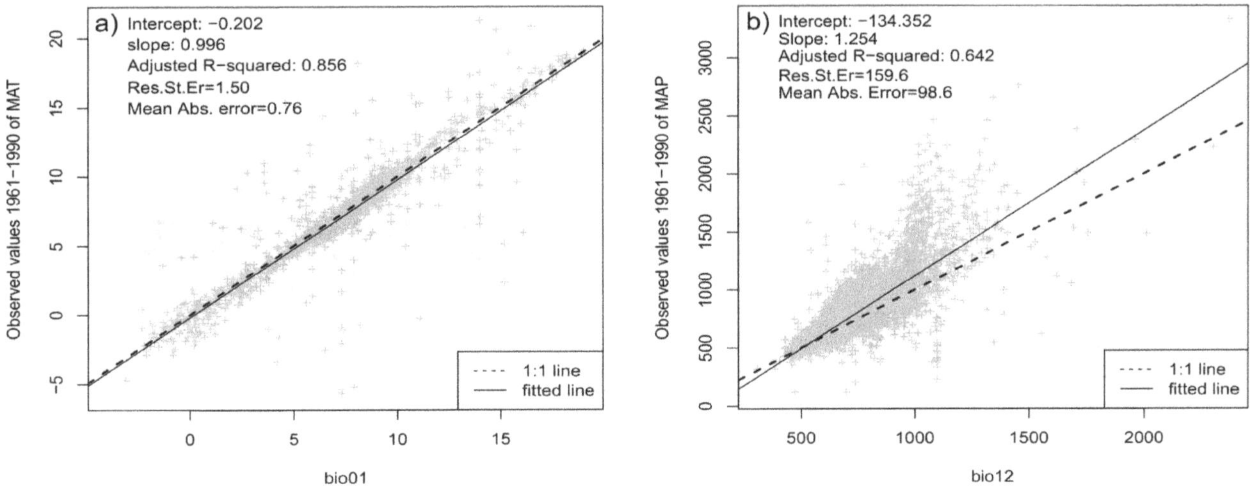

Figure 3. Results of the regression analysis for: (**a**) temperature represented by Bio1 variable of WorldClim database on x-axis and (**b**) precipitation represented by Bio12 variable of WorldClim database versus observed values on the y-axis. Regression coefficients at top-left of each figure. The 1:1 line of a perfect match shown dashed.

No relationship was found between the modelling error (ER) detected for MAT and MAP and the environmental predictors used for spatial interpolation across Europe. Modelled linear regressions explained less than 5% of variation, with one exception (Table 4). This lack of correlation can also be observed in Figure 4 where ABIAS is plotted against the average spatial distance of the "observed" meteorological station from the five WorldClim stations.

Table 4. Linear regression parameters when modelling error (ER) and the environmental predictors. Each predictor was tested separately (ADF5NM=Average distance from the five nearest meteorological stations).

Variable	Predictor	Intercept	Slope	Explained Variance	*p*-Value
MAT	Latitude	0.29	0.000000	0.56%	0.00092
	Longitude	−0.34	0.000000	1.20%	0.00000
	Elevation	0.54	−0.001025	4.95%	0.00000
	ADF5NM	0.09	0.000002	0.18%	0.04138
MAP	Latitude	−45.16	0.000014	0.05%	0.04492
	Longitude	−202.28	0.000061	4.33%	0.00000
	Elevation	8.15	−0.206690	10.26%	0.00000
	ADF5NM	−107.02	0.001059	1.61%	0.00000

The spatial distribution of BIAS in the two most represented countries is shown in Figure 5 for the two investigated variables. Spatial aggregation is especially evident in Sweden, where most of the "large dots" are clustered in the south of the country. For Sweden and Germany, variograms of the MAP variable were fitted by means of an exponential variogram model and revealed a clear spatial autocorrelation, especially for Germany (Figure 6).

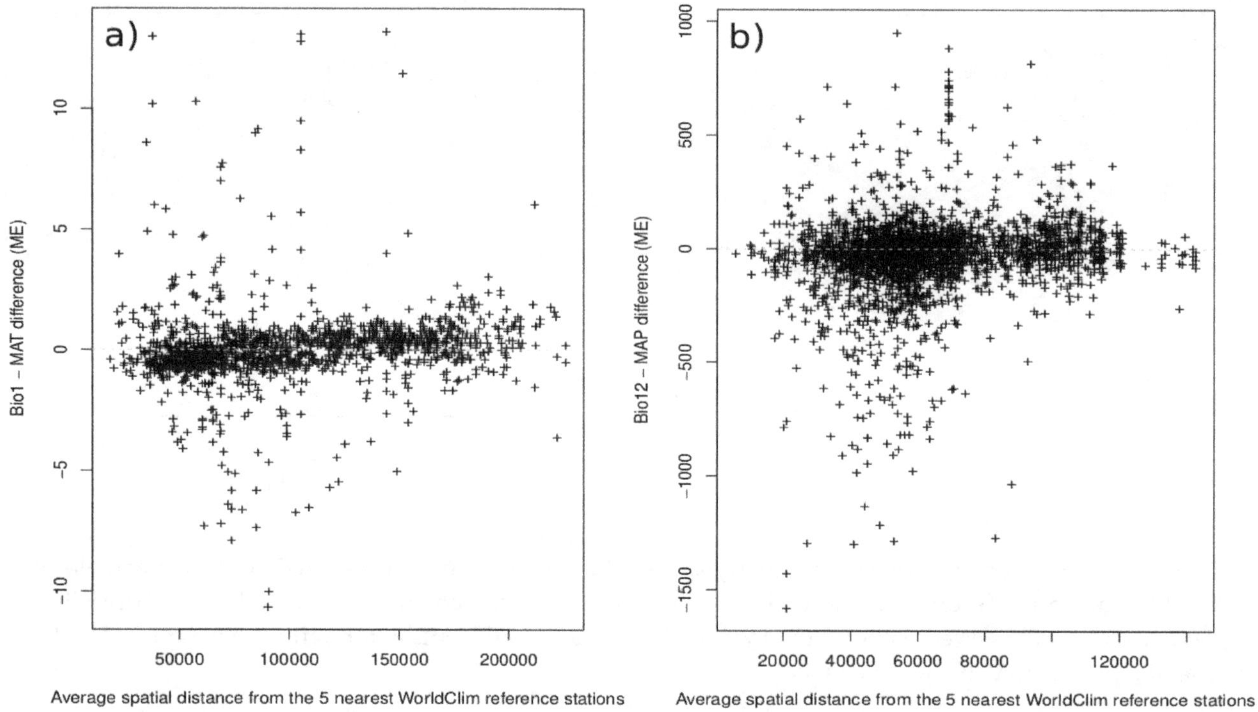

Figure 4. Relationship between the detected differences (WorldClim value, observed) and the average spatial distance of the observed record (new database) from the five nearest WorldClim reference stations for (**a**) temperature and (**b**) precipitation.

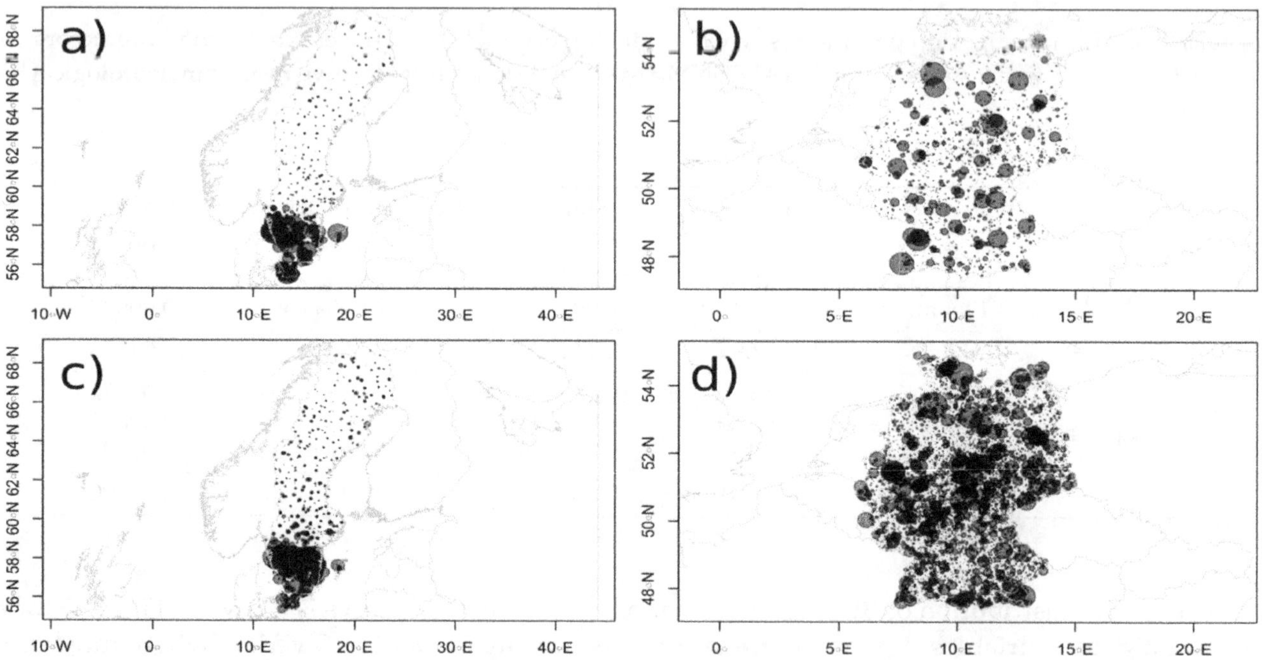

Figure 5. Spatial distribution of ABIAS for temperature (**a** and **b**) and precipitation (**c** and **d**) across Sweden and Germany, the two most sampled countries in the database. The larger the gray dot, the greater the difference between WorldClim and the independent dataset.

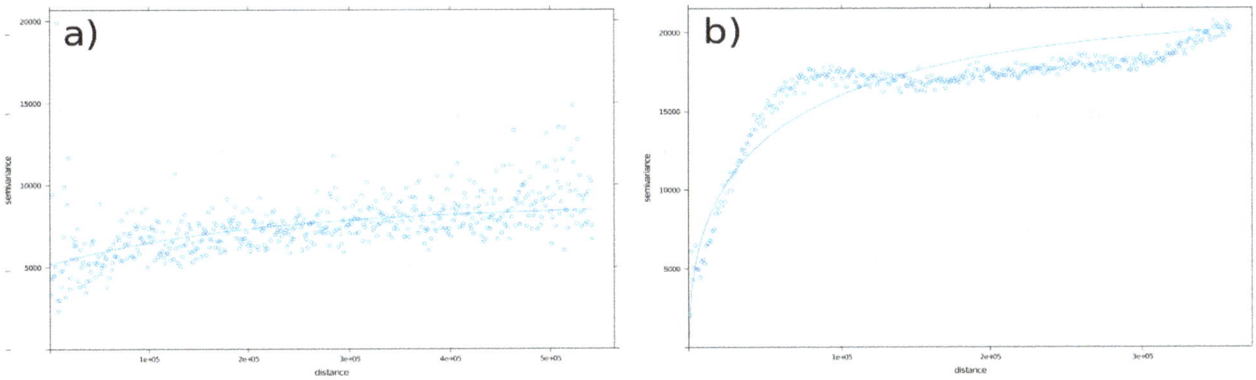

Figure 6. (Semi)variograms for precipitation calculated by the gstat package in R for Sweden (**a**) and Germany (**b**). There is a rather clear spatial structure that might be used as input for further geostatistical procedures (i.e., kriging) in order to adjust WorldClim raster maps.

4. Discussion

The quality of the reference baseline climate has a fundamental role in predictions of the potential impact of climate change on organisms and natural ecosystems. The stability and reliability of the estimated projections calculated from species distribution models [14,45,46], management simulators [18,47], or the estimation of the geographical shift of climate zones [48] rely on the differences between current and future climate. While the representativeness of WorldClim is adequate concerning air temperatures, large differences were found in the precipitation surfaces. Our results demonstrate a systematic difference of 0.76 °C between observed and interpolated values. According to the most recent IPCC report [49], the observed increase in temperature has been around 0.2 °C per decade. As a consequence such difference might affect future projections of WorldClim dataset adding uncertainties on the further modelling efforts [22,50]. In this case a likelihood analysis should be more adequate than deterministic ones and in order to include a sensitivity analysis and evaluate the probability of success of empirical models based on phenotypic plasticity and applied future projections [51]. Precipitation, by contrast, proved to be the main weakness of WorldClim surfaces in Europe. Despite its relatively small spatial extent, the European environment is characterized by many different forest systems that reflect broad climate variability, spanning from the Mediterranean to the Arctic.

The 1961–1990 baseline period is a fundamental dataset for ecological modelling because records from earlier periods were often affected by different instrumentation or changes in observational practice [30,37]. Therefore, numerous studies from climatology to biology, ecology and forestry [36,48,52] have used this baseline period, and WorldClim has been used extensively. We can expect a further warming trend in the next two decades at a rate of about 0.1 °C per decade, due mainly to the slow response of the oceans. As a consequence, even though the linear regression analysis showed a good match between observed and interpolated data (adjusted $R^2 = 0.856$), the difference is higher than the expected rate of change, which could heavily affect model predictions, adding uncertainties on future projections and smoothing results (i.e., land suitability projections) in an uncontrolled way [14,53–55]. This issue is then amplified when analysing MAP, where higher differences were found in combination with a poor regression analysis result. As a consequence, important biases may be introduced when using WorldClim's precipitation dataset. This is particularly true when WorldClim is used as the reference line and climate projections are locally downscaled and added to the WorldClim surfaces, as in the "Delta method" [56]. As a result, the calculation of climate indices might be difficult. For example, many studies used reference evapotranspiration [57–59] as the main predictor in statistical models [3,60,61]. In this case, the mathematical combination of differences in MAT and MAP might introduce uncontrolled biases through the study area. These biases could represent a critical issue, especially in the Mediterranean and anywhere else that moisture deficit is identified as the most relevant climate driver.

The main advantage of our compiled dataset might be its representativeness at small scale. The authors of WorldClim themselves warn that the high resolution of the climate surfaces does not imply high data quality in all places as this depends on local climate variability, quality and density of observations and the degree of the fitted spline [31]. In a similar study, when compared with PRISM and Daymet datasets for the continental United States, many concerns were expressed, especially regarding the quality of WorldClim's precipitation grids in mountainous areas [34,35,62,63]. For this reason, several studies at regional or national scale at higher resolution (e.g., 100–250 m) preferred the use of meteorological variables obtained at nearby observational sites [28,64–66]. Regardless of the distances of the investigation sites from the locations where meteorological datasets were gathered, orography and land use, and the surrounding area and variable characteristics, must be considered. At small scale their variability may be a strong driver of frequently overlooked heterogeneities, leading to significant discrepancies in transferred datasets used for otherwise appropriate processing methods [67,68].

The lack of any relationship between BIAS and the main physiographic parameters (i.e., latitude, longitude, and elevation) does not allow for any statistical adjustment (e.g., downscaling, locally calibrated lapse rate, etc.) for either temperature or precipitation. However precipitation regimes are very difficult for meteorological stations to record properly and this issue has often been found in other databases [59,69]. Many more data are required, especially in the case of forest monitoring, as a result of the lack of temporal autocorrelation during the timeframe [70].

The need for a freely available and representative global climate dataset is large and growing, as evidenced by WorldClim's citation statistics. These goals can be achieved with local up-to-date monitoring networks, which could play a key role in evaluating global grids at small scale [71] as well as providing data for the construction of additional global climate datasets. Harmonization efforts, as well as increased representativeness of the established networks, are paramount for construction of more accurate climate surfaces. Enhanced data recovery with regular spatial coverage may overcome the lack of dense environmental or climatological sampling [28,70,72,73]. Derived surfaces are fundamental in order to plan future management strategies. For instance, and concerning forestry, additional strata, such as homogeneous climate zones, are needed as a fundamental tool to plan the transfer of genetic resources and reproductive materials across specific geographic areas [74,75]. WorldClim grids were interpolated with spline functions, a fast method known to yield results similar to polynomial functions but without mathematical instability. Such methods do not consider the spatial autocorrelation between observations, only partially achieved by more complex models where latitude and longitude are included as predictive variables [28,76]. Therefore, the exhibited spatial aggregation of the BIAS in the case of denser observations of our dataset (i.e., Sweden and Germany) may be relevant for research activities and improvements of the climate surfaces.

5. Conclusions

A new updated beta version of WorldClim has recently been released for the 1971–2000 time period. This "Version 2" (http://worldclim.org/version2), along with the need for carefully evaluating the quality of records used for modelling and keeping climate databases up-to-date, is an essential requirement for the adequate development of tools and informative systems. The lack of reliability on MAP values can be seen as the main shortcoming of the WorldClim database in Europe and elsewhere. However, precipitation is much more difficult to interpolate, given its low spatial and temporal autocorrelation as well as the lack of statistical relationships with some of the main physiographic parameters, such as elevation. Further research should focus on this parameter, seeking more significant determinants of MAP, given its importance in climate change scenarios where drought stresses are predicted to be the most relevant issue.

Author Contributions: Conceptualization, M.W., M.B. and M.M.; methodology, M.M. and M.W.; software, M.M.; validation, M.M., I.S. and M.W.; formal analysis, M.M.; investigation, M.M., M.B. and M.W.; resources, M.B.; data curation, M.M., I.S. and M.W.; writing—original draft preparation, M.M. and M.W.; writing—review and editing, M.M., I.S., M.B. and M.W.; visualization, M.M.; supervision, M.B. and M.W.; project administration, M.W. and M.B.; funding acquisition, M.M., M.W. and M.B.

Acknowledgments: We thank Gregor Vertačnik from National Meteorological Service of Slovenia and Athanasios D. Sarantopoulos from the Hellenic National Meteorological Service Division of Climatology for contributing data.

References

1. Fridman, J.; Holm, S.; Nilsson, M.; Nilsson, P.; Ringvall, A.H.; Ståhl, G. Adapting national forest inventories to changing requirements—The case of the Swedish national forest inventory at the turn of the 20th century. *Silva Fenn.* **2014**, *48*, 1095. [CrossRef]
2. Bussotti, F.; Pollastrini, M. Traditional and novel indicators of climate change impacts on european forest trees. *Forests* **2017**, *8*, 137. [CrossRef]
3. Ray, D.; Petr, M.; Mullett, M.; Bathgate, S.; Marchi, M.; Beauchamp, K. A simulation-based approach to assess forest policy options under biotic and abiotic climate change impacts: A case study on Scotland's National Forest Estate. *For. Policy Econ.* **2019**, *103*, 17–27. [CrossRef]
4. Marchi, M.; Ferrara, C.; Biasi, R.; Salvia, R.; Salvati, L. Agro-forest management and soil degradation in mediterranean environments: towards a strategy for sustainable land use in vineyard and olive cropland. *Sustainability* **2018**, *10*, 2565. [CrossRef]
5. Mouchet, M.A.; Paracchini, M.L.; Schulp, C.J.E.; Stürck, J.; Verkerk, P.J.; Verburg, P.H.; Lavorel, S. Bundles of ecosystem (dis)services and multifunctionality across European landscapes. *Ecol. Indic.* **2017**, *73*, 23–28. [CrossRef]
6. Dyderski, M.K.; Paź, S.; Frelich, L.E.; Jagodziński, A.M. How much does climate change threaten European forest tree species distributions? *Glob. Chang. Biol.* **2018**, *24*, 1150–1163. [CrossRef] [PubMed]
7. Lindner, M.; Maroschek, M.; Netherer, S.; Kremer, A.; Barbati, A.; Garcia-Gonzalo, J.; Seidl, R.; Delzon, S.; Corona, P.; Kolström, M.; et al. Climate change impacts, adaptive capacity, and vulnerability of European forest ecosystems. *For. Ecol. Manag.* **2010**, *259*, 698–709. [CrossRef]
8. Littell, J.S.; McKenzie, D.; Kerns, B.K.; Cushman, S.; Shaw, C.G. Managing uncertainty in climate-driven ecological models to inform adaptation to climate change. *Ecosphere* **2011**, *2*, art102. [CrossRef]
9. Morton, J.F. The impact of climate change on smallholder and subsistence agriculture. *Proc. Natl. Acad. Sci. USA* **2007**, *104*, 19680–19685. [CrossRef]
10. Marchetti, M.; Vizzarri, M.; Lasserre, B.; Sallustio, L.; Tavone, A. Natural capital and bioeconomy: Challenges and opportunities for forestry. *Ann. Silvic. Res.* **2014**, *38*, 62–73.
11. Imeson, A. *Desertification, Land Degradation and Sustainability*; John Wiley & Sons, Inc.: New York, NY, USA, 2012; ISBN 9781119977759.
12. Ferrarini, A.; Selvaggi, A.; Abeli, T.; Alatalo, J.M.; Orsenigo, S.; Gentili, R.; Rossi, G. Planning for assisted colonization of plants in a warming world. *Sci. Rep.* **2016**, *6*, 28542. [CrossRef]
13. Chakraborty, D.; Wang, T.; Andre, K.; Konnert, M.; Lexer, M.J.; Matulla, C.; Schueler, S. Selecting populations for non-analogous climate conditions using universal response functions: The case of douglas-fir in central Europe. *PLoS ONE* **2015**, *10*, e0136357. [CrossRef]
14. Marchi, M.; Ducci, F. Some refinements on species distribution models using tree-level national forest inventories for supporting forest management and marginal forest population detection. *iForest Biogeosci. For.* **2018**, *11*, 291–299. [CrossRef]
15. Millar, C.I.; Stephenson, N.L.; Stephens, S.L. Climate change and forests of the future: Managing in the face of uncertainty. *Ecol. Appl.* **2007**, *17*, 2145–2151. [CrossRef]
16. Fady, B.; Aravanopoulos, F.A.; Alizoti, P.; Mátyás, C.; von Wühlisch, G.; Westergren, M.; Belletti, P.; Cvjetkovic, B.; Ducci, F.; Huber, G.; et al. Evolution-based approach needed for the conservation and silviculture of peripheral forest tree populations. *For. Ecol. Manag.* **2016**, *375*, 66–75. [CrossRef]
17. Zhang, Q.; Wei, H.; Zhao, Z.; Liu, J.; Ran, Q.; Yu, J.; Gu, W.; Zhang, Q.; Wei, H.; Zhao, Z.; et al. Optimization of the fuzzy matter element method for predicting species suitability distribution based on environmental data. *Sustainability* **2018**, *10*, 3444. [CrossRef]
18. Castaldi, C.; Vacchiano, G.; Marchi, M.; Corona, P. Projecting nonnative Douglas fir plantations in southern europe with the forest vegetation simulator. *For. Sci.* **2017**, *63*, 101–110. [CrossRef]

19. Subedi, N.; Sharma, M. Climate-diameter growth relationships of black spruce and jack pine trees in boreal Ontario, Canada. *Glob. Chang. Biol.* **2013**, *19*, 505–516. [CrossRef]
20. Garzón, M.B.; Robson, T.M.; Hampe, A. ΔTraitSDM: Species distribution models that account for local adaptation and phenotypic plasticity. *New Phytol.* **2019**, *222*, 1757–1765. [CrossRef] [PubMed]
21. Valladares, F.; Matesanz, S.; Guilhaumon, F.; Araújo, M.B.; Balaguer, L.; Benito-Garzón, M.; Cornwell, W.; Gianoli, E.; van Kleunen, M.; Naya, D.E.; et al. The effects of phenotypic plasticity and local adaptation on forecasts of species range shifts under climate change. *Ecol. Lett.* **2014**, *17*, 1351–1364. [CrossRef] [PubMed]
22. O'Neill, G.A.; Hamann, A.; Wang, T.L. Accounting for population variation improves estimates of the impact of climate change on species' growth and distribution. *J. Appl. Ecol.* **2008**, *45*, 1040–1049. [CrossRef]
23. Noce, S.; Collalti, A.; Valentini, R.; Santini, M. Hot spot maps of forest presence in the Mediterranean basin. *iForest Biogeosci. For.* **2016**, *9*, 766–774. [CrossRef]
24. Goberville, E.; Beaugrand, G.; Hautekèete, N.C.; Piquot, Y.; Luczak, C. Uncertainties in the projection of species distributions related to general circulation models. *Ecol. Evol.* **2015**, *5*, 1100–1116. [CrossRef] [PubMed]
25. Noce, S.; Collalti, A.; Santini, M. Likelihood of changes in forest species suitability, distribution, and diversity under future climate: The case of Southern Europe. *Ecol. Evol.* **2017**, *7*, 9358–9375. [CrossRef] [PubMed]
26. Pecchi, M.; Marchi, M.; Giannetti, F.; Bernetti, I.; Bindi, M.; Moriondo, M.; Maselli, F.; Fibbi, L.; Corona, P.; Travaglini, D.; et al. Reviewing climatic traits for the main forest tree species in Italy. *iForest Biogeosci. For.* **2019**, *12*, 173–180. [CrossRef]
27. Attorre, F.; Alfo, M.; De Sanctis, M.; Bruno, F. Comparison of interpolation methods for mapping climatic and bioclimatic variables at regional scale. *Int. J. Climatol.* **2007**, *1843*, 1825–1843. [CrossRef]
28. Marchi, M.; Chiavetta, U.; Castaldi, C.; Ducci, F. Does complex always mean powerful? A comparison of eight methods for interpolation of climatic data in Mediterranean area. *Ital. J. Agrometeorol.* **2017**, *1*, 59–72.
29. Azupura, M.; Dos Ramos, K. A comparison of spatial interpolation methods for estimation of average electromagnetic field magnitude. *Prog. Electromagn. Res. M* **2010**, *14*, 135–145. [CrossRef]
30. Mbogga, M.S.; Hamann, A.; Wang, T. Historical and projected climate data for natural resource management in western Canada. *Agric. For. Meteorol.* **2009**, *149*, 881–890. [CrossRef]
31. Hijmans, R.J.; Cameron, S.E.; Parra, J.L.; Jones, G.; Jarvis, A. Very high resolution interpolated climate surfaces for global land areas. *Int. J. Climatol.* **2005**, *25*, 1965–1978. [CrossRef]
32. Molyneux, N.; Soares, I.; Neto, F. Modeling current and future climates using worldclim and diva software: case studies from Timor Leste and India. *J. Crop Improv.* **2014**, *28*, 619–640. [CrossRef]
33. Vacchiano, G.; Motta, R. An improved species distribution model for Scots pine and downy oak under future climate change in the NW Italian Alps. *Ann. For. Sci.* **2015**, *72*, 321–334. [CrossRef]
34. Schueler, S.; Falk, W.; Koskela, J.; Lefèvre, F.; Bozzano, M.; Hubert, J.; Kraigher, H.; Longauer, R.; Olrik, D.C. Vulnerability of dynamic genetic conservation units of forest trees in Europe to climate change. *Glob. Chang. Biol.* **2014**, *20*, 1498–1511. [CrossRef]
35. Metzger, M.J.; Bunce, R.G.H.; Jongman, R.H.G.; Sayre, R.; Trabucco, A.; Zomer, R. A high-resolution bioclimate map of the world: A unifying framework for global biodiversity research and monitoring. *Glob. Ecol. Biogeogr.* **2013**, *22*, 630–638. [CrossRef]
36. Isaac-Renton, M.G.; Roberts, D.R.; Hamann, A.; Spiecker, H. Douglas-fir plantations in Europe: A retrospective test of assisted migration to address climate change. *Glob. Chang. Biol.* **2014**, *20*, 2607–2617. [CrossRef] [PubMed]
37. Wang, T.; Campbell, E.M.; O'Neill, G.A.; Aitken, S.N. Projecting future distributions of ecosystem climate niches: Uncertainties and management applications. *For. Ecol. Manag.* **2012**, *279*, 128–140. [CrossRef]
38. The R Development Core Team. *R: A Language and Environment for Statistical Computing*; R Foundation for Statistical Computing: Vienna, Austria, 2019.
39. KNMI Climate Explorer. Available online: https://climexp.knmi.nl (accessed on 21 April 2016).
40. Deutscher Wetterdienst. Available online: ftp://ftp-cdc.dwd.de/pub/CDC/observations_germany/climate/multi_annual/ (accessed on 13 April 2016).
41. Hellenic National Meteorological Service. Available online: http://www.emy.gr/emy/en/ (accessed on 21 April 2016).
42. National Meteorological Service of Slovenia. Available online: https://meteo.arso.gov.si/met/en/ (accessed on 21 April 2016).

43. Swedish Meteorological and Hydrological Institute. Available online: http://www.smhi.se/klimatdata/meteorologi/temperatur/dataserier-med-normalvarden-1.7354 (accessed on 30 March 2016).
44. Pebesma, E.J. Multivariable geostatistics in S: The gstat package. *Comput. Geosci.* **2004**, *30*, 683–691. [CrossRef]
45. Zhang, L.; Liu, S.; Sun, P.; Wang, T.; Wang, G.; Zhang, X.; Wang, L. Consensus forecasting of species distributions: the effects of niche model performance and niche properties. *PLoS ONE* **2015**, *10*, e0120056. [CrossRef]
46. Hamann, A.; Roberts, D.R.; Barber, Q.E.; Carroll, C.; Nielsen, S.E. Velocity of climate change algorithms for guiding conservation and management. *Glob. Chang. Biol.* **2015**, *21*, 997–1004. [CrossRef]
47. Mairota, P.; Leronni, V.; Xi, W.; Mladenoff, D.J.; Nagendra, H. Using spatial simulations of habitat modification for adaptive management of protected areas: Mediterranean grassland modification by woody plant encroachment. *Environ. Conserv.* **2013**, *41*, 144–156. [CrossRef]
48. Belda, M.; Holtanová, E.; Kalvová, J.; Halenka, T. Global warming-induced changes in climate zones based on CMIP5 projections. *Clim. Res.* **2016**, *71*, 17–31. [CrossRef]
49. IPCC. *Climate Change 2014: Impacts, Adaptation, and Vulnerability. Part A: Global and Sectoral Aspects. Contribution of Working Group II to the Fifth Assessment Report of the Intergovernmental Panel on Climate Change*; Field, C.B., Barros, V.R., Dokken, D.J., Eds.; Cambridge University Press: Cambridge, UK; New York, NY, USA, 2014; p. 1132.
50. Wood, S.N. On confidence intervals for generalized additive models based on penalized regression splines. *Aust. N. Z. J. Stat.* **2006**, *48*, 445–464. [CrossRef]
51. Fréjaville, T.; Fady, B.; Kremer, A.; Ducousso, A.; Garzón, M.B. Inferring phenotypic plasticity and local adaptation to climate across tree species ranges using forest inventory data. *bioRxiv* **2019**, 1–34. [CrossRef]
52. Ray, D.; Bathgate, S.; Moseley, D.; Taylor, P.; Nicoll, B.; Pizzirani, S.; Gardiner, B. Comparing the provision of ecosystem services in plantation forests under alternative climate change adaptation management options in Wales. *Reg. Environ. Chang.* **2015**, *15*, 1501–1513. [CrossRef]
53. Bellucci, A.; Gualdi, S.; Masina, S.; Storto, A.; Scoccimarro, E.; Cagnazzo, C.; Fogli, P.; Manzini, E.; Navarra, A. Decadal climate predictions with a coupled OAGCM initialized with oceanic reanalyses. *Clim. Dyn.* **2013**, *40*, 1483–1497. [CrossRef]
54. Heikkinen, R.K.; Luoto, M.; Araújo, M.B.; Virkkala, R.; Thuiller, W.; Sykes, M.T. Methods and uncertainties in bioclimatic envelope modelling under climate change. *Prog. Phys. Geogr.* **2006**, *30*, 751–777. [CrossRef]
55. Perdinan, P.; Winkler, J.A. Changing human landscapes under a changing climate: Considerations for climate assessments. *Environ. Manag.* **2014**, *53*, 42–54. [CrossRef]
56. Ramirez-Villegas, J.; Jarvis, A. *CIAT Decision and Policy Analysis Working Paper*; CIAT: London, UK, 2010; p. 18.
57. Hargreaves, G.H.; Samani, Z.A. Reference crop evapotranspiration from temperature. *Appl. Eng. Agric.* **1985**, *1*, 96–99. [CrossRef]
58. Maca, P.; Pech, P. Forecasting SPEI and SPI drought indices using the integrated artificial neural networks. *Comput. Intell. Neurosci.* **2016**, *2016*, 3868519. [CrossRef]
59. Vicente-Serrano, S.M.; Beguería, S.; López-Moreno, J.I.; Angulo, M.; El Kenawy, A. A new global 0.5° gridded dataset (1901–2006) of a multiscalar drought index: Comparison with current drought index datasets based on the palmer drought severity index. *J. Hydrometeorol.* **2010**, *11*, 1033–1043. [CrossRef]
60. Svenning, J.C.; Fitzpatrick, M.C.; Normand, S.; Graham, C.H.; Pearman, P.B.; Iverson, L.R.; Skov, F. Geography, topography, and history affect realized-to-potential tree species richness patterns in Europe. *Ecography (Cop.)* **2010**, *33*, 1070–1080. [CrossRef]
61. Colantoni, A.; Ferrara, C.; Perini, L.; Salvati, L. Assessing trends in climate aridity and vulnerability to soil degradation in Italy. *Ecol. Indic.* **2015**, *48*, 599–604. [CrossRef]
62. Hannachi, A.; Jolliffe, I.T.; Stephenson, D.B. Empirical orthogonal functions and related techniques in atmospheric science: A review. *Int. J. Climatol.* **2007**, *27*, 1119–1152. [CrossRef]
63. Daly, C.; Halbleib, M.; Smith, J.I.; Gibson, W.P.; Doggett, M.K.; Taylor, G.H.; Curtis, J.; Pasteris, P.P. Physiographically sensitive mapping of climatological temperature and precipitation across the conterminous United States. *Int. J. Climatol.* **2008**, *28*, 2031–2064. [CrossRef]
64. Blasi, C.; Chirici, G.; Corona, P.; Marchetti, M.; Maselli, F.; Puletti, N. Spazializzazione di dati climatici a livello nazionale tramite modelli regressivi localizzati. *Forest* **2007** *4*, 213–219. [CrossRef]
65. Wang, T.; Hamann, A.; Spittlehouse, D.; Carroll, C. Locally downscaled and spatially customizable climate data for historical and future periods for North America. *PLoS ONE* **2016**, *11*, e0156720. [CrossRef]

66. Harris, I.; Jones, P.D.; Osborn, T.J.; Lister, D.H. Updated high-resolution grids of monthly climatic observations—The CRU TS3.10 Dataset. *Int. J. Climatol.* **2014**, *34*, 623–642. [CrossRef]
67. Falk, W.; Mellert, K.H. Species distribution models as a tool for forest management planning under climate change: Risk evaluation of Abies Alba in Bavaria. *J. Veg. Sci.* **2011**, *22*, 621–634. [CrossRef]
68. Trivedi, M.R.; Berry, P.M.; Morecroft, M.D.; Dawson, T.P. Spatial scale affects bioclimate model projections of climate change impacts on mountain plants. *Glob. Chang. Biol.* **2008**, *14*, 1089–1103. [CrossRef]
69. Marchi, M.; Castaldi, C.; Merlini, P.; Nocentini, S.; Ducci, F. Stand structure and influence of climate on growth trends of a Marginal forest population of Pinus nigra spp. nigra. *Ann. Silvic. Res.* **2015**, *39*, 100–110.
70. Ferrara, C.; Marchi, M.; Fares, S.; Salvati, L. Sampling strategies for high quality time-series of climatic variables in forest resource assessment. *iForest Biogeosci. For.* **2017**, *10*, 739–745. [CrossRef]
71. Way, R.G.; Oliva, F.; Viau, A.E. Underestimated warming of northern Canada in the Berkeley Earth temperature product. *Int. J. Climatol.* **2017**, *37*, 1746–1757. [CrossRef]
72. Bhowmik, A.K.; Costa, A.C. Representativeness impacts on accuracy and precision of climate spatial interpolation in data-scarce regions. *Meteorol. Appl.* **2014**, *22*, 368–377. [CrossRef]
73. Wong, D.W.; Yuan, L.; Perlin, S. A comparison of spatial interpolation methods for the estimation of air quality data. *J. Expo. Anal. Environ. Epidemiol.* **2004**, *14*, 404–415. [CrossRef] [PubMed]
74. Metzger, M.J.; Brus, D.J.; Bunce, R.G.H.; Carey, P.D.; Gonçalves, J.; Honrado, J.P.; Jongman, R.H.G.; Trabucco, A.; Zomer, R. Environmental stratifications as the basis for national, European and global ecological monitoring. *Ecol. Indic.* **2013**, *33*, 26–35. [CrossRef]
75. De Dato, G.; Teani, A.; Mattioni, C.; Marchi, M.; Monteverdi, M.C.; Ducci, F. Delineation of seed collection zones based on environmental and genetic characteristics for *Quercus suber* L. in Sardinia, Italy. *iForest Biogeosci. For.* **2018**, *11*, 651–659. [CrossRef]
76. Bachmaier, M.; Backes, M. Variogram or Semivariogram? Variance or Semivariance? Allan Variance or Introducing a New Term? *Math. Geosci.* **2011**, *43*, 735–740. [CrossRef]

10

The Institutional Structure of Land use Planning for Urban Forest Protection in the Post-Socialist Transition Environment: Serbian Experiences

Marija Maruna, Tijana Crnčević and Milica P. Milojević

[1] Faculty of Architecture, University of Belgrade, 11000 Belgrade, Serbia
[2] Institute of Architecture and Urban & Spatial Planning of Serbia, 11000 Belgrade, Serbia
* Correspondence: m.ma@sezampro.rs.

Abstract: In recent decades, Serbia has been undergoing a period of post-socialist transition that has significantly altered the value system underlying spatial development due to alteration of ownership frameworks and land use rights. In consequence, issues have arisen of how to strike a balance between the various interests involved in the distribution of spatial resources and how to control the outcomes of public policies. Land use planning has been identified as an efficient instrument for implementing the public policy value framework. The objective of this paper is to identify the key points of land use planning in relation to urban forest management of significance for the maintenance of urban forests in the environment of post-socialist institutional transformation in Serbia. Seen as an institutional structure, the practice of land use planning in Serbia is the product of a stable interaction between the set of interrelated rules, procedures and organisational units that allows spatial development outcomes that take into account and safeguard land resources and, ultimately, urban forests. The research was carried out in relation to the concept of institutional transformation across three scales: macro/governance, meso/coordination and micro/agency: (a) components of the regulatory framework; (b) procedures for cooperation between stakeholders; and (c) specific activities of land use planning practice. As a result, the concept of Land use Planning for Urban Forest Protection (LUPUFP) in Serbia was established. It identifies components of institutional structure of importance for regulating system changes in the post-socialist transition environment and steering them towards the establishment of a value framework that allows the agenda of saving urban forests to be implemented.

Keywords: urban forest; institutional design; land use planning; Serbia; governance

---

1. Introduction

The research objective of this paper is to provide a critical overview of the institutional framework of the development planning system in relation to urban forest management, particularly land use planning, in the post-socialist environment in Serbia, in order to highlights components of importance to be used by institutional design for re-establishing a stable interaction inside the institutional structure that promotes a value system aimed at saving urban forests. As a result, this paper presents an improvement of the current land use planning system in Serbia by proposing the concept of Land Use Planning for Urban Forest Protection (LUPUFP), focusing on three major components: urban forests, land use planning and post-socialist environment in Serbia.

The problems of urban growth in the 21st century emphasise the importance of managing land resources in order to achieve sustainable development [1–4]. As a limited resource, land in cities, especially urban forests as compact large greenfield land, is particularly affected by new construction requirements that are directly reflected on environmental problems such as "heat islands", flooding and air pollution. Therefore, the network of woodlands, groups of trees and individual trees located in a city that include forests, street trees, trees in parks and gardens and trees in derelict corners [5], in terminology known as urban and peri-urban forests, play a crucial role in meeting global commitments on sustainable development as well as adaptation to climate change and mitigation of its impacts. These requirements represent a particular challenge for local governments that are expected to make land use more efficient for planning compact cities and mixed-land use [1].

As it was indicated in recent research, urban forest services enhance nine of the seventeen Sustainable Development Goals (SDGs) [6,7]. These SDGs are 1: No poverty; 2: Zero hunger; 3: Good health and well: being; 6: Clean water and sanitation; 7: Affordable and clean energy; 8: Descent work and economic growth; 11: Sustainable cities and communities; 13: Climate action; and 15: Life on land.

These analyses included five categories that were considered to be urban forest (peri-urban forest and woodlands, city parks and urban forests > 0.5 ha, pocket parks and gardens with trees, trees on streets or in public squares and other green spaces with trees) where the category of peri-urban forest and woodlands was scored with the highest score, suggesting that it contributes to human health and well-being, climate change mitigation, climate change adaptation, biodiversity and landscapes, economic benefits and a green economy, land and soil degradation, watershed protection, resilience to flooding events, food and nutrition security, wood security, recreation, education, social cohesion and social security and equity. Further, it should be noted that the positive effects of urban forests ecosystems are confirmed within numerous studies [8–11] where it is specifically stressed that in regard to the public's perceptions of the effectiveness of the ecosystems services "environmental knowledge plays a key role in fostering pro-environmental behaviours" [12] (p.171).

Land use planning is a key mechanism for reducing pressure on land resources, which facilitates the implementation of environmental protection policies and steers choices in the direction of nature protection. This intention has been affirmed in the Curitiba Declaration on Cities and Biodiversity [13], where the issue of integrating biodiversity into urban planning is placed within the context of establishing the appropriate regulatory mechanisms and implementing tools, as well as decision-making instruments that will ensure the integration of nature/biodiversity and the urban structure. Land use planning directly serves the green agenda for ecological health and management of natural ecosystems on the local level, which aims at preserving green open spaces in cities for biodiversity protection and recreation [14]. The importance of integrating green structure planning in city development planning is also borne out by the findings of a survey of 23 European cities [15] that have met green policy targets by employing land use planning as the key instrument.

The achievement of the saving urban forests agenda has been challenged in post-socialist transition Serbia. Namely, after the year 2000, with the transition of the socioeconomic system from a socialist orientation to that of the market and democracy, and the start of the process of joining the European Union, significant institutional changes were initiated in Serbia. Following major challenges are of particular importance for this research.

Firstly, ownership relations over land have changed, thus establishing a new relationship of strengths of power in society and the consequent plurality of interests in the process of creating public policies concerning the use of land resources.

Secondly, with entering into the process of joining the European Union, the principle of subsidiarity, which transfers competencies to the local level, is an important principle in decision-making on public policies, thus giving the highest responsibility and commitment to local authorities in selecting and promoting the agenda. This is particularly challenging for post-socialist transition countries that do not have developed expert capacities at the local level.

Thirdly, global demands for the implementation of sustainable development promote the concept of governance, which implies the development of horizontal and vertical coordination mechanisms at different levels of the institutional structure. These demands entail a fundamental change in the concept of decision-making in post-socialist countries, which up to now was hierarchical rather than collaborative.

Harmonization of the above components of institutional changes represents a huge challenge for the transition society, in the selection of a value framework for operation and the regulation of all domains of public policies. This is, beyond doubt, a highly sensitive institutional redesign process that will create instability and insecurity. Models that have proven successful in other situations cannot be applied, and locally-specific answers are expected to be created. In this context, the country must develop its own path of institutional transformation that will satisfy the demands of the new environment of post-socialist transition and the EU accession process, while, at the same time, retaining the positive experiences of the past. That is why this paper will investigate the current institutional framework of the development planning system in Serbia, which is in the function of achieving the agenda 'saving urban forests'. According to the concept of institutional transformation [16–25] the values are embedded in the institutional system and are a result of a balance between its components. With the disturbance of the stability of the institutional system, due to the transition process, which is induced by the influences that came into the system from the outside, an institutional transformation and consequently a distortion of the value framework of public policies and related agendas are inevitable.

Following an introductory section, the paper will first outline the theoretical framework in three important domains: land use planning, urban forest management and the concept of institutional transformation. The methodological section that follows relies on the key conclusions of the previous section to be used as the basis for the selection of methodological instruments and definition of key research questions. The analysis section comprises a detailed presentation of the institutional system of land use planning in relation to urban forest management from the national to local level, according to the concept of institutional transformation, illustrated by examples of the land use planning practice of two medium-sized cities. Within the discussion and conclusion section, as the first step in institutional design, the components of the presented institutional system are highlighted as the basis for the development of the Land use Planning for Urban Forest Protection (LUPUFP) concept.

1.1. Theoretical Background

1.1.1. Land Use Planning

In the last decades it has been emphasised that land use planning represents a significant management tool for dealing with unprecedented challenges that lie ahead of the accelerated urbanization process of cities [14,26]. Land use planning is a local development instrument that is complemented by regulations introduced hierarchically within vertical power relations. In general, land use planning is regulatory in character and is used by governments as a legal instrument intended to achieve public/common interest or public policy [27]. It essentially aims at controlling the use of and rights to land, both public and private [28], by applying various regulatory measures, such as protected areas, building codes and rezoning. More specifically, the purpose of the land use plan is to regulate the land as a category of usage, while the purpose of regulatory measures is to change ownership rights [29].

Contemporary development planning, and consequently land use planning, is a political and democratic process that mediates the conflicts over land use, not just a technical tool [30]. Contemporary planning has abandoned the traditional position of technical activity, where plans are understood as documents, to become a collaborative practice jointly undertaken by planners and local communities. This is consistent with the concept of safeguarding green areas, where traditional land use planning is seen as a passive, technical, regulative and rigid planning instrument that is lacking the capacity to protect spatial resources [15,31]. In contrast, the collaborative planning practice predominantly rooted in strategic spatial planning is considered to allow the articulation of a more coherent and coordinated long-term spatial logic for land use regulation [17]. Collaborative strategic planning practice looks at the distribution of spatial resources through process orientation, social inclusion and a multilevel approach. From this point of view, the distribution of land use is seen as a policy conditioned by the design of appropriate governance institutions and proactive management activities [16,32]. In this context, decision-making is the key mechanism for establishing the value framework for public policy, and consequently selecting objectives relevant for the protection of spatial resources. The practice of strategic planning has proven to be a highly successful instrument for redistributing land in favour of larger open green areas, such as the development of a new urban forest, 'Parkbos', in Ghent, Belgium [33].

1.1.2. Urban Forest Management

Contemporary global development policies, which prioritize climate change issues in addition to sustainability, assign particular importance to urban forests in the context of adaptation to climate change and mitigation of its impacts. In addition to the Sustainable Development Goals (SDGs) the EU's current strategic framework [34,35] promotes an integrated landscape approach, which links protected spaces with other sectoral interests.

The Framework for Assessing and Monitoring Forest Governance [36] suggests six principles of good governance as cornerstones for a stable institutional structure for managing forest land: accountability, effectiveness, efficiency, fairness, participation and transparency. These principles intersect three key pillars: (1) policy, legal, institutional and regulatory frameworks; (2) the planning and decision-making process; and (3) implementation, enforcement and compliance. The second pillar is of immediate significance for development planning processes, emphasising as it does how important it is to "examine the extent, characteristics and quality of participation of a range of stakeholders in forest governance and the capacity of different stakeholder groups to engage in governance processes. Components under this pillar also consider the transparency of forest-related decision-making and resource allocation and the degree of accountability of governance mechanisms and processes" [36] (p. 13).

Similarly, in the domain of multi-disciplinary research into urban forest governance, Lawrence et al. [37] defined the analytical framework as a research tool. This framework is proposed to comprise four variables for researching the institutional system: (1) policies including national, regional or local policies, plans and programmes that affect urban forestry; (2) planning and regulations, which comprise planning and legal requirements specific to the case; (3) ownership of the land; and (4) access and use rights, such as the right to walk/cycle and/or make use of products from the urban forest. The proposed framework is only a starting model for studying urban forest governance which, according to the authors, should be elaborated on to achieve a deeper understanding of the ways to ultimately make our cities more sustainable.

The importance of stakeholder involvement in policy making to protect natural areas is also acknowledged by the Natura 2000 network, the core pillar in the EU's biodiversity conservation policy [34] and one of the largest networks of protected areas in the world. Although member states are free to align national regulations with EU instruments [34,35], the expected changes reach deep into the institutional foundations of public policy decision-making by legitimising a broad range of stakeholders and, consequently, introducing a multitude of varied interests that need to be harmonised [38–41]. This primarily affects the local decision-making level, the most effective forum for exercising the concept of public interest, as the issues at hand are specific and easy to operationalise.

1.1.3. Concept of Institutional Transformation

According to the concept of institutional transformation, values are embedded in institutional systems and are the result of the established balance between system components [18,19]. From a normative perspective, values and commitments generated in interaction shape undermine and augment formal and official regimes [23]. The concept of institutional transformation additionally describes in detail the procedures and interactions within processes of interest that can be used to identify challenges in public administration, the formation of interests, the development of policies and links, and the implementation of administrations' programmes in the planning process [21–23].

The orientation of planning practice depends on the values built into the institutional structure. In other words, the dominant ideology informs beliefs, values and systems, which in turn shape institutions, which, ultimately, result in policies [42]. As such, land use planning, the regulation of the use of space, reflects the ideology of defining and using space.

Socioeconomic changes under the influence of external factors initiate the process of institutional transformation [24], which inevitably alters the behaviour of political and economic actors [22] and, consequently, leads to changes in public policy. In essence, institutional transformation takes place in the domain of changes to the value framework, and as such directly shapes the development of society over time. Institutional transformation disrupts what is termed 'stable interaction' within the institutional structure as the guarantee of the established value framework. Stable interaction reduces uncertainty in decision-making by securing stable outcomes, dubbed 'equilibria'. Institutional transformation, therefore, must aim at establishing these stable interactions within the institutional system and so reduce uncertainty.

Institutional design, as a normative aspect of institutional transformation, is an integral and essential part of the planning practice [22]. Institutional design determines the success and quality of interventions by agents (individuals and organisations), and so also determines the success of the planning practice [22,25]. The operation of institutions, from the national to the local level, affects the role and success of planning. Institutions create elements of order and predictability, impose orderliness on societal relationships, reduce flexibility and variability in behaviour, and limit options for one-sided exercise of personal interests [19,24]. Striking a balance between differences in land use interests is a matter of policy choice and the design of appropriate institutions and proactive management activities [22]. Consequently, it can be considered that institutional design is a useful method for changing planning practice.

The behaviour of political and economic actors is conditioned by sets of regularised practices with a rule-like quality [24] that also affect policy outcomes. The so-called regulatory regime of land use management is of crucial importance for the application of planning practice as the source of environmental protection policies [25]. The regulatory regime provides a framework for action within which agencies enjoy autonomy in choosing their modes of operation and create room for exerting influence on the value framework.

From a planning perspective, three general concepts should be viewed as the key elements to be addressed by the institutional design [22]:

(a) The first is the concept of governance, which is the most appropriate at the macro-level as it involves society as a whole and is linked to constitutional and legal amendments. This is the level of institutional design to which belong requirements for institutional change encountered by Serbia in its EU accession process.
(b) The second concept is that of coordination, at the meso-level, which pertains to the domain of planning and comprises procedures which facilitate the development and implementation of policies, programmes, projects and plans associated with professional planners' fields of practice.
(c) The third is the concept of agency, which occurs at the micro-level and involves intra-organisational design, the ordering of smaller working units and groups and the processes and interactions within and between them. This level directly entails managing planning processes and policy, the plan or project implementation [22].

That being said, institutional analysis does not focus on norms, rules and practices as integral elements of institutions, but, rather, on their mutual interaction within the context that conditions action [25]. The emphasis here is on understanding relationships between activities and the institutional context that generate practices and a power structure that subsequently determine relationships and changes. Institutional analysis, therefore, observes relationships between systems within an institution, individual processes within and between those systems, what the constituent units of an institution are, which rules or norms govern their relationships, etc. [19,20]. According to Alexander [22] (p. 213), "institutional design means designing institutions: the devising and realisation of rules, procedures and organisational structures that will enable and constrain behaviour and action so as to accord with held values, achieve desired objectives or execute given tasks". Institutional analysis, as an initial step of institutional design, will be in this research applied to the land use planning practice related to urban forest management in Serbia.

2. Materials and Methods

In accordance with the presented theoretical framework, the critical re-assessment of the institutional structure of land use planning in the post-socialist environment of Serbia related to urban forest management, will highlight the following two aspects: (a) on the one hand, understanding the institutional and regulatory context of the planning process and (b) the other on discerning the relationships between dominant stakeholders in this process that reflect value-based approaches.

Furthermore, the research is based on qualitative assessment on three scales that correspond to the levels of institutional design [22]:

1. The macro-level entails assessing the regulatory framework for the development planning system related to urban forest management in Serbia from two aspects:
   (a) Overview and assessment of the institutional and legal framework for urban forest protection standards at all levels of administrative organisation: national, regional and local. In this assessment the position of urban forest protection standards within basic land use planning documents is also included.
   (b) Overview and assessment of the value framework for urban forest protection in national policy documents. This phase of the research relies on reviewing and analysing both the primary literature (laws, strategies and other public documents) and secondary sources dealing with issues of the planning system for urban forest protection.
2. The meso-level looks at the procedures for cooperation between institutions of the land use planning process on the local level, related to urban forest management in Serbia. Institutions are observed from the national to the local levels through the lens of multilevel governance. The focus is on the procedures for collaboration between institutions in the process of the 'General zoning plan' production, as well as policy-making procedures, and is observed in two aspects:
   (a) Identification of the organisational structure, with particular attention to the position, powers and roles of the relevant institutions.
   (b) Arrangements for collaboration that includes insight into both the horizontal and the vertical levels. This segment of the research relies primarily on a review of primary literature that sets out organisational powers while also considering secondary documents devoted to how land use planning and urban forest management policies are made.

   The micro-level entails a detailed review and assessment of the land use planning process in relation to urban forest protection in selected examples of land use planning practice on the local level. It is focused on analysis of activities of concrete cases of planning practices, where formal and informal institutional arrangements in the planning process are observed. The cases of planning practice are concrete solutions for the use of space viewed through the enacted policies and regulations. The accent here is on the description and analysis of the procedure of land use planning in terms of decision-making by the stakeholders (public, private and civil sector) involved that employ various mechanisms, instruments and actions. Both policy-related and regulatory planning solutions are analysed equally. This part of the research relies on both a critical analysis of the actual land use plan development process (particularly the planning procedure for the 'General Zoning Plan') and the concrete examples of the general zoning plans (a result of the planning process) as undertaken by planners. Examples include the land use planning processes of two Serbian cities: Bor and Vrnjačka Banja. The criteria for the selection of the cases were the following.

(a) Medium-sized cities (in order to avoid the overly complex problems that are characteristic of large cities).
(b) The dominant planning agenda is 'saving urban forests'. Bor is a town where the urban environment is exposed to the impacts of copper and gold mines situated in the immediate vicinity, and Vrnjačka Banja is an urban environment purposefully developed as a spa.
(c) The planning processes started after the year 2000, following the introduction of Serbia's new socioeconomic framework.

The methodological procedure shown is seen as suitable for institutional assessment as it emphasises case development factors linked to the context, in the same manner as institutional theory links norms and procedures to the broader institutional landscape [43].

3. Results

In socialist Yugoslavia, green open spaces were considered a public resource and were, as such, accorded particular attention by urban planners. After World War II, with the institutionalisation of socialism in Yugoslavia, land policy that determined land use was based on the ideological belief in common or 'social' property, as opposed to the private ownership of land. The fundamental political, social and economic reforms pursued at the time, accompanied by the establishment of a new constitutional and legal order by the communist regime, declared which cities and urban settlements stood to be social property and excluded them from legal transactions. In consequence, any extension of the urban territory automatically made new land socially owned. Land use plans served as direct instruments for these transformations and were employed to put public interest into effect in actual space [44,45].

Although the trend under state socialism was to make forest land socially owned as well, forests could be owned by the state, cooperatives or private individuals. Nevertheless, all forests, regardless of ownership, were declared to be of general interest to the community and were placed under government protection [46]. This made the preservation of forests and forest land a matter of public interest, and a system of safeguards was designed and implemented across all levels of governance to attain this objective.

Difficulties encountered by post-socialist countries in transitional processes [47] are inseparably linked to the crucial issues of changes to the value system and established norms [48,49]. Private property, instituted by the changes as a new form of land ownership, has brought about a major shift in traditional patterns of land use planning. Private interests, needs and expectations of how land is to be used have gained legitimacy and so become major factors in land-related policy-making. Consequently, development land has come to be a fundamental resource for a city's economic growth. In these circumstances, pressure has increased to allow construction on greenfield land, where developers do not incur additional costs when investing. Therefore, in a democratising society facing privatisation and the construction of market institutions, land use planning has become both a tool to safeguard property rights and interests of various land use stakeholders and an instrument to correct for market failure [44].

Institutional changes characteristic of post-socialist transition altered the value basis for planning, which also caused a shift in the planning paradigm [50]. Lacking a common planning system model they could employ, post-socialist countries have developed their own approaches to institutional transformation [51]. The experiences of post-socialist countries have shown that no changes were possible to planners' modes of operation that would allow them to protect the public interest without an institutional foundation being laid first [52].

The issue of public interest in planning has remained ill-defined in Serbia following the democratic changes of 2000. Urban plans have endeavoured to protect interests by defining public land, public areas and public buildings, but the protection of other, privately owned land remained subject to political decision-making and mechanisms intended to safeguard public interest. As such, issues including protection of public spaces, the environment, public health and security, energy efficiency, etc., topics that the public sector is not interested in addressing [53,54], remained within the remit of regulatory regimes of land use governance, primarily at the national level.

By contrast, the EU accession process has placed a number of new demands before the practice of environmental planning. Serbia formally became a candidate country for EU membership in March 2012, starting a negotiations procedure to align the country's regulatory framework with EU law through 34 chapters. Chapter 27 envisages the creation of a sustainable environmental management system, which cuts across all policy sectors and constitutes a value framework for their formulation [55]. In addition to the requirements of the negotiations process, other instruments pertaining more directly to the preservation of green open spaces also affect the harmonisation of the Serbian regulatory framework with the European context. One major such document is the European Landscape Convention [56], ratified by Serbia in 2011 [57]. An innovation introduced in the Convention is the understanding of landscape as a dynamic category that evolves with societal change. This approach means that landscape-related planning activity can no longer be subject only to deliberation by specialised technical bodies, but that landscape development policies must be enacted through democratic dialogue wherein all stakeholders are able to present their perceptions and views of the future of landscape [56].

On the other hand, the changes brought about by transition commenced a decentralisation process in which the local level became involved in decision-making about environmental protection policies. The EU accession process explicitly requires the adoption of standards to allow equal participation of all the various stakeholders in decision-making and reduce the scope for conflict between interests and preferences for protected spaces [58], as is confirmed in the ARHUS convention ratified by Serbian law [59]. This accords with the concept of governance, which promotes the establishment of diverse forms of cooperation, partnership agreements, delegation of authority and greater powers of the local community. Good governance entails the management of protected areas pursuant to the principles and values chosen by all stakeholders. As part of societal and cultural heritage, these principles are modified in accordance with globally recognised requirements and become integral parts of constitutions, laws and other legal enactments that regulate nature protection. However, most powers and responsibilities remain within the remits of governments and their agencies [60].

3.1. Macro-Level: Regulatory Framework for Development Planning System Related to Urban Forest Management in Serbia

3.1.1. Institutional and Legal Framework for Urban Forest Protection Standards

As yet, Serbia has not enacted legislation that specifically supports planning for the system of green spaces as a separate and autonomous domain. The Law on the Protection and Improvement of Green Spaces has remained at the drafting stage for a number of years [61,62]. Furthermore, a project by the Serbian Association of Landscape Architects, supported by the Ministry of Environmental Protection, that besides requiring spatial and urban plans to acknowledge and recognise existing greenery, green spaces, spaces close to nature and ecosystems, stressed the importance of the institutional framework at the local government level related to management, maintenance and reconstruction of urban green spaces [63].

The lack of an appropriate statutory and planning basis is compounded by the absence in Serbia of guidelines and recommendations for planning green spaces. Plans and regulations do not recognise the expression 'urban forest', but rather define forest land in urban contexts as 'town forests', an echo of the German *Stadtwald* better suited to a general understanding of the urban forestry concept [64,65]. Terms of importance for urban forests formally employed in the green spaces planning system are defined in the Forests Law [66]; these are (a) 'forest', an area of land in excess of 5 acres (500 square metres) covered by forest trees and (b) 'forest land', the land on which are located structures facilitating the attainment of the generally beneficial effects of forests.

In Serbia, the powers for managing, safeguarding and improving forests in urban areas (see Table 1) in essence reside predominantly within two policy departments: the environment and planning. At the national level, the responsible institutions are the Government of Serbia and the Nature Conservation Institute, tasked with conservation activities, as well as the Ministry of Environmental Protection and the Ministry of Construction, Transportation and Infrastructure. These ministries are charged with the development of the national statutory framework for planning and protecting urban forests. The key regulations for urban forests protection are a set of planning laws that govern norms for establishing land use balance in the context of the regulation of property rights to land (i.e., public, private and cooperatively owned property). The relevant nature conservation laws govern standards for the protection, management and use of urban forests.

There is no regional governance level in Serbia, so only the Provincial Nature Conservation Institute (with powers in the Province of Vojvodina) is the only formal regional body. At the local level, urban forests are managed by local state-owned enterprises, established independently by local authorities depending on their size, status and resources. However, powers are often dispersed amongst different organisations and departments, as well as between various levels of governance. So, for instance, in Belgrade, the capital city, the state-owned enterprise manages 32.322,70 hectares of forests, while an additional 611 hectares of forests in the territory of Belgrade are owned and managed by the city itself [65] (p. 339).

Nevertheless, local land use planning has the greatest impact on urban forest protection. Land use planning, one of the most important components of planning in Serbia, is regulated chiefly by the Planning and Construction Law [67], which envisages two spatial governance instruments: Spatial Plans, more focused on the strategic orientation of development, and Urban Plans, more land use oriented with some elements of an integrated approach. Urban Plans are the most common instruments of local land use planning and are divided into three categories: (a) General Urban Plans, mostly oriented towards strategic aspects; (b) General Zoning Plans; and (c) Detailed Zoning Plans, mainly devoted to technical aspects. Land use maps and technical parameters, such as rules of planning and rules of construction, are integral parts of Urban Plans. These land use maps are effective instruments for designating land of public interest, as they formally distinguish between development land (land designated for construction) and other land, which is as a rule publicly held [67]. The Serbian planning system is characterised by a tradition of land use planning [68] that is based exclusively on regulation and where plans are rigid instruments that set out long-term land use, architectural and aesthetic standards, and landscape and natural resource protection rules [69].

Table 1. Institutional and legal framework for standards of urban forest protection in Serbia.

Territorial Organisation	Institutions	Laws	Standards of Protection
National level	Government of Serbia	Government Order on the Environmental Network [70] Government Order on Safeguards [71]	Environmentally significant areas Area safeguards
	Ministry of Construction, Transportation and Infrastructure	Law on the Spatial Plan of the Republic of Serbia [72]	Land use balance at the national level
		Planning and Construction Law [67]	Change in intended use of forest land to development land
		Expropriation Law [73]	Determination of land of public interest
		Public Property Law [74]	Resources of general interest (national level)
			Resources in general use (local level)
			Right to use public property
		Utilities Law [75]	Management and maintenance of public green surfaces
	Ministry of Environmental Protection	Nature Conservation Law [76]	Natural resources and natural values
			Protected areas
		Environmental Protection Law [77]	Forest development plans and forest management rules
			Safeguards and options for use as defined by plans
		Forests Law [66]	Ban on sale of publicly-owned forests
			Prohibited activities
			Conditions for change in intended use
		Environmental Impact Assessment Law [78]	Preventive protection measures
		Strategic Environmental Impact Assessment Law [79]	Preventive protection measures
	Serbian Nature Conservation Institute		Safeguards and conditions for their implementation
Regional level*	Vojvodina Urban Planning Institute		Vojvodina Regional Spatial Plan
	Vojvodina Nature Conservation Institute		Protection programmes Safeguards and conditions for their implementation
	Special-purpose areas (no administrative powers)		Spatial plans for special purpose areas
Local level	Local authorities' departments	Local authority spatial plan	Land use balance at the local level
		General Urban Plan	Land use Zoning Building codes
		General Zoning Plan	
		Detailed Zoning Plan	
	State-owned enterprises tasked with developing natural resources		Development and maintenance programmes

* Serbia is not formally divided into administrative regions. The only part of the country with territorial and administrative autonomy is the Province of Vojvodina.

3.1.2. Value Framework for Urban Forest Protection in Serbia

Achieving land use balance is a core task of the planning process in Serbia. So, for instance, the first Spatial Plan of the Republic of Serbia provided for three key categories of land use as the bases for striking balance in spatial development: agriculture, forests and land for other uses [80]. There are many benefits to preserving forest land and, consequently, the biodiversity of Serbia's ecosystems. With a total surface of 670.598,81 hectares of protected area in Serbia [46] (p. 88), forest and forest land account for more than 40%.

Table 2 presents an overview of the basic value framework for urban forest protection in Serbia. It reveals the extent to which strategic documents address issues of planning and managing urban forests. The key words used in this assessment were: 'urban forests', 'forest parks', 'forests', 'town forests' and 'forest land'. Clearly, the expression 'urban forests' is not recognised in any of the documents considered. The phrase 'town forests' was employed in only one of the instruments, the National Strategy for Sustainable Use of Natural Resources [81], while most of the remaining documents made use of the phrases 'forests' and 'forest land'. The highlight here is that forests are recognised as notable finite natural resources that are important for preserving biodiversity, amongst other considerations. As such, the policy documents provide significant frameworks that guide forest development and protection, in particular to ensure alignment with EU rules and institutional strengthening. Indirectly, it is noteworthy for urban forests protection that the value framework acknowledges the social and cultural functions of forests (in addition to protective and regulatory ones).

3.2. Meso-Level: Formal Cooperation Procedures between Institutions for Land Use Planning Related to Urban Forest Management in Serbia

3.2.1. Organisational Structure

The organisational structure of institutions for land use planning related to urban forest management in Serbia is analysed through three aspects: position, powers and roles.

Position: Institutions of formal importance for land use planning related to urban forest management reside at the national and the local level (Figure 1). There are no institutional powers at the regional level, except for the Autonomous Province of Vojvodina, and as such these cannot be considered to be a general rule. The Serbian Government and Parliament are the supreme institutions, and both are formally the establishing entities for specialised expert organisations responsible for urban forest protection. As such, the Government of Serbia is responsible for the Nature Conservation Institute, while Parliament has responsibility over 'Srbijašume', the Socially Owned Enterprise (SOE) charged with forest management. In addition, the central public administration includes a number of ministries responsible for sectoral duties in relation to land use planning related to urban forest management, such as the Ministry of Construction, Transportation and Infrastructure; the Ministry of Environmental Protection; and other government departments whose remits include urban forest protection issues.

There are also two key groups of institutions at the local level, city administrations and city assemblies. The administrations include a number of departments tasked with urban forest management, while city assemblies formally establish local SOEs that directly perform urban forest management duties.

Table 2. The value framework for urban forest protection in Serbia.

Policy Document	Values: Urban Forest
National Sustainable Development Strategy [82]	Sets out strategic objectives for management and use of forests and forest land, mandates an institutional framework for safeguarding the protective functions of forests, and provides a model for inter-sectoral cooperation in the development of plans.
National Strategy for Sustainable Use of Natural Resources [81]	Defines the concept of 'forests' and 'forest land'; highlights the significance of forests as finite biological resources used, amongst other purposes, for sports, recreation and tourism; and cites the overexpansion of tourism capacity and infrastructure as a threat. Lays out the ultimate objective of sustainable development—balance between the use of all forest functions to ensure lasting multifunctionality in the provision of material goods and other ecosystem services. Advocates the introduction of institutional and economic measures to preserve and advance the recreational and health-related functions of forests and forest ecosystems. Envisages the creation of 5000 hectares of new town and suburban forests (by 2020).
Forestry Development Strategy [83]	Introduces the fundamental objective of safeguarding and enhancing forests and developing forestry as an industry. Particularly advocates the preservation, advancement, sustainable use, and acknowledgment of the protective, social, cultural and regulatory functions of forests and reform and advancement of institutions in the forestry sector.
Biodiversity Strategy, 2011–2018 [84]	Provides an overview of the state of biodiversity, safeguards and the legal, institutional and financial framework for preserving biodiversity; defines strategic areas, goals and activities and includes an action plan. Divides forests by how mixed they are and provides recommendations to achieve the objectives of reducing loss of habitat, including forests, by 2020, and instituting protection of 17% of all land and water areas subject to safeguards. The strategy sets out a framework for measures to prevent adverse impacts of genetically modified species of trees and allochthonous and invasive species on forests and biodiversity. Also advocates development of forest certification programmes and best sustainable forestry practices based on an ecosystem-wide approach.
National Environmental Protection Programme [85]	Advocates the preservation, improvement and extension of existing forests and enhanced monitoring in line with international frameworks.
Law Ratifying the Convention on Biological Diversity [86]*	Advocates the preservation and sustainable use of biological diversity and all its components. Requires biodiversity issues to be considered when any national decisions are being made to preserve and sustainably use biological resources and measures to be adopted to avoid and minimise adverse impacts on biodiversity.
Law Ratifying the European Landscape Convention [57]*	Sets out the principles that each country should adjust to suit its national law and incorporate into spatial development policies. The Convention defines 'landscape' as a dynamic category that comprises areas of action of both natural and human resources, and advocates dialogue in enacting landscape development policies, especially at the local level to facilitate practical implementation.
Law Ratifying the Convention on Access to Information, Public Participation in Decision-Making and Access to Justice in Environmental Matters [60]*	This convention, also referred to as the Aarhus Convention, guarantees the right to access information, participate in decision-making, and access justice in environmental matters. It establishes principles that public administration should adhere to when communicating with the public on environmental issues so as to safeguard the right of everyone, whether belonging to present or future generations, to live in an environment adequate to his or her health and well-being.

* Although these documents have legal force, they in effect ratify international instruments and are not binding in urban forest management practice.

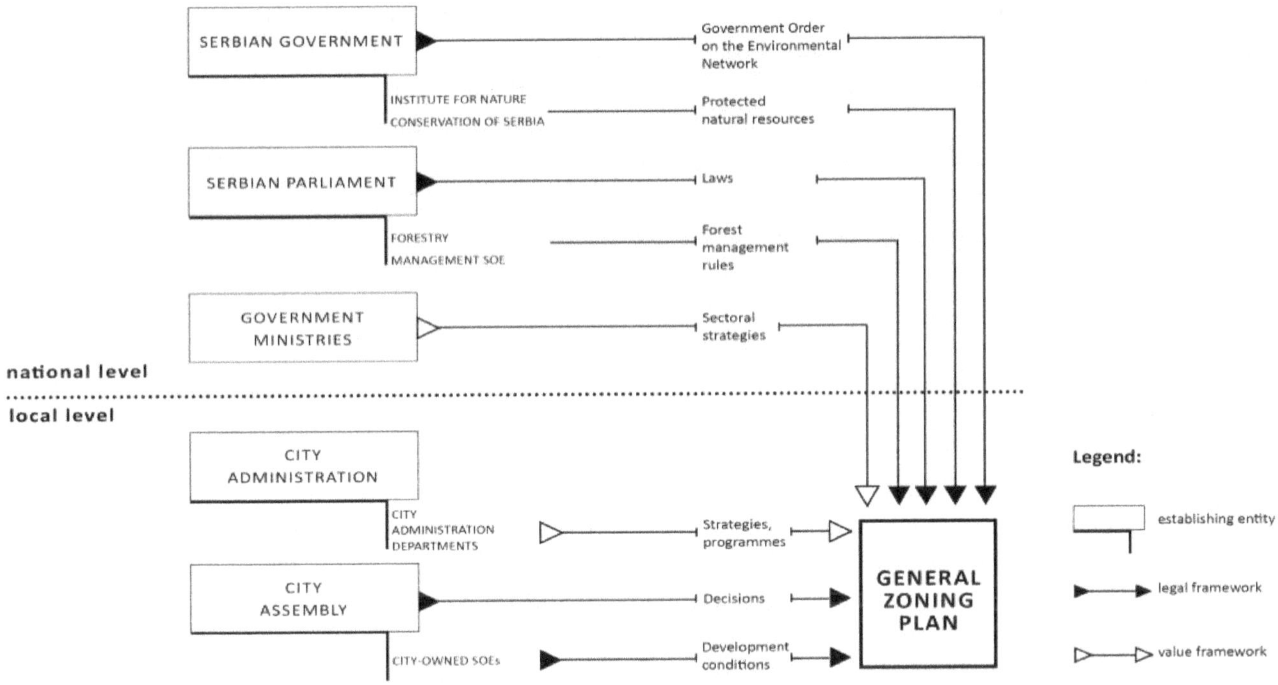

Figure 1. Formal cooperation procedures between institutions for land use planning related to urban forest management in Serbia (source: Authors).

Powers: National-level institutions are responsible for the regulatory framework binding on all levels of governance. Here, the Parliament enacts legislation proposed by the line ministries, while the Government is able to adopt 'Government orders', which carry the force of secondary legislation that elaborate on existing regulatory arrangements. The Nature Conservation Institute is specifically charged with assessing the value of natural resources and assigning protected status to 'natural resources of national importance', while the forest management SOE is responsible for regulating forest management at the national level. Documents designating 'protected natural resources' and setting out 'forest management rules' constitute a legally binding framework for land use planning at all territorial levels. Apart from legally binding decisions, the various ministries are responsible for enacting strategic documents that set out the value framework.

At the local level, it is the city assembly that is responsible for decision-making. Decisions are drafted by the city administration through its sectoral departments. In addition to these, the city assembly may incorporate local SOEs responsible for preparing 'development conditions', which regulate norms for the use of space by sector. Apart from legally binding documents, departments of city administration also prepare documents setting out values, such as strategies, action plans and programmes.

Roles: The key role in the governance system is played by Parliament, the Serbian National Assembly, which is the country's legislature. Its local counterpart is the City Assembly. The executive power is vested in Cabinet Ministries at the national level, and in Departments of City Administrations at the local level. Institutions important for enacting legally binding documents specifically aimed at urban forest management are the national Nature Conservation Institute and forest management SOE and local SOEs.

3.2.2. Arrangements for Collaboration

At the horizontal level, collaboration takes place between legislative and executive institutions responsible for land use planning related to urban forest management that are at the same time the key pillars of democratic society. National arrangements in this regard are mirrored at the local level. Particularly important for land use planning related to urban forest management is close horizontal collaboration between national and local parliaments with specialised nature and forest conservation institutions, which these legislative bodies have formal responsibility for as their incorporating entities. Their horizontal cooperation results in sectoral standards that safeguard land covered by urban forests.

Vertical collaboration between institutions of land use planning related to urban forest management is defined by the national regulatory framework which is legally binding for all lower levels of governance. Within the land use planning system itself, legislation mandates compliance with plans enacted at higher levels (i.e., plans are vertically conditioned).

3.3. Micro-Level: Activities in Land Use Planning Practice

Amendments to the Planning and Construction Law [87] established the General Zoning Plans as the binding planning documents for central built-up areas at the local level. These plans are defined as operational instruments that allow direct implementation, meaning that they set out conditions for construction. They define the intended use and status of land (publicly-owned and other); typology of construction, regulation, capacities and infrastructure; and set out safeguards and development rules. The General Zoning Plans are hierarchically linked with higher-level plans, the National Spatial Plan and the Local Spatial Plan, which provide strategic guidelines for urban development.

The procedure for preparing urban plans is governed by the latest Planning and Construction Law [67]. This is identical for all planning levels in this category (Figure 2). The planning process is preceded by the local parliament's 'Initiative for planning'. The actual process of producing the planning document is entrusted to an expert organisation, which may be a public enterprise or a private company. Various public institutions become involved at various stages of the development of the plan, as do civil or private-sector actors. At the outset of the process, when the 'Concept of the plan' is defined, binding conditions are obtained from the relevant national-level SOEs and institutions, which, for urban forest management, are the Nature Conservation Institute, the forest management SOE and local SOEs tasked with public utilities and environmental protection. Public participation is ensured at two points in the planning process: once at the very beginning, to verify the overall concept, and once at the end, to vet the proposed plan. Immediately before the proposed plan is put up for 'Public viewing', it is at the stage termed 'Inspection of the plan' by the Planning Commission, an expert body established by the local government. After the 'Inspection of the plan' and 'Public viewing' phases, the 'Plan finalization' follows, after which it is sent to the local assembly for 'Adoption of the plan'. The plan may be amended by the local assembly before it is enacted, and the amendments may significantly impact some aspects of the plan.

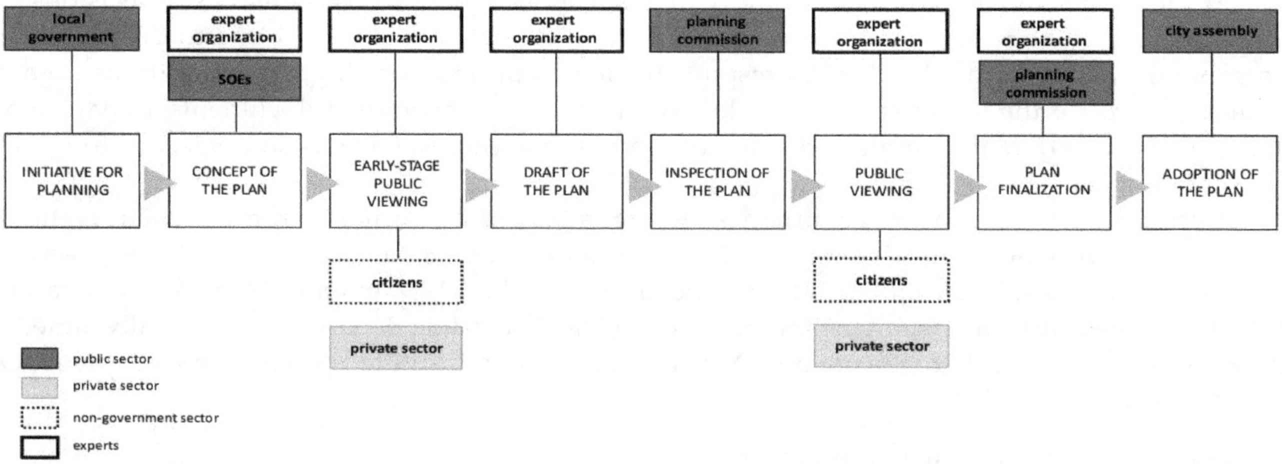

Figure 2. Formal and informal arrangements in the planning procedure (Source: Authors, based on [88]).

Strategic Environmental Impact Assessments (SEA) is yet another key tool for nature conservation. These are defined by the relevant legislation [77], and assessment is mandatory if a decision is made to produce it at the same time as the Initiative to develop a plan. Even though these assessments were first introduced as early as 2004 as a new environmental protection tool, until 2010 this was only a parallel procedure. The new regulations made them a mandatory and integral component of spatial plans; for urban plans, depending on the scope of each document, formal decisions are made to prepare or not to prepare impact assessments. The SEA aims at describing, evaluating and assessing the likely impact of the General Zoning Plan on the environment and envisages measures to mitigate adverse effects. It is developed in parallel with the draft of the plan and is subject to expert verification by the Planning Commission.

In terms of the transparency and collaborative nature of planning procedures, the practice of public participation does not differ much from that employed in the later stages of the socialist period. Until the enactment of the Planning and Construction Law [87], members of the public could become involved only at the end of the process, once the plan had already been developed, which meant that public comments could pose an issue and draw out the planning process. The 2014 the Planning and Construction Law introduced the option of 'Early-stage public viewing' at the very outset of the planning process.

Within the presented planning model, we can identify formal and informal positions that influence the issue of forest management and preservation:

(a) The formal influence is implemented through

- plans of higher order,
- conditions of institutes for environmental protection and
- conditions of SOEs for managing forest land;

(b) The informal influence is implemented through

- capacity of experts in relation to urban forest management and preservation in various positions of the planning procedure—local councillor, expert responsible for a plan creation, expert in the planning commission, expert from the civil sector and expert from the private sector; and
- capacity of nonexpert stakeholders in relation to urban forest management and preservation in various positions of the planning procedure—local councillor, civil sector and private sector.

3.3.1. Examples of the Land Use Planning Practice Related to Urban Forest Management in Serbia

The General Zoning Plans of two medium-sized Serbian cities were chosen for this assessment of the institutional structure of land use planning related to urban forest management at the micro-level. These are the General Zoning Regulation Plan for Vrnjačka Banja [89] and the General Zoning Plan for the Town of Bor [90].

General Zoning Plan for the Town of Vrnjačka Banja

Vrnjačka Banja is one of Serbia's major spas and a tourist resort of key national importance. Tourism is the city's chief industry, and as such development strategies focus on improving tourism capacities and the quality of the tourist offering.

The General Zoning Plan is based on the Spatial Plan of Serbia [72] and the Spatial Plan of the Municipality of Vrnjačka Banja [91]. A specific feature of this plan is the Government Order Establishing the Area of the Vrnjačka Banja Spa [92], which defined the boundaries of the 182-hectare spa zone, which contains a number of hot and mineral water springs harnessed for use in public baths and medical establishments. In that sense, it should be stressed that forests and forest land are recognised as critical for protection of mineral water springs. A number of the objectives set out in the higher-level plans are of strategic importance for the development of Vrnjačka Banja, including preventing continued degradation of space, addressing threats to natural resources, combating unpermitted construction and use of space contrary to intended purposes and revitalising areas, in particular those with perspectives for development.

The immediate reason for enacting the General Zoning Plan for Vrnjačka Banja was the need to take stock of new structures due to wide-ranging changes in the field resulting from not just legal but also illicit construction. The primary objectives of the Plan are to safeguard the city's character as a spa town, increasing the extent of green spaces in the broader territory of the city, especially in naturally green areas and plots of agricultural land that divide commercial zones from housing and central facilities.

The SEA is an integral part of the General Zoning Plan. The SEA concluded that the plan placed substantial emphasis on the sustainability of urban planning, zoning and construction. It also concluded that environmental protection considerations were complied with and incorporated into all aspects of the plan so as to allow the necessary development to proceed with minimum consequences for the environment.

The General Zoning Plan for Vrnjačka Banja covers an area of 2.318,97 hectares (see Table 3). One of the Plan's objectives is to protect forests, agricultural land and biodiversity in general. The summary of the current state of publicly-owned green spaces identifies only two uses: forests and forest parks. The General Zoning Plan introduces a forest park with an overall area of 150,3 hectares (see Figure 3). The purpose of this specific type of land use is to preserve forest configurations, ensure they can receive the required care and maintenance and permit them to be used for purposes of tourism and recreation.

Table 3. General Zoning Plan for Vrnjačka Banja: aspects relevant to land use planning related to urban forest management.

General Zoning Plan	**Vrnjačka Banja**
Area covered	2.318,97 hectares
Existing use	Forests (public use)
Planned use	Forest park
Area/percentage of total area covered by plan	150,3 hectares/6.5%
Objective of change of use to forest park	Forest configurations entailing additional attention in terms of maintenance, care and protection with a minimum of park facilities. The primary objective is to maximise the protection of forests and greenery in general, safeguard autochthonous vegetation, landscape configurations, and characters of areas. These zones most commonly integrate recreational and tourism-related facilities of central and boundary areas. These areas are designed for tourism and/or meeting the needs of the residential population of all ages.
Development rules	Construction of appropriate hydraulic engineering structures to provide protection from torrential flooding and floods; Provision of discreet lighting and street furniture as designed; Scheduled maintenance as part of park care projects; Provision of cafe and restaurant facilities (construction of 1 to 3 buildings of up to 150 square metres)
Restrictions	Change in intended use of space, construction of structures, tree felling and unplanned removal of vegetation, earth moving works, vehicle movements, waste disposal
Implementation conditions/instruments	Urban Planning Design required. General Zoning Plan required for Forest Park 1 (10,46 ha), Forest Park 2 (17,31 ha) and Forest Park 3 (34,91 ha)
Protection zones	Sanitary Protection Zone 2
Protection conditions	Greenery of major importance for the character of the area (Borjak) Conditions issued by the Kraljevo Cultural Heritage Institute (Forest Park 1 as whole)

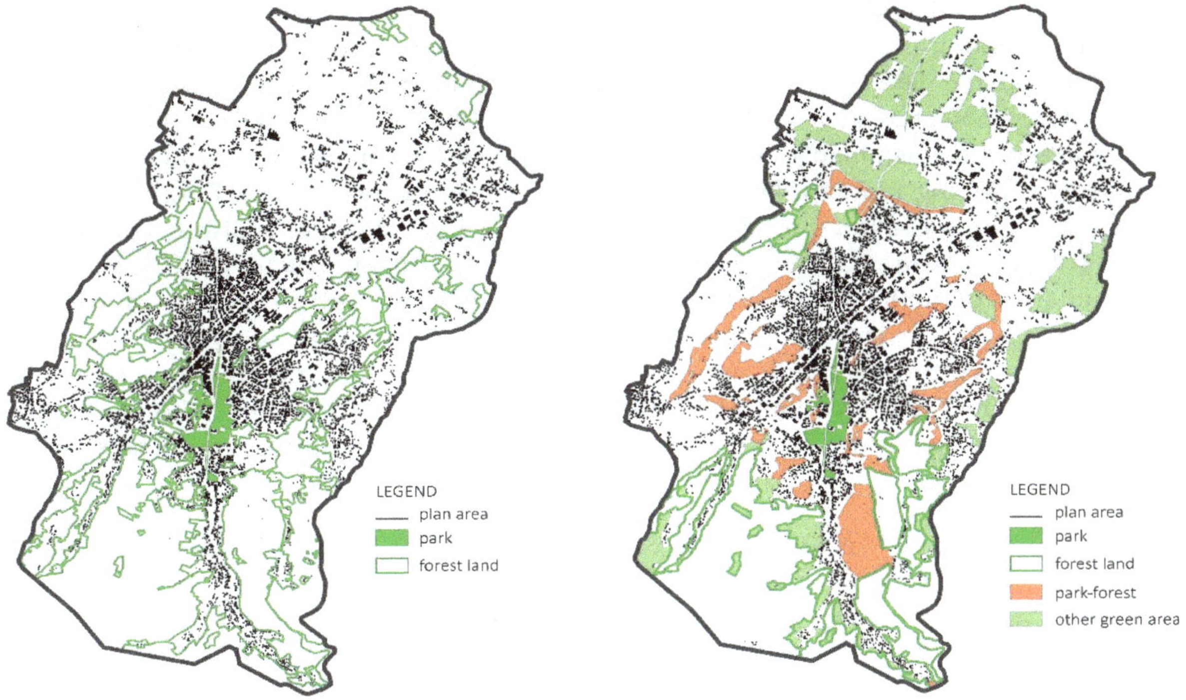

Figure 3. General Zoning Plan for Vrnjačka Banja: green spaces land use plan, current vs. planned state (source: Authors).

The plan also lays down rules for development in the forest parks and sets out conditions for particular types of constructions (catering and hydraulic engineering facilities). The plan bans any changes to the intended use of space, construction of structures, tree felling and unplanned removal of vegetation (except where required to protect plant health, as assessed by the relevant local SOE), earth moving works (except where envisaged by the project designs) and unrestricted vehicle movements and waste disposal. Detailed Zoning Plans are mandated for the three forest parks (Forest Park 1, of 10,46 ha; Forest Park 2, 17,31 ha; and Forest Park 3, 34,91 ha), whereas urban planning designs are required for any interventions within the planned forest parks. Apart from the development rules applicable to all planned forest parks, special requirements were introduced for the Borjak Forest Park (34,91 ha) to take account of the greenery of major importance for the character of the area there, as well as for the forest park near the central city park due to the need to protect cultural heritage monuments.

General Zoning Plan for the Town of Bor

Bor is located close to a major copper and gold mining and smelting facility. Rapid industrialisation in the latter half of the 20th century made the city an important centre for Eastern Serbia. Current and future development of the area is based on mining and industry and the accompanying manufacturing and services sectors.

The General Zoning Plan is based on the Spatial Plan for the Municipality of Bor [93] and the General Urban Plan of Bor [94]. Specific issues are regulated by the local Development Land Decision [95] and the Decision on Public Development Land [96], which determine which development land can be owned publicly or held otherwise: (1) development land intended for public use, comprising infrastructure and buildings (utilities facilities, urban greenery and other public structures and areas of general interest, such as those devoted to education, child protection, healthcare, social welfare, culture, sports and recreation, etc.) and (2) areas intended for other uses, meaning all other structures and areas (housing, businesses, services, etc.).

Higher-level plans designate Bor as a (sub)regional hub with a catchment area covering a number of surrounding municipalities. Bor is an industrial centre with well-developed industry and significant prospects for continuing development of the nonferrous metals sector based on the mining and processing of copper and gold ores. The Spatial Plan of Serbia envisages increasing the forested area of the Municipality of Bor from 45% to 49.2% of the total land surface, or to 3.570 hectares. The new forests are planned to take the place of poorer quality agricultural land. The General Zoning Plan sets out priorities for development in all areas; for environmental protection, they entail the implementation of the Municipality of Bor Sustainable Development Strategy and revitalisation of degraded land.

The General Zoning Plan applies to the entire territory of the town of Bor, which is divided into seven spatial units and does not comprise the mining and smelting facilities. The area covered amounts to 1312,20 hectares (Table 4). The plan aims at providing an urban planning framework for buildings and areas of public interest. It sets out the requirements for the reconstruction of the town core and other spaces, construction of public, commercial and other facilities, improved protection of the environment, cultural heritage, natural and man-made settings and other issues. Green areas are a major consideration of the Plan, especially in the town core and as part of housing complexes containing multi-dwelling units. These green areas are planned to account for 0.85% of the entire area covered (or 11,2 hectares).

Table 4. General Zoning Plan for the Town of Bor: aspects relevant to land use planning related to urban forest management.

General Zoning Plan	Town of Bor
Area covered	1312,20 ha
Existing use	Urban greenery (public use)
Planned use	Forest park
Area/percentage of total area covered by plan	11,2 ha/0.85%
Objective of change of use to forest park	Conversion of a forest, including a zoo, into a forest park. Development of recreation and tourism facilities
Development rules	Basic natural characteristics to be retained Minimum development: - Passive recreation zone: basic equipment, walking paths, lawns to be used for recreation purposes - Active recreation zone: maximum percentage of paved and built-up areas: 2.5% of total - Activity zone: additional equipment (cycle paths, running paths, miniature golf course, children's playground, restaurants, etc.); maximum percentage of paved and built-up areas: 5% of total Part of forest may be developed as park Only natural materials to be used for all paths, minimum lighting Only natural materials (wood and stone) to be used for benches and rest areas Parking spaces to be sited at the main approaches to the forest Endeavour to restrict movement to pedestrians only Provide signage and development and maintenance programmes
Restrictions	-
Implementation conditions/instruments	Appropriate technical documentation must be developed for newly planned green areas and reconstruction of existing ones. The Plan repealed the Zoning Plan for 'Section 3'—Forest Park (Municipal Official Journal Nos. 19/94, 4/01 и14/03).
Protection zones	-
Protection conditions	-

The SEA as an integral part of the Plan concludes that attaining the objectives of the plan will not seriously threaten natural and environmental values, and recommends close adherence to the guidelines of the Plan and the SEA with regard to the environment, spatial development and use of natural resources.

The objectives of the General Zoning Plan for Bor include safeguarding and improving the state of green areas and protective greenery, developing sports and recreation spaces and making the urban environment more attractive with the aim of attaining public values and interests of the urban area. The plan identifies two intended uses of publicly-owned green areas: Urban greenery (within the limits of urban development land) and Other greenery (for areas other than urban development land), which also includes forests (Figure 4). In Section 4, the General Zoning Plan introduces forest parks as a separate intended use of land, with a total area of approximately 11.2 hectares. The existing forest, which includes a zoo, is planned to be converted into a forest park designed for recreation. The plan mandates that the fundamental purpose of green areas be respected and that the key natural characteristics of the space be retained in their entirety (including vegetation, elevations, bodies of water and the like). This includes preserving autochthonous vegetation and minimum interventions in terms of introducing additional developments.

Figure 4. General zoning plan for Bor: green spaces land use plan, current vs. planned state (source: Authors).

The plan sets out development rules for the forest parks and provides conditions for development of passive and active recreation and activity zones, indicates which facilities can be constructed, regulates materials to be used in covering paths and walkways, stipulates how vehicular access is to be controlled and gives pedestrians priority within the forest parks. The creation of the newly-planned green areas and reconstruction of existing ones (which applies to the Section 4 forest park) requires the development of the appropriate technical documentation. This location is not subject to any specific restrictions.

4. Discussion and Conclusions

Land is a fundamental yet finite resource for urban development, and, as such, is directly exposed to the influence of complex socioeconomic factors. In these circumstances, urban planning must rise to the challenge of addressing the manifold development priorities that stem from public policies. Land use planning consequently becomes an efficient instrument for implementing the value framework of public policies through the definition of ways in which land can be utilised. According to the concept of institutional transformation, values are inherent to the institutional system and are the result of a balance struck between its elements. With the disturbance of the stability of the institutional system, induced by the influences outside the system, there is an institutional transformation and a distortion of the value framework of public policies and accompanying agendas, such as "saving urban forests".

The key factors that influence the transformation of the institutional system in Serbia are as follows.

(a) The shift in the economic system to embrace open market principles, directly leading to altered ownership of land and, thereby, to a new balance of power in society. This has brought about a fundamental change in perspectives of public property and promoted the diversification of interests related to land use. While it was formerly beyond question, public interest, as a category defined by ideological norms, is now a matter of political agreement that reflects the balance of power in society.

(b) Acceptance of the global value framework and the principle of sustainability as the dominant development concept, which occasioned the development of horizontal and vertical coordination mechanisms at all levels of institutional structure. This is significantly different in relation to top-down decision-making practice that was dominant in a socialist society, where communication mechanisms were exclusively in the function of carrying out the decisions made at the highest level.

(c) Acceptance of European integration, where a primary value concept is the principle of subsidiarity, whereby responsibility for decision-making on shared issues is transferred to the lowest possible tier of social organisation. It has introduced democratic dialogue as a means of determining the value orientation of future spatial development. This presents a major challenge for local authorities that should demonstrate the ability to carry out democratic dialogue within the community and the choice of development goals.

Despite a sound and well-developed tradition of nature and forest land protection under socialism, these key factors listed above, have significantly eroded the stability of the previous institutional structure and initiated a process of institutional transformation. As a result, the position of land use planning as a robust mechanism for mediating conflicts over land use and a regulatory instrument for policy implementation is highlighted.

Given the objective of this paper, to provide a critical overview of the institutional framework of land use planning in relation to urban forest management in the post-socialist environment of Serbia, we have defined the components of importance, presented below, for establishing a stable interaction inside the institutional structure for promotion of a value system aimed at saving urban forests. As a first step of institutional design, these components represent key aspects of the concept of land use planning for urban forest protection (LUPUFP) in the Serbian post-socialist transition environment.

The analysis was conducted across three scales: macro/governance, meso/coordination and micro/agency. This included the analysis of system components from the national to local level, illustrated by examples of the land use planning practice of two medium-sized cities. Accordingly, the analysis has identified the major institutional changes regarding:

(a) Regulatory structure and the value framework of public policies. This aspect of institutional analysis is aimed at examining the macro-level, which is in Serbia determined by significant macro-societal processes that take place due to the adoption of national and supranational constitutions [22]. It includes two aspects: (a) an institutional and legal framework for urban forest protection standards and (b) a value framework for urban forest protection. Serbia's planning system is hierarchically organised, from higher to lower levels of governance. Legislative changes have aimed at reducing the number of planning levels to promote efficiency and effectiveness in implementing plans. However, planning has failed to keep up with the pace of legislative change, which has in practice led to unclear planning procedures and misalignment between the outcomes of planning at various spatial levels. These circumstances have caused confusion between the national, regional and local levels as to their respective powers and roles. Further, the practice of land use planning related to urban forest management is subject to a variety of laws enacted by administrative authorities in numerous sectors. One issue here is the lack of alignment between urban forest protection standards introduced by the various regulations, which has caused problems with interpretation and implementation at the local level. On the other hand, the value framework for urban forest protection is formally implemented through the legislative framework and the standards for protection envisaged by it. A major issue here is the set of policy documents the implementation of which is not formalised and is therefore not mandatory. The multitude of formal and informal policy documents at the national level, not sufficiently aligned with one another, prevent both the establishment and the implementation of a clear value framework. As a basic drawback, the absence of a terminological framework and the identification of urban forests as a separate category of urban green land are observed, leaving at the local level a space for different interpretations, as is shown in the cases of the General zoning plans of Vrnjačka Banja and Bor. In addition, the underdeveloped capacities of local SOEs, due to the lack of expert profiles in the formation of employees, as well as the burden on the public service of many utilities, represent an obstacle in the formulation of requirements as well as the implementation of protection measures. Thus, as was illustrated in both of the General Zoning Plans of Vrnjačka Banja and Bor, standards defined on the national level serve as guidelines for particular land use planning processes; however, LUPUFP is not yet recognised as a concept. Consequently, the key components of the regulatory framework for the establishment of the LUPUFP system are

- Retaining the hierarchy of the planning system;
- Setting clear planning procedures and defining expected outcomes of planning at various spatial levels;
- Harmonising different regulations that envisage urban forest protection standards;
- Establishing a clear relationship between formal and informal policy documents;
- Mutual alignment of the multitude of policy documents;
- Formalising relationships between legally binding and nonbinding policies at the national and local governance levels.

(b) Procedures for cooperation between institutions. The next level of institutional analysis (the meso-level) involves planning and implementation structures and processes [22]. Serbia's traditional hierarchical planning system, which entails complex inter-organisational networks, requires cooperation at the horizontal and vertical levels aimed at the development and implementation of policies, programmes, projects and plans. The top-down approach, which emphasised the national decision-making level and an expert-driven approach to policy-making, is slowly opening up to bottom-up initiatives and the acknowledgment of particular interests in decision-making. This has been accompanied by a new set of regulatory reforms that aim at decentralising public administration and placing responsibility for making spatial planning decisions at the local level. This type of institutional transformation entails a reform process wherein the regulatory system is carefully harmonised both horizontally and vertically. The preconditions for these changes are a clear political orientation and the provision of appropriate professional capacity. As such, institutional design must be based on firm foundations, including institutions and regulations, which both define policies for urban forest protection and ensure decision-making procedures aimed at safeguarding the public interest. From the urban forest management perspective, the institutional structure is strictly divided between the national and the local level of governance. Each institutional level possesses a distinct unit charged with issues of nature conservation, including forests, whereby the communication between them is very weak. Also, the strict sectoral division between governance units at the same level poses a problem for horizontal communication. As was illustrated in cases of the General Zoning Plans of Vrnjačka Banja and Bor, there is a noticeable absence of horizontal communication between the sectors dealing with the "saving urban forest" agenda, as the requirements for defining planning measures such as "restrictions", "protection zones" and protection conditions" are not obligatory. This clearly shows that, for example, climate change issues, drinking water protection, energy efficiency, healthy environment, etc. are irreconcilable, and therefore they are dependent of the expertise of the organisations involved in the development of the plan as well as the knowledge of the local community. The value framework for the agenda of saving urban forests requires firm regulations for stakeholder involvement in making decisions on urban forests, indicating that various control mechanisms are necessary. The weaknesses of such a system lie in the rigidity of its mechanisms and their uncritical application in locally specific situations. Implementation of the public policies and safeguarding the adopted value framework is contributed by units specialised in nature and forest protection at all levels. As was illustrated in Vrnjačka Banja and Bor, bottom-up initiatives for forest protection and development from the local level that are recognised within land use planning processes, such as particular local decisions, reflect the adjustment of the institutional structure in order to promote the concept of LUPUFP. Consequently, the key components for the establishment of the LUPUFP system related to the procedures for cooperation between institutions are

- Strengthening vertical coordination between specialised nature and forest protection units at the national as well as local levels;
- Establishing procedures and mechanisms for horizontal communication between sectors at the same level of governance;
- Establishing procedures and mechanisms for bottom-up communication by decision-makers;
- Creating preconditions for efficient multi-stakeholder cooperation;
- Establishing firm regulations to control the impact of market forces;
- Defining legal procedures that acknowledge control mechanisms;
- Ensuring more flexibility in the application of control mechanisms in locally specific situations;
- Retaining specialised units and their instruments for implementing nature and forest protection instruments;
- Establishing mechanisms for horizontal and vertical coordination of policy implementation instruments.

(c) Activities in land use planning practice. This level of analysis pertains to intra-organisational design, addressing organisational subunits and small semiformal or informal social units, processes and interactions [22]. Also, it directly examines the extent of stakeholder participation related to the legal framework, the effectiveness of processes, and the space for the involvement of civil society [36] in relation to urban forest management. Land use planning at the local level in Serbia in general is noticeably top-down oriented, with strict control conducted by public sector, and mainly subordinate to the attainment of public sector interests. The participation of stakeholders from the private and civil sectors is partial and insufficient. The role of expert organisations does not enjoy a sufficiently clear position in the decision-making system. Substantial responsibility—and power—is given to the planning commission as an expert body of the local government. Accordingly, their position is sensitive to the influence of various interests. Furthermore, the structure of the commission does not include experts from the domain of urban forest management. As was illustrated in the example of the General zoning plan of Vrnjačka Banja, the formal institutional framework, particularly in the domain of top-down coordination and standards for protection, serves as a base for urban forest protection that was recognised as a crucial resource for further spa protection and development. In the example of Bor, where urban forest protection is not specially required outside of the formal standards, the informal institutional structure gives space for informal institutional actions for urban forest protection and bottom-up initiatives that are in line with the requirement for 'fostering pro-environmental behaviours'. Consequently, the key components for the establishment of the LUPUFP system related to the activities in the land use planning practice are

- Establishing collaborative planning, which entails informed decision-making about the directions of urban development at key stages of plan production;
- Clarifying the roles of experts in the decision-making system and ensuring their independence from political decision-making;

- Clearly defining policies and regulatory mechanisms at the national level;
- Standardising the various categories of land use at the national level;
- Retaining mechanisms that acknowledge the regulatory norms and hierarchy of the planning system;
- Harmonising regulations across various sectors;
- Strengthening the positions and capacities of local public sector experts;
- Establishing a clear methodology for the development and content of urban plans.

These three groups of components constitute possible guidelines for preserving the robust tradition of land use planning related to urban forest protection by the establishment of the concept of LUPUFP in Serbia. The complexity of the subject, rooted as it is in differing sectors of expertise, certainly calls for deeper consideration of the myriad components of the system in the future. The findings presented in this paper have no ambition to include all the components of the system, but to provide some valuable insight into the practice of land use planning as one of the most efficient instruments for protecting green land in cities and that adheres to the agenda of saving urban forests.

The concept of LUPUFP is in line with current recommendations for the safeguarding and sustainable management of forests and other green areas in cities as crucial components for the health and well-being of citizens, promoted by the most influential documents such as Agenda 2030, the Paris Agreement and the New Urban Agenda. Related to that, the main contribution of this research is in the promotion of the relevance of the concept of LUPUFP in accordance to the importance of ecosystem services especially, as outlined by the FAO report regarding the nine SDGs.

The conducted research also contributes to the concept of institutional transformation, which is verified through the system of the land resource management of society in the environment of post-socialist transition. The results of this research presents the specific, practical and applicable path of institutional redesign that leads to the establishment of a concept of LUPUFP as an experience that may assist other countries in the region seeking answers in the process of developing their own models.

Author Contributions: Conceptualization, M.M. and T.C.; Methodology, M.M.; Investigation, M.M., T.C. and M.P.M.; Writing—Original Draft Preparation, M.M., T.C. and M.P.M.; Writing—Review & Editing, M.M.; Visualization, M.P.M. and M.M.

References and Notes

1. United Nations (UN). *Transforming Our World: The 2030 Agenda for Sustainable Development*; Resolution A/RES/70/1; United Nation: New York, NY, USA, 2015.
2. United Nations Framework Convention on Climate Change (UNFCCC). Paris Agreement. 2015. Available online: https://unfccc.int/sites/default/files/english_paris_agreement.pdf (accessed on 22 June 2018).
3. United Nations Human Settlements Programme (UN HABITAT). New Urban Agenda. A/RES/71/256. United Nations, 2017. Available online: http://habitat3.org/wp-content/uploads/NUA-English.pdf (accessed on 22 April 2018).
4. World Commission on Environment and Development (WCED). *Our Common Future*; University Press: Oxford, UK, 1987.
5. Food and Agriculture Organization of the United Nations (FAO UN). Urban and Peri-Urban Forestry-Definition. Available online: http://www.fao.org/forestry/urbanforestry/87025/en/ (accessed on 22 December 2018).

6. Food and Agriculture Organization of the United Nations. *Guidelines on Urban and Peri-Urban Forestry*; FAO Forestry Paper No. 178; Salbitano, F., Borelli, S., Conigliaro, M., Chen, Y., Eds.; Food and Agriculture Organization of the United Nations: Rome, Italy, 2016.
7. Edreny, T. Strategically growing the urban forest will improve our world. *Nat. Commun.* **2018**, *9*, 1160. [CrossRef] [PubMed]
8. Hurokawa, H.K. Sustainability and the Urban Forest: An Ecosystem Services Perspective. *Nat. Resour. J.* **2011**, *51*, 233–259. [CrossRef]
9. Zhou, W.; Cao, F.; Wang, G. Effects of Spatial Pattern of Forest Vegetation on Urban Cooling in a Compact Megacity. *Forests* **2019**, *10*, 282. [CrossRef]
10. Kim, G. Assessing Urban Forest Structure, Ecosystem Services, and Economic Benefits on Vacant Land. *Sustainability* **2016**, *8*, 679. [CrossRef]
11. Jim, C.Y.; Chen, W. Ecosystem services and valuation of urban forests in China. *Cities* **2009**, *26*, 187–194. [CrossRef]
12. Gungor, B.S.; Chen, J.; Wu, S.R.; Zhou, P.; Shirkey, G. Does Plant Knowledge within Urban Forests and Parks Directly Influence Visitor Pro-Environmental Behaviors. *Forests* **2018**, *9*, 171. [CrossRef]
13. Curitiba Declaration on Cities and Biodiversity; Curitiba, Brazil. 2007. Available online: https://www.cbd.int/doc/meetings/city/mayors-01/mayors-01-declaration-en.pdf (accessed on 22 December 2018).
14. United Nations Centre for Human Settlements (UNCHS). *An Urbanizing World: Global Report on Human Settlements*; Oxford University Press for UNCHS: Oxford, UK, 1996.
15. Baycan-Levent, T.; Nijkamp, P. Planning and Management of Urban Green Spaces in Europe: Comparative Analysis. *J. Urban Plan. Dev.* **2009**, *135*, 1–12. [CrossRef]
16. Rivolin, U.J. Planning Systems as Institutional Technologies: A Proposed Conceptualization and the Implications for Comparison. *Plan. Pract. Res.* **2012**, *27*, 63–85. [CrossRef]
17. Albrechts, L.; Alden, J.; Pires, A. (Eds.) *The Changing Institutional Landscape of Planning*; Ashgate: Aldershot, UK, 2001.
18. Selznik, F. Institutionalism 'Old' and 'New'. *Adm. Sci. Q.* **1996**, *2*, 270–277. [CrossRef]
19. March, J.G.; Olsen, J.P. Elaborating the new institutionalism. In *The Oxford Handbook of Political Institutions*; Rhodes, R.A.W., Binder, S.A., Rockman, B.A., Eds.; Oxford University Press: Oxford, UK, 2006; pp. 1–20.
20. Healey, P. Institutionalist analysis, communicative planning, and shaping places. *J. Plan. Educ. Res.* **1999**, *19*, 111–121. [CrossRef]
21. Pierre, J. Models of Urban Governance: The Institutional Dimension of Urban Politics. *Urban Aff. Rev.* **1999**, *34*, 372–396. [CrossRef]
22. Alexander, E.R. Institutional Transformation and Planning: From Institutionalization Theory to Institutional Design. *Plan. Theory* **2005**, *4*, 209–223. [CrossRef]
23. Richard Scott, W. Institutional theory. In *Encyclopedia of Social Theory*; Ritzer, G., Ed.; Sage Publications: Thousand Oaks, CA, USA, 2005; pp. 408–414.
24. Hall, P.A. Historical Institutionalism in Rationalist and Sociological Perspective. In *Explaining Institutional Change: Ambiguity, Agency and Power*; Mahoney, J., Thelen, K., Eds.; Cambridge University Press: New York, NY, USA, 2009; pp. 204–225.
25. Healey, P. An institutional model of the development process. *J. Prop. Res.* **1992**, *9*, 33–44. [CrossRef]
26. United Nations Human Settlements Programme (UN HABITAT). *Planning Sustainable Cities: Global Report on Human Settlements*; Earthscan: London, UK, 2009.
27. European Commission (EC). *The EU Compendium of Spatial Planning Systems and Policies*; Office for Official Publications of the European Community: Luxembourg, 1997.
28. United Nations Economic Commission for Europe (UNECE). *Spatial Planning: Key Instrument for Development and Effective Governance with Special Reference to Countries in Transition*; United nations: Geneva, Switzerland, 2008.
29. Hopkins, L.D. *Urban Development: The Logic of Making Plans*; Island Press: Washington, DC, USA, 2001.
30. Organisation for Economic Co-operation and Development (OECD). *Land-Use Planning Systems in the OECD: Country Fact Sheets*; OECD Publishing: Paris, France, 2017.
31. European Environment Agency (EEA). *Urban Sprawl in Europe: The Ignored Challenge*; EEA Report No 10/2006; European Commission Joint Research Centre/European Environment Agency: Copenhagen, Denmark, 2006.

32. Sandström, C.; Lindkvist, A.; Öhman, K.; Nordström, E.M. Governing Competing Demands for Forest Resources in Sweden. *Forests* **2011**, *2*, 218–242. [CrossRef]
33. Buizer, M.; Van Herzele, A. Combining deliberative governance theory and discourse analysis to understand the deliberative incompleteness of centrally formulated plans. *For. Policy Econ.* **2012**, *16*, 93–101. [CrossRef]
34. European Commission (EC). *The EU Biodiversity Strategy to 2020*; Publications Office of the European Union: Luxembourg, 2011.
35. Commission of the European Communities (CEC). White Paper, Adapting to Climate Change: Towards a European Framework for Action. 2009. Available online: https://ec.europa.eu/health/ph_threats/climate/docs/com_2009_147_en.pdf (accessed on 8 March 2019).
36. Food and Agriculture Organization of the United Nations (FAO UN). *Framework for Assessing and Monitoring Forest Governance*; The Program on Forests (PROFOR) and Food and Agriculture Organization of the United Nations (FAO): Rome, Italy, 2011.
37. Lawrence, A.; De Vreese, R.; Johnston, M.; van den Bosch, C.C.K.; Sanesi, G. Urban forest governance: Towards a framework for comparing approaches. *Urban For. Urban Green.* **2013**, *12*, 464–473. [CrossRef]
38. Laktić, T.; Pezdevšek Malovrh, Š. Stakeholder Participation in Natura 2000 Management Program: Case Study of Slovenia. *Forests* **2018**, *9*, 599. [CrossRef]
39. Tacconi, L. Developing environmental governance research: The example of forest cover change studies. *Environ. Conserv.* **2011**, *38*, 234–246. [CrossRef]
40. Faggin, J.M.; Behagel, J.H.; Arts, B. Sustainable Forest Management and Social-Ecological Systems: An Institutional Analysis of Caatinga, Brazil. *Forests* **2017**, *8*, 454. [CrossRef]
41. Young, O.R. *The Institutional Dimensions of Environmental Change. Fit, Interplay, and Scale*; Cambridge University Press: Cambridge, MA, USA, 2012.
42. Campbell, H.; Marshall, R. Utilitarianism's Bad Breath? A Re-Evaluation of the Public Interest Justification for Planning. *Plan. Theory* **2002**, *1*, 163–187. [CrossRef]
43. Suddaby, R., Lefsrud, L. Institutional theory, old and new. In *Encyclopedia of Case Study Research*; Mills, A.J., Durepos, G., Wiebe, E., Eds.; Sage Publications: London, UK, 2010.
44. Zeković, S. Urban land planning in Serbia. *Arhitektura i Urbanizam* **2002**, *9*, 11–17.
45. Živanović Miljković, J. Urban Land Use regulation in Serbia: An analysis of its effects on property rights. In *A Support to Urban Development Process*; Bolay, J.C., Maričić, T., Zeković, S., Eds.; EPFL & IAUS: Belgrade, Serbia, 2018; pp. 129–147.
46. Nonić, D. *Organisation and Operation of the Forestry Service*; Univerzitet u Beogradu, Šumarski fakultet: Beograd, Serbia, 2010.
47. Committee on Spatial Development (CSD). *European Spatial Development Perspective: Towards a Balanced and Sustainable Development of the Territory of the European Union*; Office for the Official Publications of the European Communities: Luxembourg, 1999.
48. Beauregard, R.A. Epilogue: Globalization and the city. In *Change and Stability in Urban Europe*; Anderson, H., Jorgensen, G., Joye, D., Ostendorf, W., Eds.; Ashgate: Aldershot, UK, 2001; pp. 251–262.
49. Pallagst, K.M.; Mercier, G. Urban and regional planning in Central and Eastern European countries–from EU requirements to innovative practices. In *The Post-Socialist City: Urban form and Space Transformations in Central and Eastern Europe after Socialism*; Stanilov, K., Ed.; Springer: New York, NY, USA, 2007; pp. 473–490.
50. Maruna, M.; Čolić, R.; Milovanović Rodić, D. A New Regulatory Framework as both an Incentive and Constraint to Urban Governance in Serbia. In *A Support to Urban Development Process*; Bolay, J.C., Maričić, T., Zeković, S., Eds.; EPFL & IAUS: Belgrade, Serbia, 2018; pp. 80–108.
51. Stanilov, K. Urban planning and the challenges of post-socialist transformation. In *The Post-Socialist City: Urban form and Space Transformations in Central and Eastern Europe after Socialism*; Stanilov, K., Ed.; Springer-GeoJournal Library: Dodrecht, The Netherlands, 2007; pp. 413–425.
52. Tsenkova, S. Urban Futures: Strategic planning in post-socialist Europe. In *The Post-Socialist City: Urban form and Space Transformations in Central and Eastern Europe after Socialism*; Stanilov, K., Ed.; Springer-GeoJournal Library: Dordrecht, The Netherlands, 2007; pp. 447–471.
53. Petovar, K. Professional Associations as an Actor in the Enactment of Spatial Planning Decisions. In *Actors of Social Changes in Space: Spatial Transformation and Quality of Life in Croatia*; SvirčićGotovac, A., Zlatar, J., Eds.; Institut za društvena istraživanja: Zagreb, Croatiam, 2012; pp. 99–114.

54. Vujošević, M.; Petovar, K. Public interest and actor strategies in urban and spatial planning. *Sociologija* **2006**, *48*, 357–382. [CrossRef]
55. Vlada RS (Vlada Republike Srbije) Pregovaračka poglavlja, Poglavlje 27: Životna sredina [Chapters of the Acquis. Chapter 27: Environment]; Pregovarački tim za vođenje pregovora o pristupanju Republike Srbije Evropskoj uniji. 2018. Available online: http://www.eu-pregovori.rs/srl/pregovaracka-poglavlja/poglavlje-27-zivotna-sredina/ (accessed on 26 December 2018).
56. Council of Europe (CE). *European Landscape Convention*; European Treaty Series No. 176; Council of Europe: Strasbourg, France, 2000.
57. Zakon o potvrđivanju Evropske konvencije o predelu [Law Ratifying the European Landscape Convention]. 2011. Available online: http://predelisrcasrbije.rs/dokumenta.html (accessed on 23 December 2018).
58. Resolution H1: General Guidelines for the Sustainable Management of Forests in Europe. Ministerial Conference on the Protection of Forests in Europe, Helsinki. 1993. Available online: https://www.foresteurope.org/docs/MC/MC_helsinki_resolutionH1.pdf (accessed on 11 January 2019).
59. Zakon o potvrđivanju Konvencije o dostupnosti informacija, učešću javnosti u donošenju odluka i pravu na pravnu zaštitu u pitanjima životne sredine [Law Ratifying the Convention on Access to Information, Public Participation in Decision-Making and Access to Justice in Environmental Matters]. 2009. Available online: https://www.poverenik.rs/sr-yu/me%C4%91unarodni-dokumenti/1735-zakon-o-potvrdjivanju-konvencije-o-dostupnosti-informacija-ucescu-javnosti-u-donosenju-odluka-i-pravu-na-pravnu-zastitu-u-pitanjima-zivotne-sredine.html (accessed on 11 September 2018).
60. Lockwood, M. Good governance for terrestrial protected areas: A framework, principles and performance outcomes. *J. Environ. Manag.* **2010**, *91*, 754–766. [CrossRef] [PubMed]
61. Crnčević, T.; Manić, B.; Marić, I. Zeleni zidovi urbanih prostora u kontekstu klimatskih promena–pregled najnovijih okvira i iskustava [Green Walls of Urban Spaces in the Context of Climate Change-An Overview of the Latest Frameworks and Experiences]. *Arhitektura i Urbanizam* **2015**, *41*, 40–48.
62. Crnčević, T.; Sekulić, M. Green Roofs in the Context of Climate Change-A review of recent experiences. *Arhitektura i Urbanizam* **2012**, *36*, 57–67.
63. Vujičić, D.; Tubić, Lj.; Todorović, D.; Šabanović, V.; Tutundžić, A.; Jeftović, A.; Jadžić, N. *Sustainability of Green Space Legislation*; Udruženje pejzažnih arhitekata Srbije: Beograd, Serbia, 2018.
64. Lukić, N. Urban Forests and Greening in the Republic of Serbia–Legal and Institutional Aspects. *South-East Eur. For.* **2013**, *4*, 51–55. [CrossRef]
65. Gudurić, I.; Tomićević, J.; Konijnendijk, C.C. A comparative perspective of urban forestry in Belgrade, Serbia and Freiburg, Germany. *Urban For. Green.* **2011**, *10*, 335–342. [CrossRef]
66. Zakon o šumama [Forests Law]. 2015. Available online: https://www.paragraf.rs/propisi/zakon-o-sumama-republike-srbije.html (accessed on 20 January 2018).
67. Zakon o planiranju i izgradnji [Planning and Construction Law]. 2018. Available online: https://www.paragraf.rs/propisi/zakon_o_planiranju_i_izgradnji.html (accessed on 22 Jun 2018).
68. Trkulja, S.; Tošić, B.; Živanović, Z. Serbian Spatial Planning among Styles of Spatial Planning in Europe. *Eur. Plan. Stud.* **2012**, *20*, 1729–1746. [CrossRef]
69. Reimer, M.; Panagiotis, G.; Blotevogel, H. *Spatial Planning Systems and Practices in Europe*; Routledge: New York, NY, USA, 2014.
70. Uredba o ekološkoj mreži [Government Order on the Environmental Network]. 2010. Available online: http://www.zzps.rs/novo/kontent/stranicy/propisi_podzakonski_akti/uredba%20o%20ekoloskoj%20mrezi.pdf (accessed on 6 February 2019).
71. Uredba o režimima zaštite [Government Order on Safeguards]. 2012. Available online: http://www.zzps.rs/novo/kontent/stranicy/zastita_prirode_o_zasticenim_podrucjima/uredba_rezimi_zastite.pdf (accessed on 6 February 2019).
72. Zakon o prostornom planu Srbije [Law on Spatial Plan of Serbia]. 2010. Available online: https://www.mgsi.gov.rs/sites/default/files/ZAKON%20O%20PROSTORNOM%20PLANU%20RS%20OD%202010%20DO%202020.pdf (accessed on 12 December 2018).
73. Zakon o eksproprijaciji [Expropriation Law]. 2016. Available online: https://www.paragraf.rs/propisi/zakon_o_eksproprijaciji.html (accessed on 6 February 2019).
74. Zakon o javnoj svojini [Public Property Law]. 2018. Available online: https://www.paragraf.rs/propisi/zakon_o_javnoj_svojini.html (accessed on 6 February 2019).

75. Zakon o komunalnim delatnostima [Utilities Law]. 2018. Available online: https://www.paragraf.rs/propisi/zakon_o_komunalnim_delatnostima.html (accessed on 6 February 2019).
76. Zakon o zaštiti prirode [Nature Conservation Law]. 2018. Available online: https://www.paragraf.rs/propisi/zakon_o_zastiti_prirode.html (accessed on 6 February 2019).
77. Zakon o zaštiti životne sredine [Environmental Protection Law]. 2018. Available online: https://www.paragraf.rs/propisi/zakon_o_zastiti_zivotne_sredine.html (accessed on 6 February 2019).
78. Zakon o proceni uticaja na životnu sredinu [Environmental Impact Assessment Law]. 2009. Available online: https://www.paragraf.rs/propisi/zakon_o_proceni_uticaja_na_zivotnu_sredinu.html (accessed on 7 February 2019).
79. Zakon o strateškoj proceni uticaja na životnu sredinu [Strategic Environmental Impact Assessment Law]. 2010. Available online: https://www.paragraf.rs/propisi/zakon_o_strateskoj_proceni_uticaja_na_zivotnu_sredinu.html (accessed on 7 February 2019).
80. Stojković, S. *Spatial Plan of the Republic of Serbia*; Službeni glasnik: Beograd, Srbija, 1996.
81. Nacionalna strategija održivog korišćenja prirodnih resursa i dobara [National Strategy for Sustainable Use of Natural Resources]. 2012. Available online: http://www.zzps.rs/novo/kontent/stranicy/propisi_strategije/S_prirodnih%20resursa.pdf (accessed on 6 February 2019).
82. Nacionalna strategija održivog razvoja [National Sustainable Development Strategy]. 2008. Available online: http://www.zurbnis.rs/zakoni/Nacionalna%20strategija%20odrzivog%20razvoja.pdf (accessed on 6 February 2019).
83. Strategija razvoja šumarstva Republike Srbije [Serbia Forestry Development Strategy]. 2006. Available online: https://www.fornetserbia.com/doc/shared/Strategija_razvoja_sumarstva.pdf (accessed on 6 February 2019).
84. Strategija biološke raznovrsnosti Republike Srbije za period 2011. do 2018. godine [Serbia Biodiversity Strategy, 2011 to 2018]. 2011. Available online: http://www.zzps.rs/novo/kontent/stranicy/propisi_strategije/strategija_bioloske_raznovrsnosti.pdf (accessed on 8 February 2019).
85. Nacionalni program zaštite životne sredine [National Environmental Protection Programme]. 2010. Available online: http://www.zzps.rs/novo/kontent/stranicy/propisi_strategije/Nacionalni_program_zastite_%20zs.pdf (accessed on 11 February 2019).
86. Zakon o potvrđivanju Konvencije o biološkoj raznovrsnosti [Law Ratifying the Convention on Biological Diversity]. 2001. Available online: http://www.vojvodinasume.rs/wp-content/uploads/2012/04/sertifikacija/Zakon%20o%20potvrdjivanju%20KONVENCIJE%20O%20BIOLOSKOJ%20RAZNOVRSNOSTI.pdf (accessed on 11 February 2019).
87. Zakon o planiranju i izgradnji [Planning and Construction Law]. 2014. Available online: https://www.mgsi.gov.rs/sites/default/files/ZAKON%20O%20PLANIRANJU%20I%20IZGRADNJI%20PRECTEKST%202015.pdf (accessed on 11 February 2019).
88. Graovac, A.; Danilović Hristić, N.; Stefanović, N. Technical and logical methods for improving the process of urban planning in Serbia. *Spatium* **2017**, *38*, 27–34. [CrossRef]
89. Plan generalne regulacije Vrnjačke Banje [General Zoning Plan of Vrnjačka Banja2016. Available online: http://vrnjackabanja.gov.rs/privreda/urbanizam/plan-generalne-regulacije?alphabet=lat (accessed on 15 March 2019).
90. Plan generalne regulacije gradskog naselja Bor [General Zoning Plan of the Town of Bor]. 2018. Available online: http://bor.rs/wp-content/uploads/2018/02/Nacrt-Bor-Knjiga-I-Planska-resenja-konacno.pdf?script=lat (accessed on 16 March 2019).
91. Prostorni plan opštine Vrnjačka Banja [Spatial Plan of the Municipality of Vrnjacka Banja]. 2011. Available online: https://vrnjcispa.biz/baze-i-registri/vazeci-planovi/prostorni-plan-opstine-vrnjacka-banja (accessed on 15 March 2019).
92. Uredba o utvrđivanju područja Banje [Government Order Establishing the Area of the Vrnjačka Banja Spa]. 1997. Available online: http://www.pravno-informacioni-sistem.rs/SlGlasnikPortal/eli/rep/sgrs/vlada/uredba/1997/26/2/reg (accessed on 15 March 2019).
93. Prostorni plan opštine Bor [Spatial Plan of the Municipality of Bor]. 2014. Available online: http://bor.rs/wp-content/uploads/2018/02/PPO-Bor-Knjiga-1_januar-2014.pdf?script=lat (accessed on 16 March 2019).

94. Generalni urbanistički plan Bora [General Urban Plan of Bor]. 2015. Available online: http://bor.rs/wp-content/uploads/2018/02/Knjiga-I_Plan.pdf?script=lat (accessed on 16 March 2019).
95. Odluka o građevinskom zemljištu [Decision on Development Land]. Službeni list opštine Bor, br. 3/1983.
96. Odluka o javnom građevinskom zemljištu [Decision on Public Development Land]. Službeni list opštine Bor, br. 2/2016.

11

Soil Organic Carbon Accumulation in Post-Agricultural Soils under the Influence Birch Stands

Tomasz Gawęda [1,*], Ewa Błońska [2] and Stanisław Małek [3]

1. State Forests National Forest Holding, Bielsko Forest District, 43-382 Bielsko–Biała, Poland
2. Department of Forest Soil Science, Faculty of Forestry, University of Agriculture, 31-425 Krakow, Poland
3. Department of Forest Ecology and Reclamation, Faculty of Forestry, University of Agriculture, 31-425 Krakow, Poland

* Correspondence: tgtomik@poczta.onet.pl

Abstract: The aim of this study was to demonstrate the effects birch renewal on the soil organic carbon accumulation and on dehydrogenase activity. We selected 12 research plots with birch stands of different ages (1–4 years, 5–8 years, 9–12 years, and 13–17 years) to determine soil texture, pH, total carbon and nitrogen levels, and base cation content. The total organic carbon stock was calculated for the soil profiles. Additionally, dehydrogenase activity was determined. Naturally regenerated birch stands on post-agricultural land facilitated carbon accumulation. Based on our results, dehydrogenase activity is useful in assessing the condition of post-agricultural soils, and its determination allowed for us to assess the processes occurring in post-agricultural soils that are associated with the formation and carbon distribution.

Keywords: afforestation; birch stands; carbon stock; dehydrogenase activity

---

1. Introduction

Poland has one of the largest forest areas in Europe; forests occupy 29% of the country's territory and cover an area of 9.1 million hectares [1]. Forest stands on arable lands occupy nearly 25% of the forest area. The soil and soil processes are crucial in maintaining the productivity of forest ecosystems [2]. Soil is an important reservoir of carbon, and it is estimated that the global soil carbon stocks amount to more than 1500 Pg C, which are significantly higher than those of the atmosphere (750 Pg C) or the biomass of terrestrial ecosystems (650 Pg C) [3]. Recently, the mechanisms that are responsible for carbon stabilisation in soils have received considerable interest due to their relevance in our understanding of the global carbon cycle [3]. The fertility and productivity of soils depend on soil organic matter (SOM), which serves as a nutrient reservoir and it therefore plays an important role in nutrient cycling [4]. Changes in soil management are the main factor that affects SOM dynamics [5]. For example, transforming natural ecosystems into arable fields generally depletes soil organic carbon (SOC) reserves by as much as 75% (mostly between 30 and 50%), depending on the climate zone and the ecosystem type [6]. However, such losses can be limited by converting arable land into grassland and forest [7]. Afforestation positively affects various soil properties, and the soil organic carbon values of afforested sites are generally higher than those of bare sites [8]. Afforestation also positively influences the physical soil characteristics, which are important for maintaining soil stability and productivity [9–11].

In recent years, the interest in soil quality has been stimulated by the growing awareness of the fact that the soil is an important component of the biosphere. It functions not only to produce food, timber, and other forest resources, but it also plays a role in maintaining the local, regional, and global quality of the environment. Karlen et al. [12] and Gil-Sotres et al. [13] state that soil quality enables the healthy functioning of an ecosystem and it maintains its biological production. One of the most important factors in determining soil fertility are the soil biological properties in relation to the activity of microorganisms and higher organisms (plants and animals), including enzymes that are secreted by them [14]. Dehydrogenase activity, as an integral part of an intact cell and soil microflora activity, can provide information regarding the biologically active population of microorganisms in a given soil [15]. Soil microbial and enzymatic activity responds relatively quickly to slight changes in soil conditions and can reflect the changes in soil quality before they can be detected by other soil analyses [16]. Dehydrogenase plays a significant role in the biological oxidation of soil organic matter by transferring hydrogen from organic substrates to inorganic acceptors [17]. In this sense, the determination of dehydrogenase activity can be used to reflect the changes in soil biology [18,19], including assessing soil quality, the influence of soil management on soil quality, and the degree of regeneration of degraded soil [13,20]. Afforestation induces a rapid increase in microbial biomass, with changes apparent within one year of tree planting [21]. In a previous study, afforestation increased bacterial PLFAs by 20–120%, whereas it had a stronger impact on the development of fungal communities (increases by 50–200%) [22].

In this context, the main aim of this research was to determine the effects of changes in soil management from agriculture to forestry on the soil organic carbon accumulation and on enzymatic activity. Dehydrogenase activity, which plays a key role in the carbon cycle, was determined, and we tested the following hypotheses: (1) natural birch regeneration has a positive effect on the soil organic carbon accumulation and (2) dehydrogenase activity reflects the changes that occurred in the soil of the studied chronosequence.

2. Materials and Methods

The soil samples were collected from 12 research plots at four locations in the Mazowieckie province of Poland (Table 1, Figure 1). The study area is characterized by the following climatic conditions: average annual rainfall of 629 mm, average annual temperature of 8.4 °C, and a growing season of 210 days. The area in which the sample plots were located was dominated by fluvioglacial and glacial sand and loam with Gleysols, Cambisols, Podzols, and Arenosols [23]. The study plots were used as cropland in the past.

The study plots were divided into four groups based on the age of the self-seeded birch trees: I—1–4 years, II—5–8 years, III—9–12 years, and IV—13–17 years. In each plot, we took three soil samples from the 0–5, 5–15, and 15–50 cm layers. The samples were air-dried, sieved through a 2-mm-mesh, and the following physicochemical properties were determined [24]: pH (potentiometrically, in 1 M KCl and H_2O solution), texture (using laser diffraction in an Analysette 22: Fritsch, Idar-Oberstein, Germany), nitrogen, and organic carbon contents (with a LECO CNS True Mac Analyser: Leco, St. Joseph, MI, USA), C/N ratio, basic cations content (in 1 M ammonium acetate, using a Thermo Scientific iCAP 6000 ICP OES analyser, Thermo Fisher Scientific, Cambridge, UK). The data presented is the mean of the three soil replicates.

The results were used to calculate the carbon stock in the soils of the chronosequences, based on bulk density (BD), which were determined using Kopecky's cylinders. The carbon stock was calculated according to the following formula:

$$\text{SOCstock} = \text{C} \times \text{BD} \times \text{T}/100 \quad (1)$$

where SOCstock is the carbon stock in the soil ($kg \cdot m^{-2}$), C is the carbon content in the soil layers ($g \cdot kg^{-1}$), BD is bulk density [$g \cdot cm^{-3}$], and T is the thickness of the soil layers (cm).

Fresh samples, with natural moisture content, were taken to determine dehydrogenase (DH) activity (DH) via the Lenhard method, according to the Casida procedure. The DH activity was expressed as μmol TPF kg^{-1} h^{-1} [25].

The biomass [kg·ha^{-1}] of the aboveground and belowground parts of the stands in groups I–IV was determined, using the trunks, branches, assimilation apparatus, bark, and roots. For this, 10 trees were randomly selected at each location and were separated into trunk, branches, assimilation apparatus, bark, and roots. All the parts of the tree were weighed in the field while using portable scales with an accuracy of 0.01 g. Samples from each of the components from each tree model were collected to determine the relationship between fresh and dry biomass. Briefly, the samples were oven-dried at 105 °C and then weighed. On the basis of appropriate fresh-to-dry mass ratios, we calculated the dry biomass of the components for each tree.

Basic statistical data were calculated (i.e., the arithmetic mean and measures to determine the degree of differentiation among the results—standard deviation). The obtained data did not show normality, the Shapiro–Wilk test was used to to check the normal distribution. Tukey's HSD multiple comparisons of means were used in post hoc analysis to assess the effect of the age of regenerated birch trees and soil depth on the studied soil properties. Principal components analysis (PCA) was used to interpret the relationships among the studied variables, while the Pearson's correlation was applied to determine the relationships between dehydrogenase activity and soil properties. By applying Ward's method, the samples were grouped according to DH activity and carbon content. Average and standard deviation (SD) were presented in tables. Differences with $p < 0.05$ were considered to be statistically significant. Statistical analyses were performed in the Statistica 10 software (StatSoft Inc., Tulsa, OK, USA).

Table 1. Location of research plots and soil type.

Study Site	GPS	Soil Type
Mińsk Maz.	52°10′ N, 21°40′ E	Brunic arenosol
Kozienice	51°24′ N, 21°26′ E	Brunic arenosol
Dobieszyn 1	51°35′ N, 21°10′ E	Brunic arenosol
Dobieszyn 2	51°33′ N, 21°09′ E	Brunic arenosol

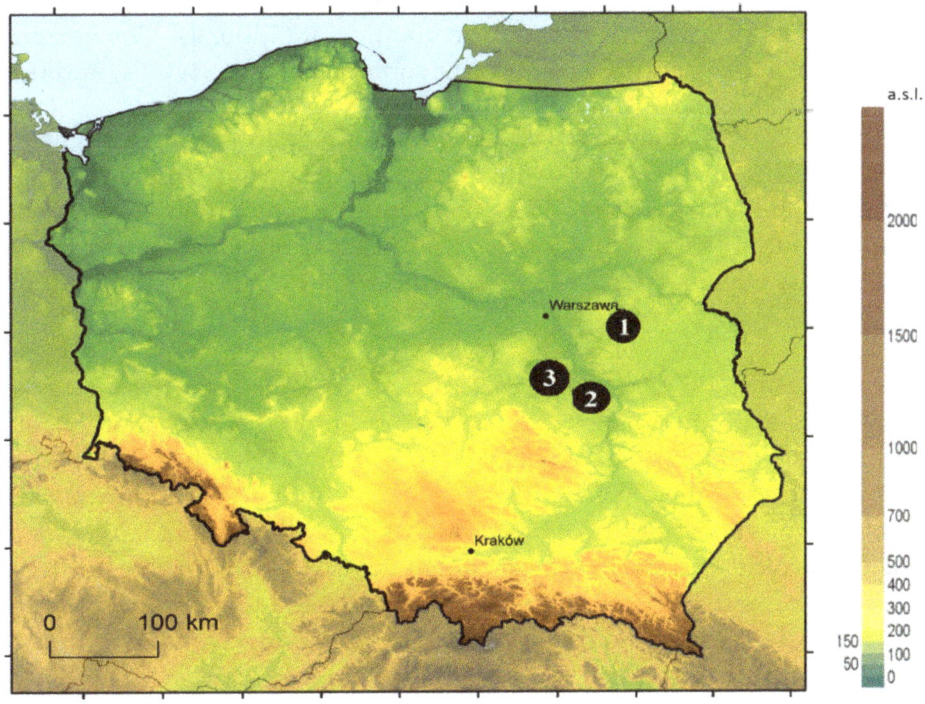

Figure 1. Localization of study plots (1—Mińsk Maz., 2—Kozienice, 3—Dobieszyn 1, and Dobieszyn 2).

3. Results

The soils differed in terms of pH values. The highest average pH was recorded in soils from the youngest birch stands (groups I and II); in the surface soil layer of these stands, the pH in H_2O was 4.52 and 4.62, respectively. The lowest pH was recorded in soils of the oldest stands (group IV) (Table 2). All of the sites were similar in terms of silt, and clay contents; there were no statistically significant differences of the studied chronosequences. Slight differences were noted in the sand content (Table 2). We also found no statistically significant differences in the C contents of the subsequent soil layers. There were no significant differences in the rate of organic matter decomposition, being expressed as the C/N ratio. The highest C/N ratio was recorded for the soil of the youngest birch stands (group I, average 22.3) and the lowest for the soil of the oldest birch stands (group IV, average 16.6). There were statistically significant differences in the Ca content (Table 2).

The total carbon stock did not significantly differ among the groups (Table 3). A slightly lower than average carbon stock was found in soils of the younger stands (groups I and II), while the values were above the average in the soils of the older stands (groups III and IV). However, these differences were not statistically significant. The average carbon stock depended on the age of the forest stand and it ranged from 4.34 to 6.16 $kg \cdot m^{-2}$. The carbon stock in the different soil layers changed with the age of the tree stands (Table 3). In the soils of the younger stands (groups I and II), a greater amount of accumulated carbon was found in the upper layer (0–5 cm) as compared to the same layer in the older stands (groups III and IV). In the soils of groups I and II, the proportion of the total carbon, which was determined to a depth of 50 cm, in the surface layer accounted for about 28%, while it accounted for 13.6% in the soils of the oldest stands. The highest amount of C in the deeper layers was recorded in soils of the oldest stands; the carbon in the 15–50-cm soil layer of group IV accounted for nearly 60% of the total carbon stock, while it did not exceed 40% in the soils of group I.

Dehydrogenase activity was used as a proxy for the biological activity of the studied soils and it varied among the sites. The highest mean value of dehydrogenase activity was recorded for group I soils and the lowest for group II–IV soils (Figure 2), which indicated a decrease in dehydrogenase activity with stand age. A strong relationship between dehydrogenase activity and the basic cation content was determined while using Pearson's correlation coefficient (Table 4). The correlation coefficients between dehydrogenase activity and K, Ca, and Mg contents were 0.81, 0.64, and 0.60, respectively.

Table 5 presents the components of the aboveground and belowground biomass of the examined stands. The biomass components significantly increased with stand age. In the youngest stand (group I), the average aboveground biomass was 2521.5 $kg \cdot ha^{-1}$ and the root biomass was 1058.7 $kg \cdot ha^{-1}$. In the oldest stand (group IV), the aboveground biomass was 30 times higher than that in the youngest stand, whereas the root biomass was more than 12 times higher in the older group IV than in the younger (group I) stand.

The first two axes of the PCA explained 46.2% of the variance of the analyzed soil properties (Figure 3). The first axis explained 31.74% of the variability and it was mainly related to the basic cation content, while the second axis explained 14.47% of the variability and it was associated with the C and N contents and with the pH. The results of the PCA analysis confirmed the dependence of dehydrogenase activity on the amounts of basic cations that are available. The C and N levels were higher in the soils of the older stands. To discriminate the distinction of the studied chronosequence of birch stands, we performed a cluster analysis, which enabled us to identify the two main groups differing in dehydrogenase activity and carbon content. The youngest stands (groups I and II) clearly differed from the oldest stands (groups III and IV) (Figure 4).

Table 2. Soil properties of the studied chronosequence of birch stands, with statistic test results.

Chronosequence	Depth	pH in H_2O	pH in KCl	C	N	C/N	Na	K	Ca	Mg	Sand	Silt	Clay
				%			cmol (+)·kg^{-1}					%	
I	0–5	4.52 ± 0.14 ^{a}	3.88 ± 0.12 ^{a}	1.57 ± 0.33 ^{a}	0.07 ± 0.02 ^{a}	22.3 ± 6.5 ^{a}	0.17 ± 0.05 ^{a}	0.95 ± 0.30 ^{a}	1.62 ± 0.21 ^{ab}	1.57 ± 0.57 ^{a}	83 ± 4 ^{a}	14 ± 3 ^{a}	2 ± 1 ^{a}
	5–15	4.73 ± 0.30 ^{ab}	3.97 ± 0.19 ^{a}	1.17 ± 0.71 ^{a}	0.07 ± 0.03 ^{a}	14.1 ± 4.9 ^{a}	0.21 ± 0.07 ^{a}	1.05 ± 0.47 ^{a}	1.60 ± 0.57 ^{ab}	1.60 ± 0.70 ^{a}	84 ± 4 ^{ab}	13 ± 3 ^{a}	2 ± 1 ^{a}
	15–50	5.03 ± 0.48 ^{a}	4.23 ± 0.16 ^{a}	0.49 ± 0.49 ^{a}	0.03 ± 0.02 ^{b}	19.0 ± 16.2 ^{a}	0.25 ± 0.07 ^{a}	1.50 ± 0.40 ^{a}	1.69 ± 0.56 ^{a}	2.60 ± 0.85 ^{a}	79 ± 11 ^{a}	18 ± 7 ^{a}	3 ± 2 ^{a}
II	0–5	4.62 ± 0.31 ^{a}	3.98 ± 0.07 ^{a}	1.30 ± 0.31 ^{a}	0.09 ± 0.07 ^{a}	15.2 ± 6.5 ^{a}	0.22 ± 0.10 ^{a}	1.98 ± 0.30 ^{a}	2.13 ± 0.71 ^{a}	2.23 ± 1.39 ^{a}	76 ± 2 ^{b}	19 ± 2 ^{a}	4 ± 1 ^{a}
	5–15	4.87 ± 0.23 ^{a}	4.13 ± 0.26 ^{a}	1.10 ± 0.47 ^{a}	0.32 ± 0.35 ^{a}	11.1 ± 2.7 ^{a}	0.21 ± 0.10 ^{a}	1.39 ± 1.34 ^{a}	2.37 ± 1.11 ^{a}	2.30 ± 1.48 ^{a}	78 ± 5 ^{b}	18 ± 4 ^{a}	4 ± 2 ^{a}
	15–50	5.29 ± 0.38 ^{a}	4.35 ± 0.39 ^{a}	0.16 ± 0.10 ^{a}	0.04 ± 0.03 ^{ab}	8.4 ± 5.3 ^{a}	0.22 ± 0.07 ^{a}	1.96 ± 0.34 ^{a}	2.25 ± 1.01 ^{a}	3.55 ± 0.31 ^{a}	80 ± 5 ^{a}	17 ± 4 ^{a}	3 ± 1 ^{a}
III	0–5	4.58 ± 0.15 ^{a}	3.79 ± 0.12 ^{a}	1.38 ± 0.58 ^{a}	0.11 ± 0.06 ^{a}	15.8 ± 7.2 ^{a}	0.20 ± 0.10 ^{a}	0.80 ± 0.25 ^{a}	1.38 ± 0.14 ^{b}	1.44 ± 0.32 ^{a}	83 ± 7 ^{ab}	14 ± 5 ^{a}	2 ± 2 ^{a}
	5–15	4.60 ± 0.20 ^{ab}	3.94 ± 0.15 ^{a}	1.19 ± 0.63 ^{a}	0.06 ± 0.04 ^{a}	19.7 ± 4.9 ^{a}	0.19 ± 0.06 ^{a}	0.81 ± 0.27 ^{a}	1.41 ± 0.18 ^{a}	1.51 ± 0.37 ^{a}	86 ± 4 ^{ab}	12 ± 3 ^{a}	2 ± 1 ^{a}
	15–50	4.91 ± 0.36 ^{a}	4.27 ± 0.05 ^{a}	0.68 ± 0.53 ^{a}	0.05 ± 0.05 ^{ab}	18.2 ± 15.1 ^{a}	0.21 ± 0.05 ^{a}	1.07 ± 0.55 ^{a}	1.35 ± 0.24 ^{ab}	1.92 ± 0.52 ^{a}	87 ± 3 ^{a}	12 ± 2 ^{a}	2 ± 1 ^{a}
IV	0–5	4.38 ± 0.13 ^{a}	3.85 ± 0.25 ^{a}	1.39 ± 0.50 ^{a}	0.09 ± 0.06 ^{a}	16.6 ± 6.2 ^{a}	0.21 ± 0.03 ^{a}	1.11 ± 0.18 ^{a}	1.66 ± 0.31 ^{ab}	1.97 ± 0.27 ^{a}	85 ± 7 ^{ab}	13 ± 5 ^{a}	2 ± 1 ^{a}
	5–15	4.35 ± 0.30 ^{b}	3.93 ± 0.09 ^{a}	1.33 ± 0.55 ^{a}	0.15 ± 0.06 ^{a}	11.5 ± 2.7 ^{a}	0.23 ± 0.03 ^{a}	1.10 ± 0.26 ^{a}	1.41 ± 0.19 ^{b}	1.81 ± 0.38 ^{a}	88 ± 1 ^{a}	11 ± 1 ^{a}	1 ± 1 ^{a}
	15–50	4.65 ± 0.59 ^{a}	4.13 ± 0.29 ^{a}	1.11 ± 0.90 ^{a}	0.19 ± 0.02 ^{a}	12.2 ± 8.9 ^{a}	0.22 ± 0.08 ^{a}	1.19 ± 0.33 ^{a}	1.54 ± 0.09 ^{a}	2.22 ± 0.58 ^{a}	88 ± 2 ^{a}	11 ± 2 ^{a}	2 ± 1 ^{a}

Mean ± SD; small letters in the upper index of the mean values mean significant differences of soils properties between chronosequence and dept.

Table 3. Total and percentage carbon storage (kg·m^{-2}—SOCstock) in soil layers of the studied chronosequence of birch stands.

Chronosequence	Depth	SOCstock (kg·m^{-2})	Total SOCstock in All Layers	% Participation SOCstock
I	0–5	0.94	4.57	27.5
	5–15	1.43		32.2
	15–50	2.20		40.2
II	0–5	0.78	4.34	24.1
	5–15	1.37		38.4
	15–50	2.19		37.5
III	0–5	0.82	5.41	18.0
	5–15	1.49		29.1
	15–50	3.10		52.8
IV	0–5	0.69	6.16	13.6
	5–15	1.54		27.7
	15–50	3.93		58.6

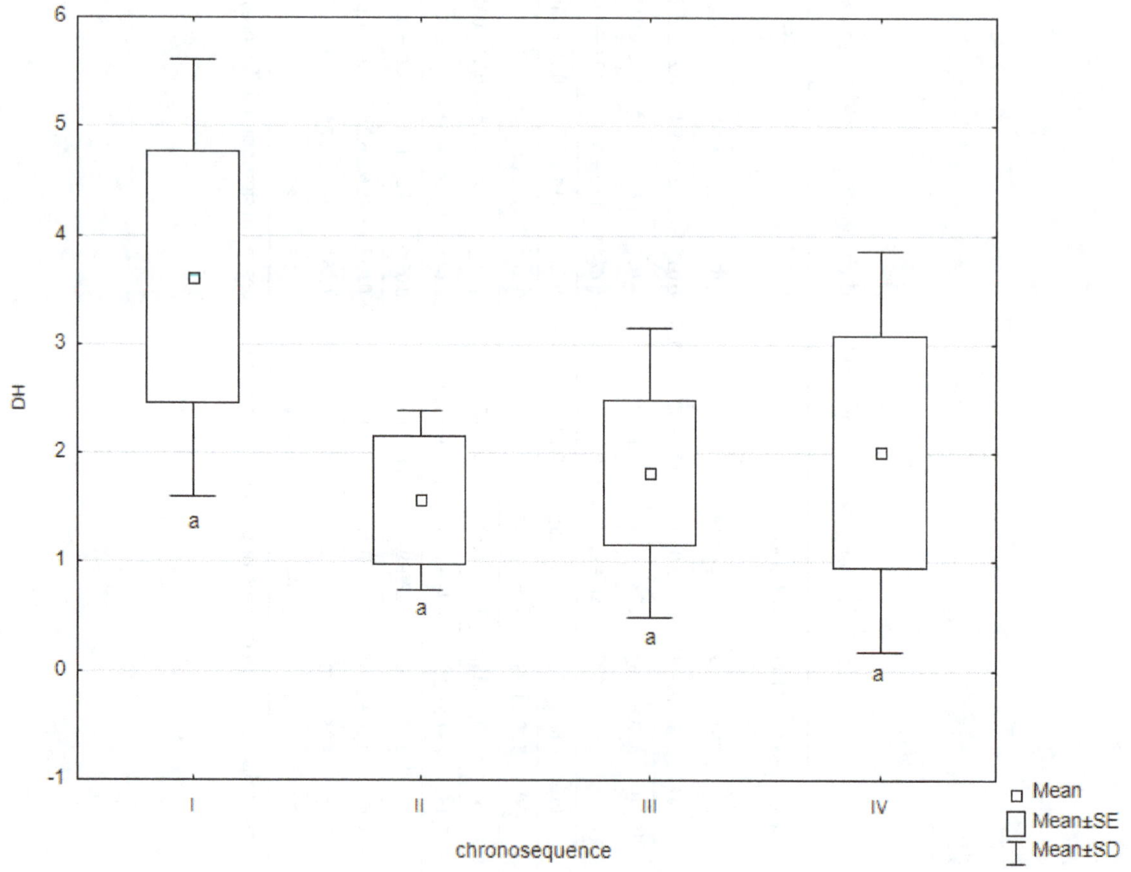

Figure 2. Dehydrogenase activity (DH) (µmol TPF·kg^{-1}·h^{-1}) in first soil layers of the studied chronosequences of birch stands.

Table 4. Correlations between dehydrogenase activity (DH) and basic soil properties.

	pH_{H2O}	pH_{KCl}	Na	K	Ca	Mg	C	N	Sand	Silt	Clay
DH	−0.04	−0.13	0.43	0.81 *	0.64 *	0.60 *	0.09	0.10	−0.20	0.20	0.11

* $p < 0.05$.

Table 5. Average biomass ($kg \cdot ha^{-1}$) of stand components in the studied chronosequence.

Chronosequence	Stem	Branches	Foliage	Bark	Roots
I	867.6 ^{b}	548.0 ^{b}	863.5 ^{b}	242.4 ^{b}	1058.7 ^{b}
II	7947.7 ^{ab}	1924.9 ^{b}	1745.9 ^{ab}	1931.4 ^{ab}	2866.8 ^{b}
III	23907.2 ^{ab}	5147.2 ^{ab}	2299.7 ^{ab}	4581.3 ^{ab}	6485.4 ^{ab}
IV	54307.9 ^{a}	11357.8 ^{a}	3242.6 ^{a}	9173.7 ^{a}	13492.7 ^{a}

Small letters in the upper index of the mean values mean significant differences.

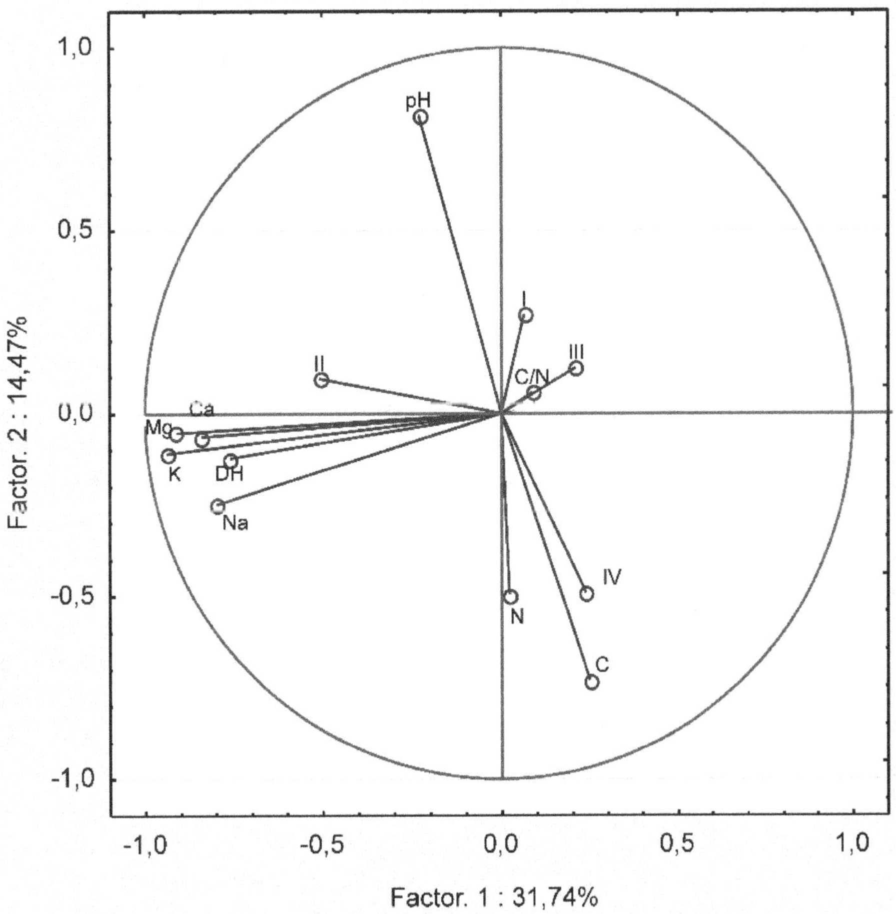

Figure 3. Projection of soil properties of birch stands chronosequence of on a plane of the first and second factors in the PCA (I—from 1 to 4 years, II—from 5 to 8 years, III—from 9 to 12 years, and IV—from 13 to 17 years; DH—dehydrogenase activity; C—carbon content; N—nitrogen content).

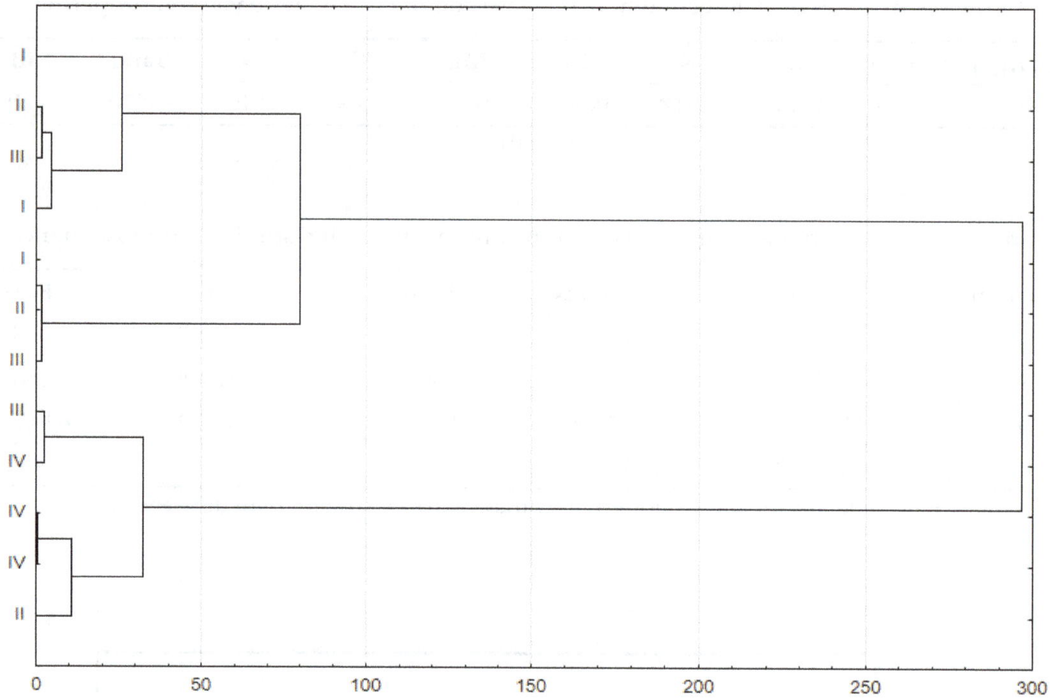

Figure 4. Dendrogram with group identified in the cluster analysis. The dehydrogenase activity and carbon content in surface layers were used for diagram preparation. I–IV—studied chronosequences of birch stands.

4. Discussion

Our results support the hypothesis that natural birch regeneration has a positive effect on the soil organic carbon accumulation. The tendency to increase carbon stocks was observed in the studied chronosequence of the birch tree stands. Several studies have shown that the decomposition of soil organic matter exceeded the input of organic matter from the trees in the initial following afforestation [26,27]. In the younger stands, the soil organic carbon accumulation was the greatest in the surface layer. In the 5-cm layer, accumulation accounted for about 30% of the total accumulation in the 50-cm deep soil column. Conversely, carbon accumulation was considerably lower than in the deeper soil layers in the 5-cm layer of the soils from the older stands. In the group I soils, accumulation in the deeper layers constituted 40% of the total stock of carbon, while it accounted for 60% in group IV. This increase in C accumulation in the deeper layers of the soils is associated not only with the processes of transporting dissolved organic compounds downwards, but also with an effect of supplying organic debris from the root systems. This is confirmed by the increase in root biomass in particular groups of stands. Subsoil soil organic carbon (SOC) storage may be promoted by the translocation of OM into deeper soil layers as DOC with the percolating water and due to bioturbation by soil animals [28]. Kotroczó et al. [29] found that plants cause greater changes in soil properties through their roots and secretions than via litter. In this sense, aboveground OM only probably has limited effects on SOM levels when compared to belowground OM [28]. Roots are a key component of the belowground part of the forest ecosystem, constituting the basic source of SOM that significantly affects soil microbiological activity [30,31]. Over time, soil organic matter input increases with the productivity of the forest stands, and the soils switch from being a C source to a C sink [32]. According to Laganiere et al. [33], the positive impact of afforestation on soil organic carbon stock is more pronounced in the cropland soils than in pastures or natural grasslands. Afforestation usually results in the establishment of higher plant biomass, and trees modify the quality and quantity of litter inputs and microclimatic conditions, such as moisture and temperature. Deng and Shangguang [34] highlight the importance of previous land use, tree species, soil depth, and forest age in determining soil C and N changes in a range of environments and land use transitions. In our study, birch stands, through aboveground and belowground biomass accumulation, had a positive effect on the quality of SOM, as expressed by the C/N ratio, which is an indicator of the extent of plant nitrogen being made available to plant residues. Li et al. [35] state that land use changes from agricultural areas to forest alter the ratios between soil C, N, and P. Springob and Krichmann [36] found that a soil C/N ratio of >20 could limit SOM mineralisation. According to Cools et al. [37], tree species are the main factor in explaining the variability of the C/N ratio. The content of better decomposed soil organic matter increases with stand age. At the same time, soil acidity and nutrient uptake increase with tree growth. Riqueiro-Rodríguez [38] note that the Pinus radiate more drastically decreases the soil pH than *Betula alba*. In another study, the acidifying affect of afforestation on mineral soil has been confirmed by a significant decrease in soil pH in the 0–5-cm layer and by a slightly weaker decrease in the 5–15-cm layer [39].

The results of the cluster analysis confirmed the distinctness in terms of C content and enzymatic activity of the soils of younger stands when compared to the soils of older stands. The soil parameters pH and soil organic carbon are important factors that shape dehydrogenase activity [15,40]. The highest pH, with the highest alkaline cation content, resulted in the highest dehydrogenase activity in soils of the younger stands (Groups I and II), reflecting the previous agricultural use of the soils and the associated intensive fertilisation and liming. According to Rousk et al. [41], pH is the main determinant of the structure of soil microbial populations. Soil pH directly determines plant growth, nutrient absorption, and the intensity of biological and chemical processes in the soil. In this work, dehydrogenase activity was positively associated with exchangeable Ca, K, and Mg contents, with a higher content of basic ions leading to an increase in pH, which results in the stimulation of soil microorganisms. Soil pH may be the major factor controlling the biomass and composition of microbial communities and their maintenance demand [42]. When assessing the properties of soils that were subjected to long-term agricultural use, several authors have considered the high plant-nutrient content as evidence of systematic fertilisation [43]. For example, Ren et al. [44] have noted that catalase, saccharase, urease, and alkaline phosphatase were significantly increased by land-use conversion from farmland to forest. According to this, significant correlations between soil enzyme activities and soil properties indicate that the soil enzyme activities are closely related to soil nutrients dynamics [18]. Dehydrogenase activity differed among the soils of the studied birch stands. The activity of this enzyme reflects that changes in the soil that are associated with the growth of the birch stands. According to previous studies, enzymatic activity is strongly stimulated by SOM [15,18], and processes that are related to organic matter transformations are carried out with the participation of soil microorganisms and their enzymes [45]. In our study, no direct relationship between dehydrogenase activity and carbon accumulation was found. Dehydrogenase activity was high in the soil of the youngest stands (first age class), and subsequently considerably decreased in class II. Our results indicate a trend to increased dehydrogenase activity in the soils of the oldest stands (IV group of stands). The highest pH, with the highest alkaline cation content, resulted in the highest dehydrogenase activity in soils of the younger stands, reflecting the previous agricultural use of the soils and the associated intensive fertilization and liming. The effects of fertilization disappear in the following years of tree stand growth. Forest stands grow and provide more litter fall to the soil, which stimulates the dehydrogenase activity. With age, greater amounts of carbon were accumulated in the surface soil layers. With increased litter input and in the absence of soil cultivation, conversion from cropland to forest could result in increased SOM stocks [46]. Similarly, Kara et al. [47] and Kang et al. [48] suggest that long-term afforestation could significantly enhance SOM contents, accumulate microbial biomass, and improve potential enzyme activities.

5. Conclusions

Our results confirm the beneficial effect of birch stand regeneration on the soil properties on post-agricultural land. We observed a clear trend of increasing carbon accumulation in the soil under the influence of birch trees. With age, greater amounts of carbon were accumulated in the surface soil layers. Dehydrogenase activity is a suitable indicator of the condition of post-agricultural soils with birch stands and, in combination with soil chemical properties, reflects historical soil management. In this sense, the determination of dehydrogenase activity allows for an assessment of the processes occurring in post-agricultural soils, which are associated with the soil organic carbon accumulation. A high nutrient content and high pH are characteristic of post-agricultural soils, facilitating a greater biochemical activity in the initial stages of stand formation.

Author Contributions: Conceptualization, T.G. and S.M.; Methodology, T.G. and E.B.; Project administration, T.G. and S.M.; Supervision, S.M.; Writing—original draft, T.G. and E.B.; Writing—review & editing, T.G.

References

1. Report 2018. *Report on the Condition of Forests in Poland*; CILP: Warszawa, Poland, 2018.
2. Schoenholtz, S.H.; Van Miegroet, H.; Burger, J.A. A review of chemical and physical properties as indicators offorest soil quality: Challenges and opportunities. *For. Ecol. Manag.* **2000**, *138*, 335–356. [CrossRef]
3. Van-Camp, L.; Bujarrabal, B.; Gentile, A.R.; Jones, R.J.A.; Montanarella, L.; Olazabal, C.; Selvaradjou, S.K. *Reports of the Technical Working Groups Established under the Thematic Strategy for Soil Protection*; EUR 21319EN/3; Office for official Publications of the European Communities: Luxembourg, 2004; p. 872.
4. Steiner, C.; Teixeira, W.G.; Lehman, J.; Nehls, T.; de Macêdo, J.L.U.; Blum, W.E.M.; Zech, W. Long term effects of manure, charcoal and mineral fertilization on crop production and fertility on a highly weathered Central Amazonian upland soil. *Plant Soil* **2007**, *291*, 275–290. [CrossRef]
5. Yang, L.; Luo, P.; Wen, L.; Li, D. Soil organic carbon accumulation during post-agricultural succession in a karst area, southwest China. *Sci. Rep.* **2016**, *6*, 37118. [CrossRef] [PubMed]
6. Lal, R. Soil carbon sequestration impacts on global climate change and food security. *Science* **2004**, *304*, 1623–1627. [CrossRef] [PubMed]
7. Guo, L.; Gifford, R. Soil carbon stocks and land use change: A meta analysis. *Glob. Chang. Biol.* **2002**, *8*, 345–360. [CrossRef]
8. Korkanç, S.Y. Effects of afforestation on soil organic carbon and other soil properties. *Catena* **2014**, *123*, 62–69. [CrossRef]
9. Mao, D.M.; Min, Y.W.; Yu, L.L.; Martens, R.; Insam, H. Effect of afforestation on microbial biomass and activity in soils of tropical China. *Soil Biol. Biochem.* **1992**, *24*, 865–872.
10. Podrázský, V.; Holubík, O.; Vopravil, J.; Khel, T.; Moser, W.K.; Prknová, H. Effects of afforestation on soil structure formation in two climatic regions of the Czech Republic. *J. For. Sci.* **2015**, *61*, 225–234. [CrossRef]
11. Bolat, I.; Kara, Ö.; Sensoy, H.; Yüksel, K. Influences of Black Locust (*Robinia pseudoacacia* L.) afforestation on soil microbial biomass and activity. *iFor. Biogeosci. For.* **2017**, *9*, 171–177. [CrossRef]
12. Karlen, D.L.; Mausbach, M.J.; Doran, J.W.; Cline, R.G.; Harris, R.F.; Schuman, G.E. Soil quality: A concept, definition, and framework for evaluation. *Soil Sci. Soc. Am. J.* **1997**, *61*, 4–10. [CrossRef]
13. Gil-Sotres, F.; Trasar-Cepeda, C.; Leiros, M.C.; Seoane, S. Different approaches to evaluating soil quality using biochemical properties. *Soil Biol. Biochem.* **2005**, *37*, 877–887. [CrossRef]
14. Kucharski, J. Relacje między aktywnością enzymów a żyznością gleby. In *Drobnoustroje w Środowisku, Występowanie, Aktywność i Znaczenie*; Barabasz, W., Ed.; AR: Kraków, Poland, 1997; pp. 327–347.
15. Wolińska, A.; Stępniewska, Z.; Pytlak, A. The effect of environmental factors on total soil DNA content and dehydrogenase activity. *Arch. Biol. Sci.* **2015**, *67*, 493–501. [CrossRef]
16. Błońska, E.; Lasota, J.; Gruba, P. Enzymatic activity and stabilization of organic matter in soil with different detritus inputs. *J. Soil Sci. Plant Nutr.* **2017**, *63*, 242–247.
17. Kaczynski, P.; Lozowicka, B.; Hrynko, I.; Wolejko, E. Behaviour of mesotrione in maize and soil system and its influence on soil dehydrogenase activity. *Sci. Total Environ.* **2016**, *71*, 1079–1088. [CrossRef] [PubMed]
18. Błońska, E.; Lasota, J.; Zwydak, M.; Klamerus-Iwan, A.; Gołąb, J. Restoration of forest soil and vegetation 15 years after landslides in a lower zone of mountains in temperate climates. *Ecol. Eng.* **2016**, *97*, 503–515. [CrossRef]
19. Pająk, M.; Błońska, E.; Frąc, M.; Oszust, K. Functional diversity and microbial activity of forest soils that are heavily contaminated by lead and zinc. *Water Air Soil Pollut.* **2016**, *227*, 348. [CrossRef]

20. Józefowska, A.; Pietrzykowski, M.; Woś, B. Tree species and soil substrate effects on soil biota during early soil forming stages at afforested mine sites. *Appl. Soil Ecol.* **2016**, *102*, 70–79. [CrossRef]
21. Van der Wal, A.; van Veen, J.A.; Smant, W.; Boschker, H.; Bloem, J.; Kardol, P.; van der Putten, W.H.; de Boer, W. Fungalbiomass development in a chronosequence of land abandon-ment. *Soil Biol. Biochem* **2006**, *38*, 51–60. [CrossRef]
22. Gunina, A.; Smith, A.R.; Godbold, D.L.; Jones, D.L.; Kuzyakov, Y. Response of soil microbial community to afforestation with pure and mixed species. *Plant Soil* **2016**, *412*, 357–368. [CrossRef]
23. *WRB (World Reference Base for Soil Resource)*; FAO: Rome, Italy, 2014.
24. Ostrowska, A.; Porębska, G.; Kanafa, M. Carbon accumulation and Distribution in Profiles of Forest Soils. *Pol. J. Environ. Stud.* **2010**, *19*, 1307–1315.
25. Alef, K.; Nannipieri, P. Enzyme activities. In *Methods in Applied Soil Microbiology and Biochemistry*; Alef, K., Nannipieri, P., Eds.; Academic Press: London, UK, 1995.
26. Richter, D.D.; Markewitz, D.; Trumbore, S.E.; Wells, C.G. Rapid accumulation and turnover of soil carbon in a reestablishing forest. *Nature* **1999**, *400*, 56–58. [CrossRef]
27. Mao, R.; Zeng, D.H.; Hu, Y.L.; Li, L.J.; Yyng, D. Soil organic carbon and nitrogen stocks in an age-sequence of poplar stands planted on marginal agricultural land in Northeast China. *Plant Soil* **2010**, *332*, 277–287. [CrossRef]
28. Lorenz, K.; Lal, R. The depth distribution of soil organic carbon in relation to land use and management and the potential of carbon sequestration in subsoil horizons. *Adv. Agron.* **2005**, *88*, 35–66.
29. Kotroczó, Z.; Veres, Z.; Fekete, J.; Krakomperger, Z.; Tóth, J.A.; Lajtha, K.; Tóthmérisz, B. Soil enzyme activity in response to long-term organic matter manipulation. *Soil Biol. Biochem.* **2014**, *70*, 237–243. [CrossRef]
30. Janssens, J.A.; Sampson, D.A.; Curiel-Yuste, J.; Carrara, A.; Cenlemans, R. The carbon cost of fine root turnover in a Scots pine forest. *For. Ecol. Manag.* **2002**, *168*, 231–240. [CrossRef]
31. Kätterer, T.; Bolinder, M.A.; Andrén, O.; Kirchmann, H.; Menichetti, L. Roots contribute more to refractory soil organic matter than above-ground crop residues, as revealed by a long-term field experiment. *Agric. Ecosyst. Environ.* **2011**, *141*, 184–192. [CrossRef]
32. Holubík, O.; Podrázský, V.; Vopravil, J.; Khel, T.; Remeš, J. Effect of agricultural lands afforestation and tree species composition on the soil reaction, total organic carbon and nitrogen content in the uppermost mineral soil profile. *Soil Water Res.* **2014**, *9*, 192–200. [CrossRef]
33. Laganiere, J.; Angers, D.A.; Parc, D. Carbon ac-cumulation in agricultural soils after afforestation: A meta-analysis. *Glob. Chang. Biol.* **2010**, *16*, 439–453. [CrossRef]
34. Deng, L.; Shangguang, Z. Afforestation drives soil carbon and nitrogen changes in China. *Land Degrad. Dev.* **2016**, *21*, 151–165. [CrossRef]
35. Li, D.; Wen, L.; Zhang, W.; Yang, L.; Xiao, K.; Chen, H.; Wang, K. Afforestation effects on soil organic carbon and nitro gen pools modulated by lithology. *For. Ecol. Manag.* **2017**, *400*, 85–92. [CrossRef]
36. Springob, G.; Kirchmann, H. Bulk soil C to N ratio as a simple measure of net N mineralization from stabilized soil organic matter in sandy arable soils. *Soil Biol. Biochem.* **2003**, *35*, 629–632. [CrossRef]
37. Cools, N.; Vesterdal, L.; De Vos, B.; Vanguelova, E.; Hansen, K. Tree species is the major factor explaining C:N ratios in European forest soil. *For. Ecol. Manag.* **2014**, *311*, 3–16. [CrossRef]
38. Riqueiro-Rodríguez, A.; Mosquera-Losada, M.R.; Férnández-Núñez, E. Afforestation of agricultural land with Pinus radiata D. Don and Betula alba L. in NW Spain: Effects on soil pH, understory production and floristic diversity eleven years after establishment. *Land Degrad. Dev.* **2012**, *23*, 227–241.
39. Ritter, E.; Vesterdal, L.; Gundersen, P. Changes in soil properties after afforestation of former intensively manager soil with oak and Norway spruce. *Plant Soil* **2003**, *249*, 319–330. [CrossRef]
40. Błońska, E.; Lasota, J.; Gruba, P. Effect of temperate forest tree species on soil dehydrogenase and urease activities in relation to other properties of soil derived from loess and glaciofluvial sand. *Ecol Res* **2016**, *31*(5), 655–664. [CrossRef]
41. Rousk, J.; Bååth, E.; Brookes, P.C.; Lauber, C.L.; Lozupone, C.; Caporaso, J.G.; Knight, R.; Fierer, N. Soil bacterial and fungal communities across a pH gradient in an arable soil. *ISME J.* **2010**, *4*, 1340–1351. [CrossRef] [PubMed]
42. Chen, C.R.; Condron, L.M.; Cavis, M.R.; Sherlock, R.R. Effects of afforestation on phosphorus dynamics and biological propertiesin a New Zealand grassland soil. *Plant Soil* **2000**, *220*, 151–163. [CrossRef]
43. Wanic, T.; Błońska, E. Zastosowanie metody SIG w ocenie przydatności terenów porolnych do hodowli lasu.

Rocz. Glebozn. **2011**, *62*, 173–181.

44. Ren, C.; Kang, D.; Wu, J.P.; Zhao, F.; Yang, G.; Han, X.; Feng, Y.; Ren, G. Temporal variation in soil enzyme activities after afforestation in the Loess Plateau, China. *Geoderma* **2016**, *282*, 103–111. [CrossRef]
45. Schimel, J.P.; Bennett, J. Nitrogen mineralization: Challenges of a changing paradigm. *Ecology* **2004**, *85*, 591–602. [CrossRef]
46. Georgiadis, P.; Vesterdal, L.; Stupak, I.; Raulund-Rasmussen, K. Accumulation of soilorganic carbon after cropland conversion to short-rotation willow and poplar. *GCB Bioenergy* **2017**, *9*, 1390–1401. [CrossRef]
47. Kara, O.; Babur, E.; Altun, L.; Seyis, M. Effects of afforestation on microbial biomass C and respiration in eroded soils of Turkey. *J. Sustain. For.* **2016**, *35*, 385–396. [CrossRef]
48. Kang, H.; Gao, H.; Yu, W.; Yi, Y.; Wang, Y.; Ning, M. Changes in soil microbial community structure and function after afforestation depend on species and age: Case study in a subtropical alluvial Island. *Sci. Total Environ.* **2018**, *625*, 1423–1432. [CrossRef] [PubMed]

12

Species Distribution Model for Management of an Invasive Vine in Forestlands of Eastern Texas

Hsiao-Hsuan Wang [1,*], Tomasz E. Koralewski [2], Erin K. McGrew [1], William E. Grant [1] and Thomas D. Byram [2,3]

1 Department of Wildlife and Fisheries Sciences, Texas A&M University, College Station, TX 77843, USA; ekmcgrew@email.tamu.edu (E.K.M.); wegrant@tamu.edu (W.E.G.)

2 Department of Ecosystem Science and Management, Texas A&M University, College Station, TX 77843, USA; tkoral@tamu.edu (T.E.K.); t-byram@tamu.edu (T.D.B.)

3 Texas A&M Forest Service, College Station, TX 77843, USA

* Correspondence: hsuan006@tamu.edu.

Academic Editor: Diana F. Tomback

Abstract: Invasive plants decrease biodiversity, modify vegetation structure, and inhibit growth and reproduction of native species. Japanese honeysuckle (*Lonicera japonica* Thunb.) is the most prevalent invasive vine in the forestlands of eastern Texas. Hence, we aimed to identify potential factors influencing the distribution of the species, quantify the relative importance of each factor, and test possible management strategies. We analyzed an extensive dataset collected as part of the Forest Inventory and Analysis Program of the United States Department of Agriculture (USDA) Forest Service to quantify the range expansion of Japanese honeysuckle in the forestlands of eastern Texas from 2006 to 2011. We then identified potential factors influencing the likelihood of presence of Japanese honeysuckle using boosted regression trees. Our results indicated that the presence of Japanese honeysuckle on sampled plots almost doubled during this period (from 352 to 616 plots), spreading extensively, geographically. The probability of invasion was correlated with variables representing landscape conditions, climatic conditions, forest features, disturbance factors, and forest management activities. Habitats most at risk to invasion under current conditions occurred primarily in northeastern Texas, with a few invasion hotspots in the south. Estimated probabilities of invasion were reduced most by artificial site regeneration, with habitats most at risk again occurring primarily in northeastern Texas.

Keywords: biodiversity; biological invasions; boosted regression trees; Japanese honeysuckle; likelihood of invasion; *Lonicera japonica* Thunb.

---

1. Introduction

Invasive species have caused dramatic economic damage and loss of biodiversity worldwide and non-native species continue to invade forest ecosystems in the United States at an accelerating rate [1]. Despite general consensus on the risks posed by alien species, the exact economic and ecological impact of invasions is an active area of debate and, in some cases, certain potential benefits are pointed out, depending on particular species, scenario, and perspective (reviewed in [2]). Nevertheless, there are an estimated 50,000 non-native species on U.S. soil [3]. According to Wilcove *et al.* [4], 57% of the 1055 imperiled plant species in the U.S., for which they managed to collect threat data, were negatively impacted by alien species. Based on the same data, Gurevitch and Padilla [5], estimated that out of 602 plant species in the U.S. imperiled by invaders, 410 (about 68%) could be linked to competition and indirect habitat effects caused by alien plants, but highlighted that invasive plants often act in parallel with other factors, such as habitat loss. Invasive plants reduce forest productivity [6] and degrade forest health [2], and pose a particularly severe threat to the economically and ecologically important forestlands in the southeastern U.S. Southeastern forests contain about 201 million acres of timberland, corresponding to about 58% of U.S. timber production and about 16% of timber production world-wide [7], and also support a high level of biodiversity, with over 3000 native plant species [8]. However, invasive plants threaten both the productivity of timber species essential to economic sustainability of the region [9] and the biodiversity equilibrium essential to the provision of ecosystem services, such as erosion control [10].

Japanese honeysuckle (*Lonicera japonica* Thunb.), which was introduced into the U.S. from China and Japan in the early 1800s [11], is one of the most aggressive invasive species in North America. It has become naturalized in 45 states of the U.S., and the Canadian province of Ontario [11,12]. Its attractive and fragrant flowers made it a valued cultivar, but its long flowering period and hardy growth characteristics allowed it to escape cultivation [11], probably during the 1890s, and spread to most of its present range within about 30 years [13].

Japanese honeysuckle prefers moist woodlands, often growing in tangled clumps along slopes, but is able to tolerate most forest conditions [14]. It grows rapidly, flowers early and for long durations [15], reproduces via animal-dispersed seeds or vegetatively [16], possesses phenotypic plasticity which reduces herbivore damage [17], and has no known predators or pathogens within its invaded range [11]. It negatively affects native vegetation [18], modifying vegetation structure [19], decreasing biodiversity [20,21], and inhibiting natural regeneration of native species, including loblolly pine [22], a predominant tree species in both artificially and naturally regenerated forests of the southeast. While natural pine forest range has been gradually declining, the acreage of pine plantations has been steadily expanding since 1950s, reaching 32 million acres in 1999 with further expected increase up to 54 million acres in 2040 [23].

The expanding plantations and increased timber yield additionally highlight economic potential of the southeastern forests. Approximately 84% of the seedlings planted annually throughout the South are loblolly pine [24]. In the forestlands of eastern Texas, largely corresponding to the western part of the Piney Woods ecoregion, where loblolly pine is of both commercial and ecological importance, Japanese honeysuckle is the most prevalent invasive vine [25]. A recent study identified areas vulnerable to invasion by Japanese honeysuckle throughout Mississippi, stretching northward across western Tennessee and western Kentucky, westward across southern Arkansas, eastward across north-central Alabama, and also in several counties scattered within Virginia [26]. These authors used logistic regression to correlate land characteristics and climatic conditions with presence/absence of Japanese honeysuckle based primarily on analysis of data collected prior to 2006.

In the present study, we calculated the recent range expansion of Japanese honeysuckle within the forestlands of eastern Texas based on analyses of an extensive set of field data collected by the U.S. Forest Service on fixed plots during the period from 2007 to 2011. We then used boosted regression trees to analyze an extensive data set collected as part of the Forest Inventory and Analysis (FIA) program of the United States Department of Agriculture (USDA) Forest Service to identify areas within the forestlands of eastern Texas that were invaded by Japanese honeysuckle during the FIA inventory period that extended from 2007 to 2011. We quantitatively assessed a suite of landscape features, climatic conditions, forest conditions, forest management activities, and disturbances as potential factors affecting the likelihood of invasion. Based on identification of the potentially most influential factors, we predicted likelihoods of future invasions by Japanese honeysuckle under a common management strategy (artificial regeneration of loblolly pine).

2. Methods

2.1. Study Area and Data Sources

We focused our investigation on eastern Texas, which is an area of vast timber resources [27] that has been invaded by Japanese honeysuckle [25]. The warm and humid climate in the area is conducive to invasion.

To document range expansion of Japanese honeysuckle in eastern Texas between the years of 2006 and 2011, we extracted data on the presence or absence of Japanese honeysuckle using the Southern Nonnative Invasive Plant data Extraction Tool [25] of the USDA Forest Service. In the early 2000s, as part of the Forest Inventory and Analysis (FIA) Program, the Forest Service's Southern Research Station began an intensive regional monitoring of nonnative plant invasions [28]. The FIA Program is a forest inventory program in which each state inventory is reported every five years in most southeastern states [29]. The basic sampling design consists of a lattice of 4047-m^2 hexagons, with one sample plot located randomly within each hexagon [29,30]. Each sample plot consists of four subplots of radius 7.32 m which form a cluster consisting of a central subplot and three peripheral subplots equidistant from each other arrayed in a circle of radius 36.58 m centered on the central plot. On each subplot, inventory crews estimate percent cover by target invasive species [28], and also record a suite of landscape conditions and forest features, as well as past disturbances and forest management activities [30].

2.2. Quantification of the Range Expansion

At present, two state inventory cycles recording the presence of nonnative plant species have been reported (in 2006 and 2011) for eastern Texas [25]. We documented range expansion by noting the presence (with cover) or absence (without cover) of Japanese honeysuckle on each subplot sampled during each of these two inventories and then mapping and counting the plots in each inventory with Japanese honeysuckle present. We also calculated an index of potential rate of spread based on the distance between each plot in which Japanese honeysuckle was first detected during the second survey and the nearest plot in which Japanese honeysuckle had been detected during the first survey, that is, the distance to the nearest known propagule source.

2.3. Identification of Potential Factors Influencing Likelihood of Invasion

Widely recognized factors influencing the likelihood of invasion of Japanese honeysuckle include landscape conditions, climatic conditions, forest features, disturbance factors, and forest management activities [26,31,32]. Landscape conditions include slope and adjacency to water bodies within 300 m. Climatic conditions include mean annual precipitation, and mean annual minimum and maximum temperature. Forest features include stand age, site productivity, basal area, and natural regeneration (growth of existing trees, natural seeding, or both, resulting in a stand at least 50% stocked with live trees of any size). Disturbance factors include distance between the plot and the nearest road, distances between the plot and the nearest known propagule source, and forest disturbance (such as those caused by animals, disease, fire, insects, and/or wind). Forest management activities such as harvesting, site preparation (including clearing, slash burning, chopping, disking, bedding, or other practices clearly intended to prepare a site for reforestation), and artificial regeneration (planting or direct seeding resulting in a stand in at least 50% stocking with trees of any size). Data on several factors in each of these broad categories are available from the traditional FIA dataset for plots within our study area (Table 1). Drawing on the literature on plant invasions, we selected a set of possible variables for plots within our study area from the PRISM database [33,34] and the FIA dataset (Table 1). We also used the same dataset to compute Shannon's index of tree species diversity, H_s, for each plot [35,36]:

$$H_s = -\sum_{i=1}^{n_s} \frac{B_i}{B} \ln(\frac{B_i}{B}) \tag{1}$$

where B and B_i are the total stand basal area and the basal area of trees of species i, respectively, and n_s is the number of tree species.

Table 1. Descriptions, possible values or units of measure, and means or counts of landscape conditions, forest features, disturbance factors, and forest management activities evaluated as potential factors of site invasion by Japanese honeysuckle in forest plots of eastern Texas.

Variable	Value or Unit of Measure	Mean (Range) for Continuous Data/Count for Categorical Data
Landscape conditions		
Slope	Degree	2.07 (0.00–32.50)
Adjacency to water bodies within 300 m	No Yes	1830 471
Climate conditions		
Mean annual precipitation	cm	127.06 (99.06–154.94)
Mean annual minimum temperature	°C	12.43 (10.56–16.67)
Mean annual maximum temperature	°C	24.89 (22.78–26.11)
Forest features		
Stand age	Year	35.76 (1–104)
Site productivity	L1: 0–1.39 $m^3 \cdot ha^{-1} \cdot year^{-1}$ L2: 1.40–3.49 L3: 3.50–5.94 L4: 5.95–8.39 L5: 8.40–11.54 L6: 11.55–15.74 L7: >15.74	9 154 660 931 541 95 4
Basal area	m^2	1.26 (0.00–7.10)
Species diversity	Shannon's species diversity	1.02 (0.00–2.53)
Natural regeneration [a]	No Yes	2348 46
Disturbance factors		
Distance to the nearest road	D1: <30 m D2: 30–91 D3: 92–152 D4: 153–305 D5: 306–805 D6: 806–1609 D7: 1610–4828 D8: 4829–8047 D9: >8047	169 382 273 492 674 239 62 3 2
Distance to the nearest known propagule source	m	14,343 (0–75,529)
Forest disturbance [a,b]	No Yes	2294 100

Table 1. *Cont.*

Variable	Value or Unit of Measure	Mean (Range) for Continuous Data/Count for Categorical Data
Forest management activities		
Harvesting [a]	No	1969
	Yes	425
Site preparation [a]	No	2309
	Yes	85
Artificial regeneration [a]	No	1802
	Yes	592

[a] Normally within the past five years; [b] A disturbance code of "yes" indicates that at least 25% of the trees in a stand are damaged by animals (from deer/ungulates, rabbits, or a combination of animals), disease, fire (from crown or ground fire, either prescribed or natural), insects, and/or wind (including, but not limited to, damages from hurricanes and tornados).

2.4. Data Analysis

We associated the data on presence or absence of Japanese honeysuckle (SNIPET) with the data on landscape conditions, forest features, disturbance factors, and forest management activities (FIA Data and Tools) using the FIA plot identification numbers. We then conducted our analysis with a process of machine learning that uses boosted regression trees, which combines decision trees and a boosting algorithm with a form of logistic regression [37]. For boosted regression trees, the probability (p) of species occurrence ($y = 1$) at a location with the potential explanatory variables (X) is given by $p(y = 1|X)$ and is modeled via the logit: $logit\ p\ (y = 1 \mid X) = f\ (X)$. We fitted our model in R (R Development Core Team 2006 version 2.14.1, R Foundation for Statistical Computing, Vienna, Austria) using the gbm package version 1.5-7 [38]. Our aim was to find the best combination of parameters (learning rate and tree complexity) that achieved the minimum predictive error (minimum error for predictions of independent samples). The learning rate, also known as the shrinkage parameter, determines the contribution of each tree to the growing model, and the tree complexity controls whether interactions are fitted. We determined the optimal model following the recommendations of Elith *et al.* [37] by altering the learning rate and tree complexity (the number of split nodes in a tree) until the predictive deviance was minimized without over-fitting, and by limiting our choice of the final model to those that contained at least 1000 trees. We included randomness into the models to reduce over-fitting and also to improve accuracy and speed of the model selection process [39]. We represented randomness using a "bag fraction" that specifies the proportion of data to be selected at each step [37]. We set the default bag fraction at 0.6, meaning that 60% of the data were drawn at random without replacement from the full training set during each iteration. When we obtained the optimal model, we evaluated model performance using a ten-fold cross-validation procedure with re-substitution. For each cross-validation trial, we randomly selected 60% of the dataset for model fitting and we used the excluded 40% for testing. We then used plot Universal Transverse Mercator (UTM) coordinates, x for distance east and y for distance north of the UTM origin, to test for spatial autocorrelation of the residuals of the model, employing Moran's I index. We calculated the response variance explained, the area under the receiver operator characteristic curve (AUC), and the overall accuracy based on the aggregated cross-validation results. We evaluated the reliability and validity of the optimal model as fair ($0.50 < \text{AUC} \leqslant 0.75$), good ($0.75 < \text{AUC} \leqslant 0.92$), very good ($0.92 < \text{AUC} \leqslant 0.97$), or excellent ($0.97 < \text{AUC} \leqslant 1.00$) based on the value of AUC [40]. We then used the gbm library to derive the relative influence of each potential explanatory variable in the optimal model and constructed partial dependence plots and fitted values plots for the most influential variables [37].

2.5. *Likelihood of Invasion*

We used the optimal model to calculate the probability of presence of Japanese honeysuckle on each of the study plots. We then re-estimated the probabilities of presence assuming that artificial regeneration occurred on 25%, 50%, 75%, or 100% of the plots. We represented these different levels of management activity by changing the value of the corresponding element of X (e.g., artificial regeneration) from "No" to "Yes." To visually display the results, we superimposed the probabilities of presence predicted under each level of management activity on maps of the study area using ArcMap^{TM} 10.2.1 (ESRI, Redlands, CA, USA).

3. Results

3.1. *Historical Trends in Range Expansion*

As indicated by the FIA records from 2011, Japanese honeysuckle spread extensively throughout eastern Texas during only a five-year period. The range expansion has been particularly dramatic in the northern part of the studied area, but the few, initially sparse, hot-spots in the south have also shown dynamism that seems to match that of the north (Figure 1). The presence of Japanese honeysuckle almost doubled from 352 plots (14.69%) in 2006 to 616 plots (25.7%) in 2011, and the number of counties invaded increased from 23 to 41, leaving only two counties un-invaded. The distance between each plot in which Japanese honeysuckle was first detected during the second survey and the nearest plot in which Japanese honeysuckle had been detected during the first survey ranged from 0 km to 75 km, with a mean (±SE) of 16 km (±0.5 km).

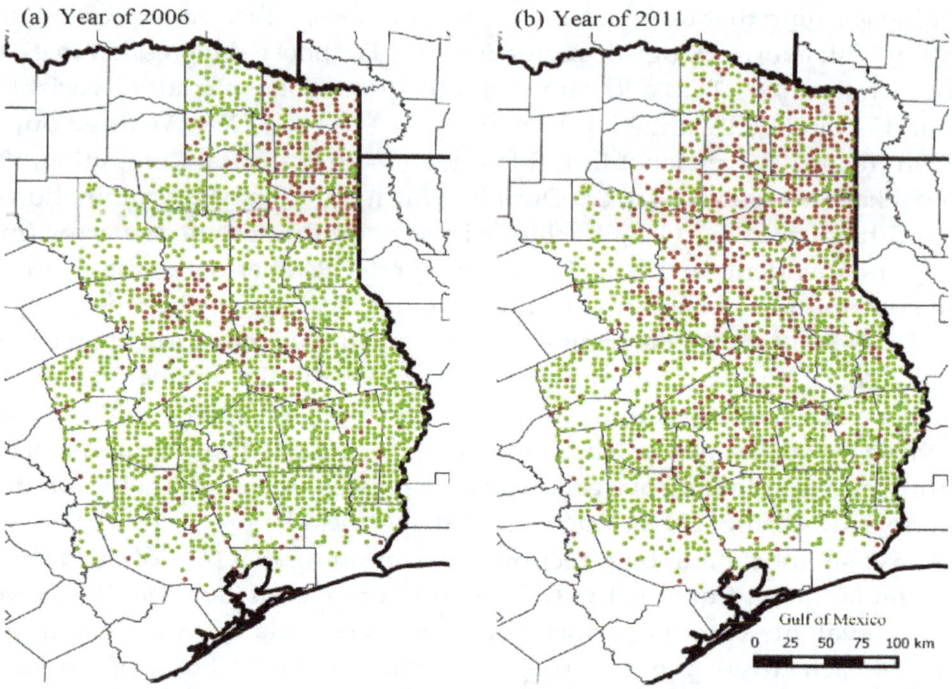

Figure 1. Presence (red dots) and absence (green dots) of Japanese honeysuckle in forested plots sampled in eastern Texas in (**a**) 2006 and (**b**) 2011 as part of the Forest Inventory and Analysis Program of the U.S. Forest Service [25].

3.2. *Potential Determinants of Invasion*

We explored 250 combinations of tree complexity (ranging from three to seven) and learning rate (ranging from 0.001 to 0.05); these produced models with between 800 and 1400 trees. The optimal model had a tree complexity of five, a learning rate of 0.01, and a total of 1100 trees. Moran's I index

(I = 0.002) indicated no statistically significant (p = 0.17) spatial autocorrelation. Model predictive deviance was 0.894 ± 0.015 with 80.6% of the total response variance explained. The AUC score was 0.855 ± 0.008 ("good" ability to discriminate between species presence and absence) and the overall accuracy was 80.8%. Recursive feature elimination tests showed that five variables could be removed from the model before the resulting predictive deviance exceeded the initial predictive deviance of the model with all variables.

Examination of the relative contribution of the predictor variables indicated that the top five accounted for approximately 65% of the contribution in the overall model (Figure 2). Of the five most influential variables, three were forest features and two were climatic conditions. Basal area was the most influential variable, contributing 14.8%. Mean annual maximum temperature, biodiversity index (Shannon's species diversity), stand age, and precipitation were the second, third, fourth, and fifth most important variables, contributing 13.7%, 12.8%, 12.0%, and 11.7%, respectively. Distances to the nearest known propagule source and distance to the nearest road were selected from disturbance factors. Landscape conditions, climatic conditions, forest features, disturbance factors, and forest management activities had total contributions of 4.0%, 32.1%, 44.8%, 17.3%, and 1.7%, respectively (Figure 2).

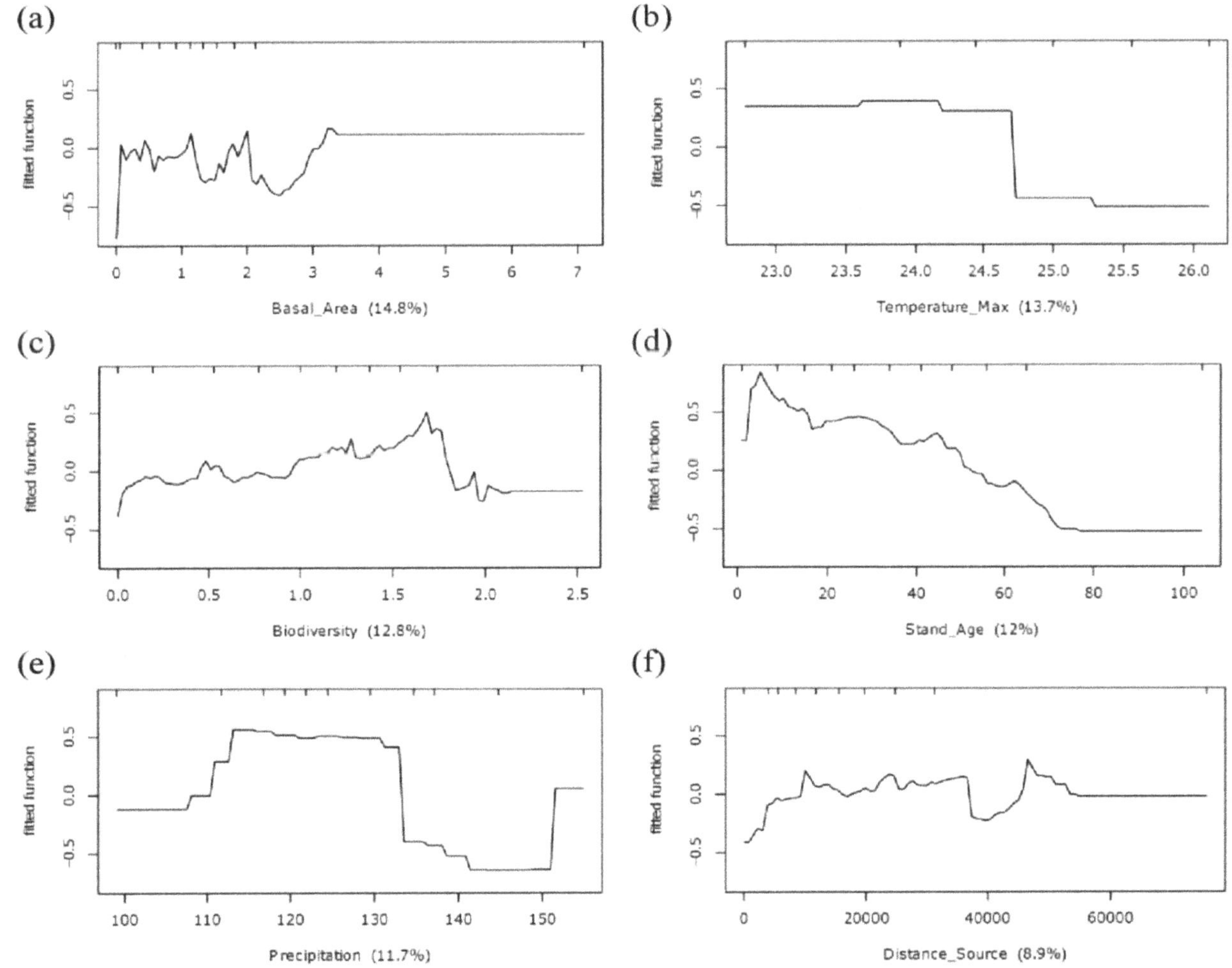

Figure 2. *Cont.*

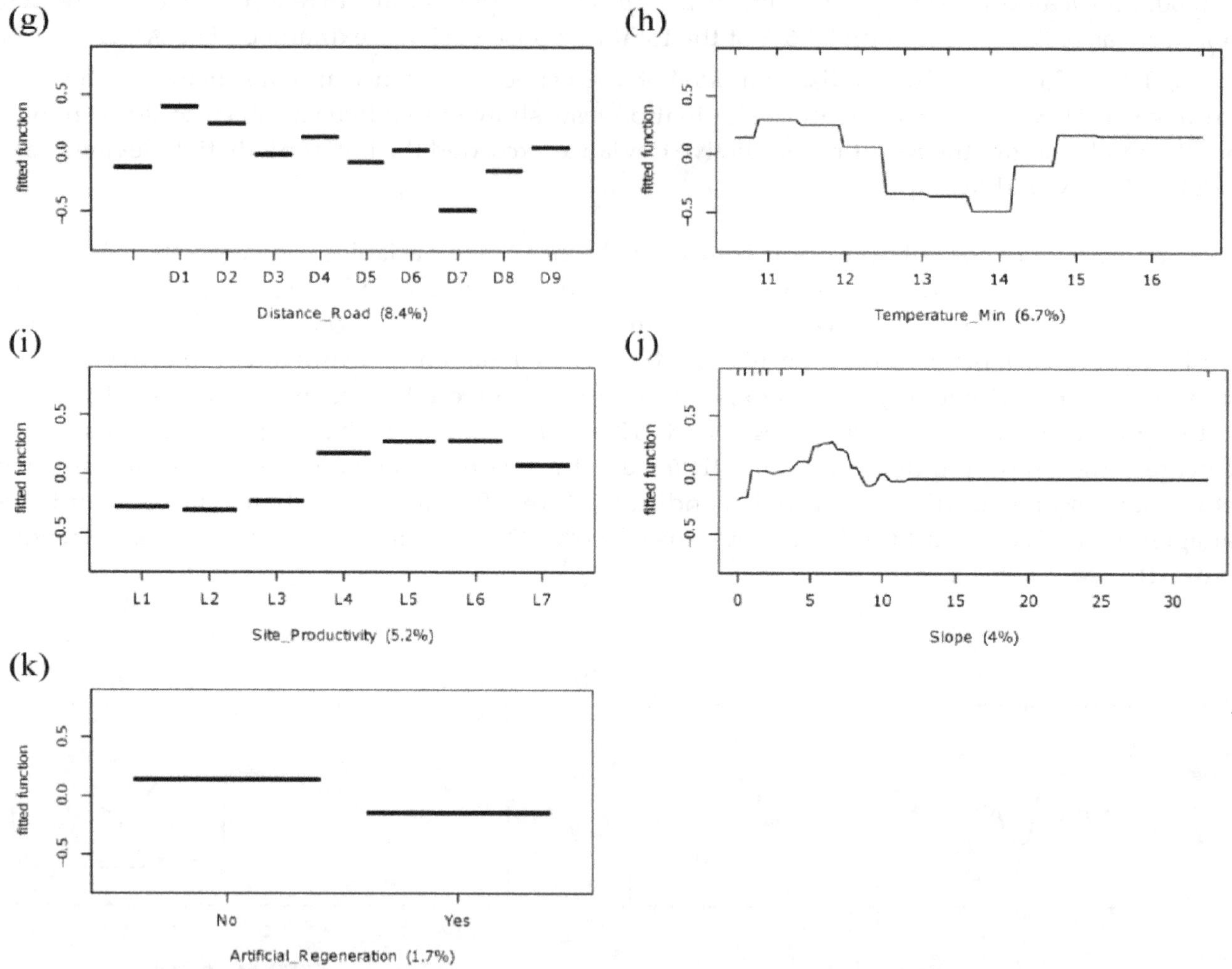

Figure 2. Partial dependence plots for the explanatory variables included in the optimal boosted regression tree models for Japanese honeysuckle based on analyses of the 11 most influential variables. Hash marks at the top of each plot indicate distribution of sample plots along the range of the indicated variable. X axis labels include the relative contributions (%) of each influential variable included in the final model (see Table 1 for the description of variables). Y axes are based on the logit scale used for the indicated variable.

Partial dependence plots indicated that Japanese honeysuckle occurrences were associated with plots with a slope ratio lower than 8% (Figure 2j). Climatic conditions included a mean annual maximum temperature below 24.7 °C (Figure 2b), an annual precipitation between 110 and 133 cm (Figure 2e), and a mean annual minimum temperature between 10.6 °C and 12.5 °C (Figure 2h). Forest features usually included a basal area less than 2 m^2 (Figure 2a), a biodiversity index between 1.0 and 1.8 (Figure 2c), stand age younger than 50 years (Figure 2d), and site productivity higher than 5.95 $m^3 \cdot ha^{-1} \cdot year^{-1}$ (above L3, Figure 2i). Occurrences were more likely in plots with distances to the nearest known propagule less than 35,000 m (Figure 2f) and to the nearest road less than 91 m (below D3, Figure 2g), and with no artificial regeneration within the past five years (Figure 2k). In addition, the fitted values plots (Figure 3) indicated all fitted values in each factor type.

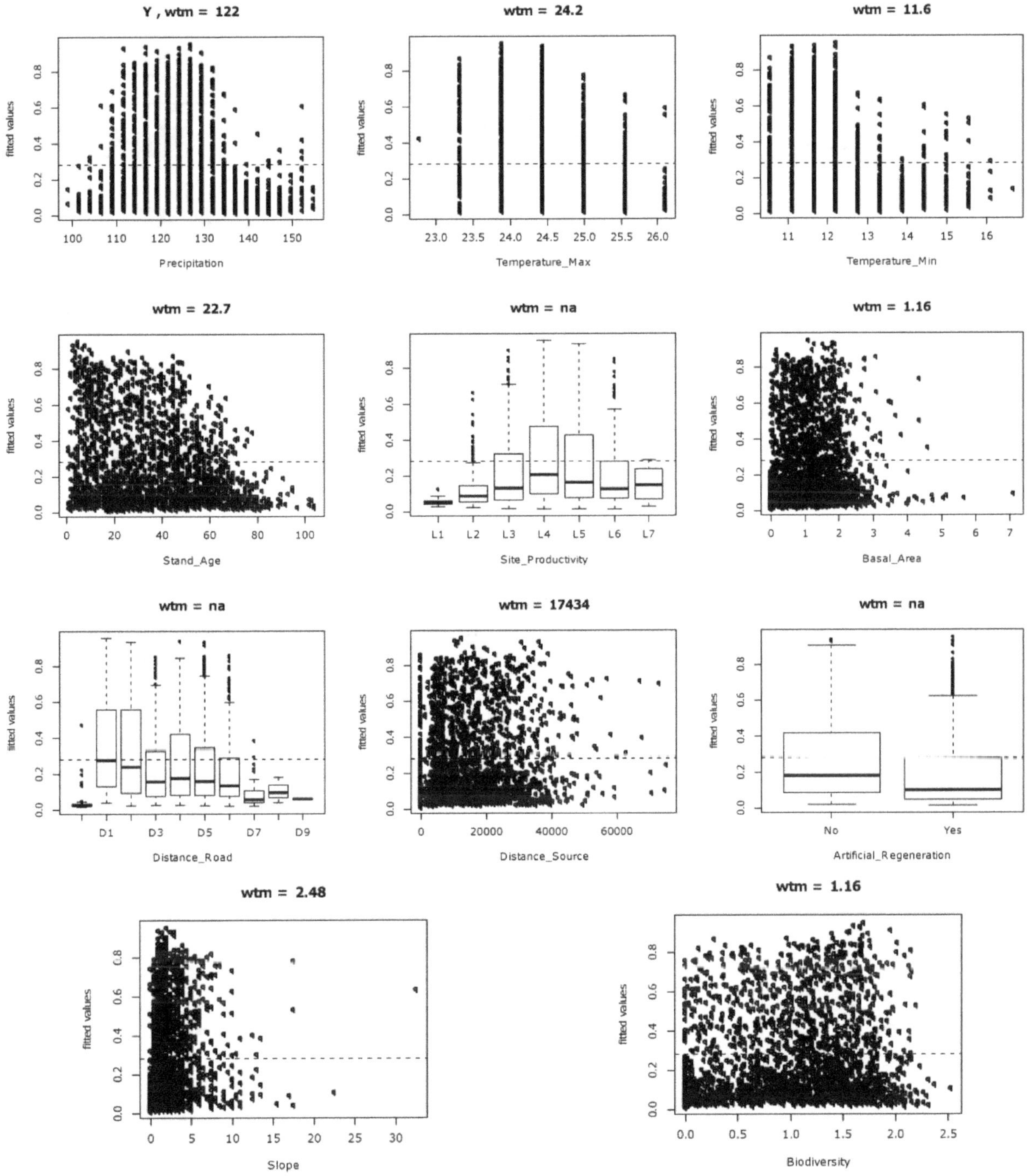

Figure 3. Fitted values plots for the explanatory variables included in the optimal boosted regression tree models for Japanese honeysuckle based on analyses of the 11 most influential variables. Values above each graph indicate the weighted mean of fitted values in relation to each non-factor explanatory variable.

3.3. Likelihood of Invasion under Current Conditions and Different Levels of Management

Estimated probabilities (p) of invasion under current conditions indicated that approximately 70% of the plots fell within the $0 < p \leqslant 0.2$ category, 24% within the $0.2 < p \leqslant 0.4$ category, 5% within the $0.4 < p \leqslant 0.6$ category, and about 1% within the $0.6 < p \leqslant 0.8$ category (Figure 4). The majority of the relatively higher estimated probabilities ($p > 0.4$) were located in northeastern Texas (Figure 5a).

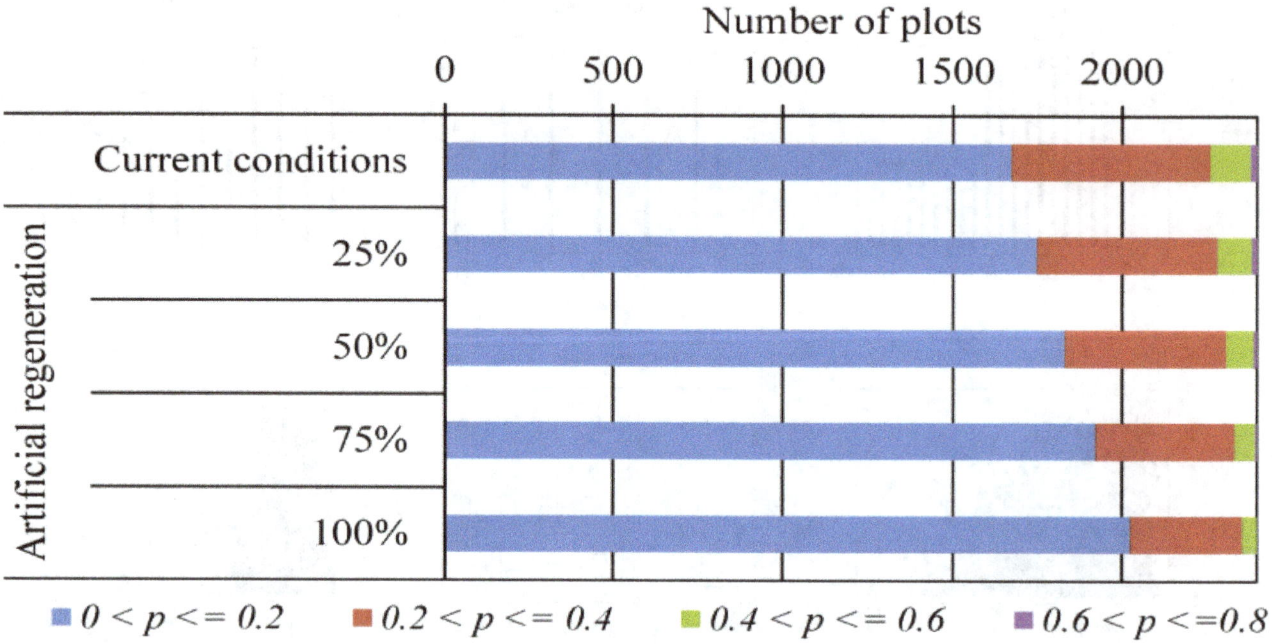

Figure 4. Number of plots within the indicated categories of estimated probability (p) of invasion (blue: $0 < p \leqslant 0.2$, red: $0.2 < p \leqslant 0.4$, green: $0.4 < p \leqslant 0.6$, purple: $0.6 < p \leqslant 0.8$) by Japanese honeysuckle assuming current conditions, and artificial regeneration on 25%, 50%, 75%, or 100% of the plots (chosen at random).

Increasing the level of artificial regeneration from zero to 100% in increments of 25% reduced estimated mean probabilities (p) of invasion from 0.17 to 0.16, 0.15, 0.13, and 0.12, respectively, and decreased the number of plots with $p > 0.2$ from 724 to 648, 564, 474, and 373, respectively, (Figure 4). Once again, the majority of the plots with relatively higher estimated probabilities ($p > 0.2$) were located in northeastern Texas (Figure 5b–e).

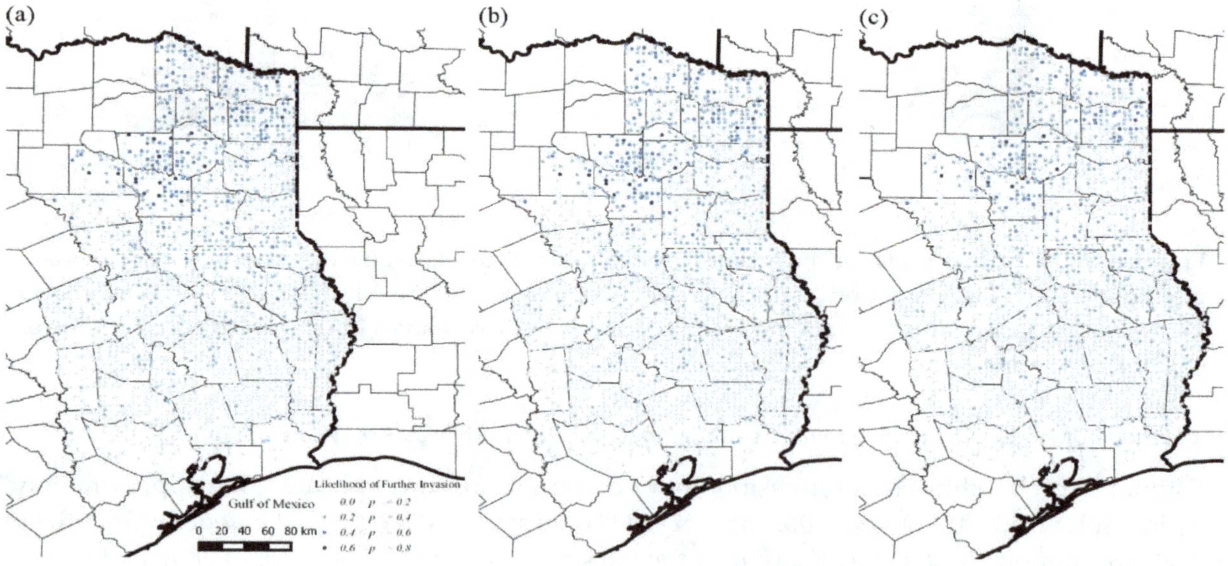

Figure 5. *Cont.*

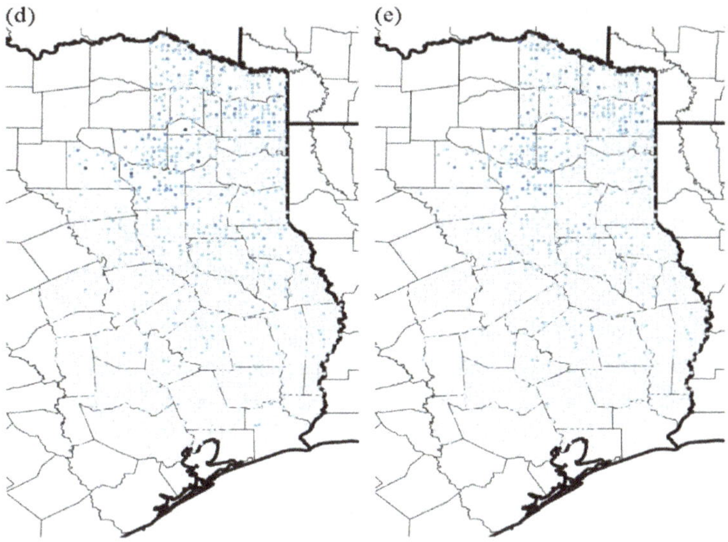

Figure 5. Estimated probabilities of invasion by Japanese honeysuckle in plots sampled in eastern Texas as part of the Forest Inventory and Analysis Program of the USDA Forest Service assuming (**a**) current conditions, (**b**) artificial regeneration on 25%, (**c**) 50%, (**d**) 75%, or (**e**) 100% of the plots (chosen at random).

4. Discussion

The number of sites invaded by Japanese honeysuckle almost doubled within only five years. Clearly, northern sites are far more frequently invaded than those in the south (Figure 1), and the species was identified in, what seems moderate, 25.7% of the sampled sites. However, for a better understanding of the species presence one should perhaps consider that in 2011 less than 5% of the forested counties in eastern Texas remained uninvaded. The species abundance throughout the region seems uneven, but given its ability to spread rapidly, as evidenced by comparison of 2006 *vs.* 2011 presence-absence records, and its current distribution, there is potential for further invasion into other currently uninvaded sites. The change in spatial patterns, as shown by our results, agrees with the earlier work by Wang *et al.* [26] who estimated probabilities of invasion of Japanese honeysuckle throughout the Southern US forestlands based primarily on analysis of data collected during the FIA inventory period that ended in 2006. In particular, they identified areas vulnerable to invasion in eastern Texas, with northeastern sites showing higher probability of invasion. Our index of potential rate of spread, assuming a mean interval of five years between samples on any given plot, suggests a mean potential dispersal velocity of 3200 m/year (16 km/5 years), which is faster than the rates used in the models of Morales and Carlo [41] and Clark *et al.* [42]. Morales and Carlo [41] used a spatially-explicit simulation model to estimate a dispersal velocity of approximately 800 m per year for plants dispersed by frugivores. Clark *et al.* [42] used a stochastic model based on information from a variety of sources to estimate a range of annual velocities of seed dispersal by birds: 90% of the seeds were dispersed 0–100 m from the parent tree, 8% were dispersed 100–500 m, 1.7% were dispersed 500–5000 m and 0.3% were >5000 m. Our index was almost surely biased by exclusion of propagule sources in relatively close, un-sampled areas; the FIA field data were collected from a lattice of 4047-m^2 hexagons, with only one sample plot located randomly within each hexagon [29]. On the other hand, given that seeds of Japanese honeysuckle are dispersed by wide-ranging animals, such as white-tailed deer, a relatively high dispersal velocity is not surprising.

While several factors affect a site's susceptibility to Japanese honeysuckle invasion, forest features and climate conditions seem to play predominant roles. The majority of the species occurrences have been noted on sites with very low basal area, although the exact impact of basal area is somewhat ambiguous considering the interlacing positive and negative effects upon the fitted function along the variable's distribution. A clear and consistent effect, however, may be observed in the case of other major contributors from the forest features group of variables. In particular, the youngest stands are most invasion-prone, and the susceptibility gradually decreases with age. High site productivity appears to favor invasions, as well. Logically, relatively young and highly-productive sites provide favorable conditions for any species, native and invasive ones alike [43]. The main factor identified as a limitation to mature forest stand invasions by Japanese honeysuckle was limited light availability, even though the species can tolerate a relatively low light environment [44,45]. Relatively open habitats facilitate early age flowering and prolific seed production. Halls [46] showed that Japanese honeysuckle in eastern Texas reproduced fruit at three years of age when plants were grown in the open, but at age five when grown in the shade. In addition, signs of stress developed after two years of growth under 8% of ambient light [47]. Even though new growth was initiated each spring, leaders would subsequently die back and a portion of the current leaf crop would abscise following maturation of the flush [47].

Japanese honeysuckle invasions were more abundant on sites with medium-high biodiversity index values. Although it seems counterintuitive that a new species would successfully establish itself in a diverse and likely balanced community (a classic view by Elton; [48]), post-invasion dynamics within this community as well as fate of the invader remain unknown. Debate is ongoing with arguments and evidence both supporting and opposing the classic perspective, however [48–51]. This disagreement also was noted between theoretical analyses and field studies (reviewed in [52]). In a recent case study, Wang and Grant [53] noted that sites with dominant loblolly pine overstory were less diverse than mixed evergreen-deciduous and mostly-deciduous stands. They suggested that the evergreen canopy was the likely factor diminishing risk of encroachment by invasive shrubs due to the decreased amount of light reaching the understory. On the contrary, the mixed canopy partially opens up from fall to early spring allowing for more light to reach the understory during that period. Indeed, the interspecies feedback loops [54,55] that characterize ecological communities challenge us to focus on the processes by which resident species affect the availability of physical resources, such as light, with which the likelihood of species invasions often are correlated.

Apart from favorable forest features, climate conditions of eastern Texas forest stands also appear to facilitate Japanese honeysuckle invasions. Leatherman [15] reported that Japanese honeysuckle in Eastern North America expands generally south of an isotherm where mean January temperature is −1 °C, north of an isotherm where only 5% of January daily low temperatures are <0 °C, and east of a mean annual precipitation of 1016 mm. Although Japanese honeysuckle may tolerate cool winter temperatures and regrow vigorously the following summer, late spring frosts that damage new growth and short growing seasons limit the northern distribution [13,56]. Sasek and Strain [57] suggested that projected future climate change may allow Japanese honeysuckle to expand its northern range; however, its southern distribution may be limited by mild winter temperatures that are insufficient for seed stratification. Our simulated intensity of invasion generally decreased with increasing mean annual temperature and decreasing annual precipitation.

Among the investigated disturbance factors, proximities of the nearest known propagule source and road were significantly correlated with the distribution of Japanese honeysuckle. Propagule pressure has been shown to affect the distribution and abundance of many invasive plants [58,59] and, like many other invasive plants, recruitment limitation tends to slow the range expansion of Japanese honeysuckle [60,61]. These results support the concept that invasions are best mitigated during the early stages of recruitment and establishment [62,63] or via control of outlier populations [64]. The control of biological invasions has become a topic discussed at the highest scientific, policy and management levels [65]. While long-term invasive plant management strategies are recognized as essential to reduce the ecological damage and economic costs associated with invasions, the development and implementation of such plans lags considerably behind [66]. Most of invasive plant management strategies, including that of Japanese honeysuckle, have emphasized controlling highly infested areas, despite this knowledge.

Roads have been shown as invasion corridors that facilitate the spread of invasive plants [67,68]. This is especially likely for an invasive vine such as Japanese honeysuckle, capable of plastic responses to heterogeneous habitat and thus of rapid growth in light gaps such as road-sides [69]. If there are no structures available for twining, Japanese honeysuckle can grow in a mat form along the ground, increasing the likelihood that roads could act as corridors for, rather than barriers to, spread. Spread of the species along roads may be additionally augmented by seed dispersal, also across the road, by wildlife. Japanese honeysuckle has been recognized as an important browse and forage species, especially for white-tailed deer, but also for other wildlife [70–72] throughout much of the eastern and southern United States, particularly during poor mast years and during winter when other food sources are scarce or inaccessible [73,74]. For example, in areas of northern Alabama managed primarily for loblolly pine production, Japanese honeysuckle constituted 49.4% of the year-round diet of white-tailed deer, whereas no other single food item accounted for more than 6% [75]. White-tailed deer can move over large distances [76–78] and have been observed utilizing roadways [79,80] and, thus, may be important vectors of Japanese honeysuckle seed dispersal.

The roles of other studied factors, namely landscape conditions and forest management activities, were rather minor. About 75% of the plots in eastern Texas currently occupied by Japanese honeysuckle occur at the northern part of the area, and about 98% occur on lands with a slope ratio less than 8%, where favorable habitats are created on well-drained sites [81]. In general, the species is noticeably absent on coarse sands and poor peat soils [82]. Extensive areas of poorly drained soils have also been linked with the absence of Japanese honeysuckle in southern Florida [15].

Anthropogenic effects on Japanese honeysuckle distribution are multifaceted. However, while roadways may facilitate the spread of the invader, artificial forest regeneration, one of the management practices we considered in this work, was shown effective in reducing the likelihood of Japanese honeysuckle invasions. A positive outcome of artificial regeneration, which generally involves some effort to control woody competition prior to planting, is that these types of sites are at least temporarily less susceptible to Japanese honeysuckle invasion. Approaches like herbicide application, mechanical treatment, or burning, aim to reduce interspecific competition and allow for successful establishment of the newly planted seedlings. However, as a side effect, this step may also reduce biodiversity on artificially regenerated sites in general (mean $H_S = 0.69$ *vs.* $H_S = 1.06$ on sites with and without history of artificial regeneration, respectively). The majority of the industrial plantations are located in southeastern Texas and, consequently, artificial regeneration is more common in this part of the state, with about 30% of the sites being classified in this category, as compared to about 21% of the sites in northeastern Texas, a region with less frequent human intervention and more favorable soil types.

We estimated invasion likelihood by applying boosted regression trees to data on landscape features, climatic conditions, forest conditions, disturbance factors, and forest management activities collected over a five-year period. Two valid criticisms of our approach are that our estimates of invasion likelihood are unique to our method of analysis [83] and to our specification of the variables included in that analysis [84]. These criticisms are generic problems related to structural uncertainty in the mathematical representation of natural systems [85]. We identified most of our potential independent variables following a recent study focused on the invasion of Japanese honeysuckle in forestlands of the southern United States [27]; however, we did not use exactly the same set of independent variables and we used a different modeling approach. The possibility remains that there might be a more powerful model [83] and/or a more useful set of explanatory variables [66]. Evaluation of the relative merits of the different methodological approaches to geographical distribution modeling currently is a topic of much debate [83,86,87], but is beyond the scope of the present study.

In conclusion, our analyses suggest that the range of Japanese honeysuckle in forestlands of eastern Texas is expanding. Even though providing reliable predictions of habitats that are most at risk, distribution limits, and efficacy of management strategies for any invasive species remain a challenge, our model showed the positive impact of selected management practices upon reducing the likelihood of invasion. Our model may also guide forest managers in development of long term monitoring and control strategies for Japanese honeysuckle by facilitating detection and eradication of newly established invasions in eastern Texas.

Acknowledgments: Acknowledgments: The study was supported by the Undergraduate Research Fund of the Department of Wildlife and Fisheries Sciences, Texas A&M University and The Pine Integrated Network: Education, Mitigation, and Adaptation Project (PINEMAP), a Coordinated Agricultural Project funded by the USDA National Institute of Food and Agriculture, Award #2011-68002-30185. We also thank three anonymous reviewers for their time and effort. The manuscript is greatly improved as a result of their comments.

Author Contributions: Author Contributions: This study was designed by Hsiao-Hsuan Wang. Erin K. McGrew organized the data. Tomasz E. Koralewski computed indexes. Hsiao-Hsuan Wang conducted statistical analyses. Hsiao-Hsuan Wang, Tomasz E. Koralewski, William E. Grant and Thomas D. Byram led the writing.

References

1. Moser, W.K.; Barnard, E.L.; Billings, R.F.; Crocker, S.J.; Dix, M.E.; Gray, A.N.; Ice, G.G.; Kim, M.-S.; Reid, R.; Rodman, S.U.; *et al.* Impacts of nonnative invasive species on US forests and recommendations for policy and management. *J. For.* **2009**, *107*, 320–327.
2. Pejchar, L.; Mooney, H.A. Invasive species, ecosystem services and human well-being. *Trends Ecol. Evol.* **2009**, *24*, 497–504. [CrossRef] [PubMed]
3. Pimentel, D.; Zuniga, R.; Morrison, D. Update on the environmental and economic costs associated with alien-invasive species in the United States. *Ecol. Econ.* **2005**, *52*, 273–288. [CrossRef]
4. Wilcove, D.S.; Rothstein, D.; Jason, D.; Phillips, A.; Losos, E. Quantifying threats to imperiled species in the United States. *BioScience* **1998**, *48*, 607–615. [CrossRef]
5. Gurevitch, J.; Padilla, D.K. Are invasive species a major cause of extinctions? *Trends Ecol. Evol.* **2004**, *19*, 470–474. [CrossRef] [PubMed]
6. Wang, H.-H.; Grant, W.E.; Gan, J.; Rogers, W.E.; Swannack, T.M.; Koralewski, T.E.; Miller, J.H.; Taylor, J.W. Integrating spread dynamics and economics of timber production to manage Chinese tallow invasions in southern US forestlands. *PLoS ONE* **2012**, *7*, e33877. [CrossRef] [PubMed]
7. McNulty, S.G.; Moore, J.A.; Iverson, L.; Prasad, A.; Abt, R.; Smith, B.; Sun, G.; Gavazzi, M.; Bartlett, J.; Murray, B.; *et al.* Application of linked regional scale growth, biogeography, and economic models for southeastern United States pine forests. *World Resour. Rev.* **2000**, *12*, 298–320.
8. Linder, E.T. *Biodiversity and Southern Forests*; General Technical Report SRS 75; US Department of Agriculture, Forest Service, Southern Research Station: Asheville, NC, USA, 2004; pp. 303–306.
9. Hayes, D. Earth day 1990: Threshold of the green decade. *World Policy J.* **1990**, *7*, 289–304.
10. Reubens, B.; Poesen, J.; Danjon, F.; Geudens, G.; Muys, B. The role of fine and coarse roots in shallow slope stability and soil erosion control with a focus on root system architecture: A review. *Trees* **2007**, *21*, 385–402. [CrossRef]
11. Schierenbeck, K.A. Japanese honeysuckle (*Lonicera japonica*) as an invasive species; history, ecology, and context. *Crit. Rev. Plant Sci.* **2004**, *23*, 391–400. [CrossRef]
12. Natural Resources Conservation Service (NRCS). The PLANTS Database. Available online: http://plants.usda.gov (accessed on 1 July 2015).
13. Hardt, R.A. Japanese honeysuckle: From "one of the best" to ruthless pest. *Arnoldia* **1986**, *46*, 27–34.
14. Andrews, E.F. The Japanese honeysuckle in the eastern United States. *Torreya* **1919**, *19*, 37–43.
15. Leatherman, A.D. Ecological Life-History of *Lonicera japonica* Thunb. Ph.D. Thesis, University of Tennessee, Knoxville, TN, USA, 1955.
16. Schierenbeck, K.A. Comparative ecological and genetic studies between a native (*Lonicera sempervirens* L.) and an introduced congener (*L. japonica* Thunb.). Ph.D. Thesis, Washington State University, Pullman, WA, USA, 1992.
17. West, N.M.; Gibson, D.J.; Minchin, P.R. Microhabitat analysis of the invasive exotic liana *Lonicera japonica* Thunb. *J. Torrey Bot. Soc.* **2010**, *137*, 380–390. [CrossRef]
18. Garrison, G.A.; Bjugstad, A.J.; Duncan, D.A.; Lewis, M.E.; Smith, D.R. *Vegetation and Environmental Features of Forest and Range Ecosystems*; Forest Service, Department of Agriculture: Washington, DC, USA, 1977; p. 68.

19. Nyboer, R. Vegetation management guideline—Japanese honeysuckle (*Lonicera japonica* Thunb.). *Nat. Areas J.* **1992**, *12*, 217–218.
20. Oosting, H.J.; Livingston, R.B. A resurvey of a loblolly pine community twenty-nine years after ground and crown fire. *Bull. Torrey Bot. Club* **1964**, *91*, 387–395. [CrossRef]
21. Yurkonis, K.A.; Meiners, S.J. Invasion impacts local species turnover in a successional system. *Ecol. Lett.* **2004**, *7*, 764–769. [CrossRef]
22. McLemore, B.F. *Comparison of Three Methods for Regenerating Honeysuckle-Infested Openings in Uneven-Aged Loblolly Pine Stands*; Southern Forest Experiment Station, Forest Service, US Department of Agriculture: Atlanta, GA, USA, 1985.
23. Wear, D.N.; Greis, J.G. Southern forest resource assessment: Summary of findings. *J. For.* **2002**, *100*, 6–14.
24. McKeand, S.; Mullin, T.; Byram, T.; White, T. Deployment of genetically improved loblolly and slash pines in the south. *J. For.* **2003**, *101*, 32–37.
25. United States Department of Agriculture (USDA). Southern Nonnative Invasive Plant data Extraction Tool (SNIPET). Available online: http://srsfia2.fs.fed.us/data_center/index.shtml (accessed on 1 July 2015).
26. Wang, H.-H.; Wonkka, C.L.; Grant, W.E.; Rogers, W.E. Potential range expansion of Japanese honeysuckle (*Lonicera japonica* Thunb.) in southern US forestlands. *Forests* **2012**, *3*, 573–590. [CrossRef]
27. Gan, J.; Smith, C.T. Co-benefits of utilizing logging residues for bioenergy production: The case for East Texas, USA. *Biomass Bioenergy* **2007**, *31*, 623–630. [CrossRef]
28. Rudis, V.A.; Gray, A.; McWilliams, W.; O'Brien, R.; Olson, C.; Oswalt, S.; Schulz, B. Regional Monitoring of Nonnative Plant Invasions with the Forest Inventory and Analysis program, Proceedings of the Sixth Annual FIA Symposium, Denver, CO, USA, 21–24 September 2006; McRoberts, R.E., Reams, G.A., Eds.; General Technical Report WO-70. USDA Forest Service: Denver, CO, USA, 2006; pp. 49–64.
29. Bechtold, W.A.; Patterson, P.L. *The Enhanced Forest Inventory and Analysis Program—National Sampling Design and Estimation Procedures*; General Technical Report SRS-80; Southern Research Station, Forest Service, US Department of Agriculture: Asheville, NC, USA, 2005; p. 14.
30. United States Department of Agriculture (USDA). *The Forest Inventory and Analysis Database: Database Description and Users Manual Version 5.1*; US Department of Agriculture Forest Service: Arlington, VA, USA, 2011; pp. 12–15.
31. Lemke, D.; Hulme, P.E.; Brown, J.A.; Tadesse, W. Distribution modelling of Japanese honeysuckle (*Lonicera japonica*) invasion in the Cumberland Plateau and Mountain Region, USA. *For. Ecol. Manag.* **2011**, *262*, 139–149. [CrossRef]
32. Lemke, D.; Schweitzer, C.J.; Tadesse, W.; Wang, Y.; Brown, J.A. Geospatial assessment of invasive plants on reclaimed mines in Alabama. *Invasive Plant Sci. Manag.* **2013**, *6*, 401–410. [CrossRef]
33. Daly, C.; Halbleib, M.; Smith, J.I.; Gibson, W.P.; Doggett, M.K.; Taylor, G.H.; Curtis, J.; Pasteris, P.P. Physiographically sensitive mapping of climatological temperature and precipitation across the conterminous United States. *Int. J. Climatol.* **2008**, *28*, 2031–2064. [CrossRef]
34. Belda, E.J.; Sánchez, A. Seabird mortality on longline fisheries in the western Mediterranean: Factors affecting bycatch and proposed mitigating measures. *Biol. Conserv.* **2001**, *98*, 357–363. [CrossRef]
35. Wills, C.; Condit, R.; Foster, R.B.; Hubbell, S.P. Strong density- and diversity-related effects help to maintain tree species diversity in a neotropical forest. *Proc. Natl. Acad. Sci. USA* **1997**, *94*, 1252–1257. [CrossRef] [PubMed]
36. Filipescu, C.N.; Comeau, P.G. Competitive interactions between aspen and white spruce vary with stand age in boreal mixedwoods. *For. Ecol. Manag.* **2007**, *247*, 175–184. [CrossRef]
37. Elith, J.; Leathwick, J.R.; Hastie, T. A working guide to boosted regression trees. *J. Anim. Ecol.* **2008**, *77*, 802–813. [CrossRef] [PubMed]
38. Ridgeway, G. Generalized Boosted Regression Models. Documentation on the R Package 'gbm', Version 1-5-7. Available online: http://www.i-pensieri.com/gregr/gbm.shtml (accessed on 4 April 2012).
39. Friedman, J.H. Stochastic gradient boosting. *Comput. Stat. Data Anal.* **2002**, *38*, 367–378. [CrossRef]
40. Hosmer, D.W.; Lemeshow, S. *Applied Logistic Regression*; John Wiley and Sons, Inc.: New York, NY, USA, 2000.
41. Morales, J.M.; Carlo, T.A. The effects of plant distribution and frugivore density on the scale and shape of dispersal kernels. *Ecology* **2006**, *87*, 1489–1496. [CrossRef]
42. Clark, J.S.; Lewis, M.; McLachlan, J.S.; HilleRisLambers, J. Estimating population spread: What can we forecast and how well? *Ecology* **2003**, *84*, 1979–1988. [CrossRef]

43. Davies, K.F.; Harrison, S.; Safford, H.D.; Viers, J.H. Productivity alters the scale dependence of the diversity-invasibility relationship. *Ecology* **2007**, *88*, 1940–1947. [CrossRef] [PubMed]
44. Baars, R.; Kelly, D. Survival and growth responses of native and introduced vines in New Zealand to light availability. *N. Z. J. Bot.* **1996**, *34*, 389–400. [CrossRef]
45. Carter, G.A.; Teramura, A.H. Vine photosynthesis and relationships to climbing mechanics in a forest understory. *Am. J. Bot.* **1988**, *75*, 1011–1018. [CrossRef]
46. Halls, L.K. *Flowering and Fruiting of Southern Browse Species*; US Department of Agriculture, Forest Service, Southern Forest Experiment Station: New Orleans, LA, USA, 1973; p. 10.
47. Blair, R.M. Growth and nonstructural carbohydrate content of southern browse species as influenced by light intensity. *J. Range Manag.* **1982**, *35*, 756–760. [CrossRef]
48. Elton, C.S. *The Ecology of Invasions by Plants and Anmials*; University of Chicago Press: Chicago, IL, USA, 1958.
49. Hector, A.; Dobson, K.; Minns, A.; Bazeley-White, E.; Hartley Lawton, J. Community diversity and invasion resistance: An experimental test in a grassland ecosystem and a review of comparable studies. *Ecol. Res.* **2001**, *16*, 819–831. [CrossRef]
50. Levine, J.M. Species diversity and biological invasions: Relating local process to community pattern. *Science* **2000**, *288*, 852–854. [CrossRef] [PubMed]
51. Lyons, K.G.; Schwartz, M.W. Rare species loss alters ecosystem function—Invasion resistance. *Ecol. Lett.* **2001**, *4*, 358–365. [CrossRef]
52. Levine, J.M.; D'Antonio, C.M. Elton revisited: A review of evidence linking diversity and invasibility. *Oikos* **1999**, *87*, 15–26. [CrossRef]
53. Wang, H.-H.; Grant, W.E. Determinants of Chinese and European privet (*Ligustrum sinense* and *Ligustrum vulgare*) invasion and likelihood of further invasion in southern US forestlands. *Invasive Plant Sci. Manag.* **2012**, *5*, 454–463. [CrossRef]
54. Richardson, D.M.; Pyšek, P.; Rejmánek, M.; Barbour, M.G.; Panetta, F.D.; West, C.J. Naturalization and invasion of alien plants: Concepts and definitions. *Divers. Distrib.* **2000**, *6*, 93–107. [CrossRef]
55. Mitchell, C.E.; Agrawal, A.A.; Bever, J.D.; Gilbert, G.S.; Hufbauer, R.A.; Klironomos, J.N.; Maron, J.L.; Morris, W.F.; Parker, I.M.; Power, A.G.; *et al.* Biotic interactions and plant invasions. *Ecol. Lett.* **2006**, *9*, 726–740. [CrossRef] [PubMed]
56. Pelczar, R. When vigor is not a virtue: Beware the beautiful but invasive Japanese honeysuckle. *Horticulture* **1995**, *73*, 64–66.
57. Sasek, T.W.; Strain, B.R. Implications of atmospheric CO_2 enrichment and climatic change for the geographical distribution of two introduced vines in the USA. *Clim. Chang.* **1990**, *16*, 31–51. [CrossRef]
58. Holle, B.V.; Simberloff, D. Ecological resistance to biological invasion overwhelmed by propagule pressure. *Ecology* **2005**, *86*, 3212–3218. [CrossRef]
59. Merritt, D.M.; Nilsson, C.; Jansson, R. Consequences of propagule dispersal and river fragmentation for riparian plant community diversity and turnover. *Ecol. Monogr.* **2010**, *80*, 609–626. [CrossRef]
60. Primack, R.B.; Miao, S.L. Dispersal can limit local plant distribution. *Conserv. Biol.* **1992**, *6*, 513–519. [CrossRef]
61. Tilman, D. Community invasibility, recruitment limitation, and grassland biodiversity. *Ecology* **1997**, *78*, 81–92. [CrossRef]
62. Moody, M.E.; Mack, R.N. Controlling the spread of plant invasions: The importance of nascent foci. *J. Appl. Ecol.* **1988**, *25*, 1009–1021. [CrossRef]
63. Sheley, R.L.; Mangold, J.M.; Anderson, J.L. Potential for successional theory to guide restoration of invasive plant-dominated rangeland. *Ecol. Monogr.* **2006**, *76*, 365–379. [CrossRef]
64. Miller, J.H.; Schelhas, J. Adaptive collaborative restoration: A key concept for invasive plant management. In *Invasive Plants and Forest Ecosystems*; Kohli, R.K., Singh, J.P., Eds.; CRC Press: Boca Raton, FL, USA, 2008; pp. 251–265.
65. Lodge, D.M.; Williams, S.; MacIsaac, H.J.; Hayes, K.R.; Leung, B.; Reichard, S.; Mack, R.N.; Moyle, P.B.; Smith, M.; Andow, D.A.; *et al.* Biological invasions: Recommendations for US policy and management. *Ecol. Appl.* **2006**, *16*, 2035–2054. [CrossRef]
66. Wang, H.-H.; Swannack, T.M.; Grant, W.E.; Gan, J.; Rogers, W.E.; Koralewski, T.E.; Miller, J.H.; Taylor, J.W. Predicted range expansion of Chinese tallow tree (*Triadica sebifera*) in forestlands of the southern United States. *Divers. Distrib.* **2011**, *17*, 552–565. [CrossRef]

67. Gelbard, J.L.; Belnap, J. Roads as conduits for exotic plant invasions in a semiarid landscape. *Conserv. Biol.* **2003**, *17*, 420–432. [CrossRef]
68. Parendes, L.A.; Jones, J.A. Role of light availability and sispersal in exotic plant invasion along roads and streams in the H.J. Andrews Experimental Forest, Oregon. *Conserv. Biol.* **2000**, *14*, 64–75. [CrossRef]
69. Schweitzer, J.A.; Larson, K.C. Greater morphological plasticity of exotic honeysuckle species may make them better invaders than native species. *J. Torrey Bot. Soc.* **1999**, *126*, 15–23. [CrossRef]
70. Pitts, T.D. Foods of eastern bluebirds during exceptionally cold weather in Tennessee. *J. Wildl. Manag.* **1979**, *43*, 752–754. [CrossRef]
71. Suthers, H.B.; Bickal, J.M.; Rodewald, P.G. Use of successional habitat and fruit resources by songbirds during autumn migration in central New Jersey. *Wilson Bull.* **2000**, *112*, 249–260. [CrossRef]
72. Halls, L.K. Managing deer habitat in loblolly-shortleaf pine forest. *J. For.* **1973**, *71*, 752–757.
73. Nixon, C.M.; McClain, M.W.; Russell, K.R. Deer food habits and range characteristics in Ohio. *J. Wildl. Manag.* **1970**, *34*, 870–886. [CrossRef]
74. Sotala, D.J.; Kirkpatrick, C.M. Foods of white-tailed deer, *Odocoileus virginianus*, in Martin County, Indiana. *Am. Midl. Nat.* **1973**, *89*, 281–286. [CrossRef]
75. Sheldon, J.J.; Causey, K. Deer habitat management—Availability use of Japanese honeysuckle by white-tailed deer. *J. For.* **1974**, *72*, 286–287.
76. Michael, E.D. Movements of white-tailed deer on the Welder Wildlife Refuge. *J. Wildl. Manag.* **1965**, *29*, 44–52. [CrossRef]
77. Inglis, J.M.; Hood, R.E.; Brown, B.A.; DeYoung, C.A. Home range of white-tailed deer in Texas coastal prairie brushland. *J. Mammal.* **1979**, *60*, 377–389. [CrossRef]
78. Thomas, J.W.; Teer, J.G.; Walker, E.A. Mobility and home range of white-tailed deer on the Edwards Plateau in Texas. *J. Wildl. Manag.* **1964**, *28*, 463–472. [CrossRef]
79. Sage, R.W., Jr.; Tierson, W.C.; Mattfeld, G.F.; Behrend, D.F. White-tailed deer visibility and behavior along forest roads. *J. Wildl. Manag.* **1983**, *47*, 940–953. [CrossRef]
80. Kilgo, J.C.; Labisky, R.F.; Fritzen, D.E. Influences of hunting on the behavior of white-tailed deer: Implications for conservation of the Florida panther. *Conserv. Biol.* **1998**, *12*, 1359–1364. [CrossRef]
81. Conner, W.; Stanturf, J.A.; Gardiner, E.S.; Schweitzer, C.J.; Ezell, A.W. Recognizing and overcoming difficult site conditions for afforestation of bottomland hardwoods. *Ecol. Restor.* **2004**, *22*, 183–193.
82. Jackson, L.W. Honeysuckles. *Shrubs and Vines for Northeastern Wildlife*; Gill, J.D., Healy, W.M., Eds.; US Department of Agriculture, Forest Service, Northeastern Forest Experiment Station: Upper Darby, PA, USA, 1974; pp. 71–82.
83. Elith, J.; Graham, C.H. Do they? How do they? WHY do they differ? On finding reasons for differing performances of species distribution models. *Ecography* **2009**, *32*, 66–77. [CrossRef]
84. Agresti, A. *An Introduction to Categorical Data Analysis*; John Wiley and Sons, Inc.: Hoboken, NJ, USA, 2007.
85. Walters, C.J. *Adaptive Management of Renewable Resources*; Macmillan Publishing Company: New York, NY, USA, 1986.
86. Barry, S.; Elith, J. Error and uncertainty in habitat models. *J. Appl. Ecol.* **2006**, *43*, 413–423. [CrossRef]
87. Elith, J.; Leathwick, J.R. Species distribution models: Ecological explanation and prediction across space and time. *Annu. Rev. Ecol. Evol. Syst.* **2009**, *40*, 677–697. [CrossRef]

13

Forestry Policy, Conservation Activities and Ecosystem Services in the Remote Misuku Hills of Malawi

Christopher Coutts, Tisha Holmes and April Jackson

[1] Department of Urban and Regional Planning, Florida State University, Tallahassee, FL 32306, USA; ttholmes@fsu.edu (T.H.); ajackson5@fsu.edu (A.J.)

[2] Department of Built Environment, Mzuzu University, Mzuzu 2, Malawi

* Correspondence: ccoutts@fsu.edu

Abstract: Research Highlights: Most of Malawi's land area has been deforested; however, expansive indigenous forests remain in the remote Misuku Hills in Malawi's northern region. Despite its conservation potential, this region of Malawi has been overlooked in forestry conservation research. Background and Objectives: The Misuku Hills is one the most floristically diverse regions in Malawi, but this region is facing similar pressures and forestry policy enforcement challenges that drive deforestation of other regions. This study therefore addresses the questions: What are the forestry policy challenges and opportunities for forest conservation in Malawi? What conservation activities are taking place in the Misuku Hills in support of these policies? What ecosystem services are residents using that are in need of protection? Materials and Methods: A comprehensive inventory and review of the national forest policies and current programs in the Misuku Hills region was compiled through document reviews and communications with governmental and non-governmental stakeholders. A Photovoice exercise was conducted with residents of Chikutu village to create an inventory of resident-identified ecosystem services. Results: While there is an impressive array of policies in place to protect the forests of Malawi, there is little institutionalization or enforcement of these policies. There have been funded conservation programs in the Misuku Hills, but these have been limited to the areas surrounding the three small public forest reserves. The Photovoice exercise revealed that residents rely on an abundance of forest ecosystem services to support their livelihoods, including food, medicine, and timber products. Conclusions: The challenges to conserving forests and their ecosystem services are being met at a local level in a variety of creative ways in the Misuku Hills (e.g., tree planting, beekeeping) that could be used as community-based models for other areas in Africa and elsewhere, where people depend directly on these services to meet daily needs.

Keywords: forests; environment; Malawi; ecosystem services; Photovoice; conservation; policy; community-based forest management; participatory forest management

---

1. Introduction

1.1. Rates of Deforestation in Malawi

Deforestation rates in the tropics are increasing after a promising lag supported by conservation efforts and good governance measures during the 1990s and early 2000s [1]. Increased deforestation is particularly concerning in Africa, where land is rapidly converted to support growing populations and expansion of small-scale and subsistence agriculture [2–4]. In the southeast African nation of Malawi, concerns are growing around the preservation of forest resources, the livelihoods dependent on these resources, and the dynamics which threaten their viability. The most methodologically forthcoming and rigorous estimate available of the forested land area in Malawi reveals that 26.8% of this small African country is forested (Figure 1), and Malawi experienced a 1.6% aggregated loss of tree cover between 1990 and 2010 (1.4% 1990–2000; 0.2% 2000–2010) [5]. The 1.6% total forest loss over this 20-year period is almost completely explained by the equivalent expansion of agricultural land [5]. Most of the agricultural expansion in Malawi has happened on customary land (defined below), outside the jurisdiction of government-controlled parks and reserves [6].

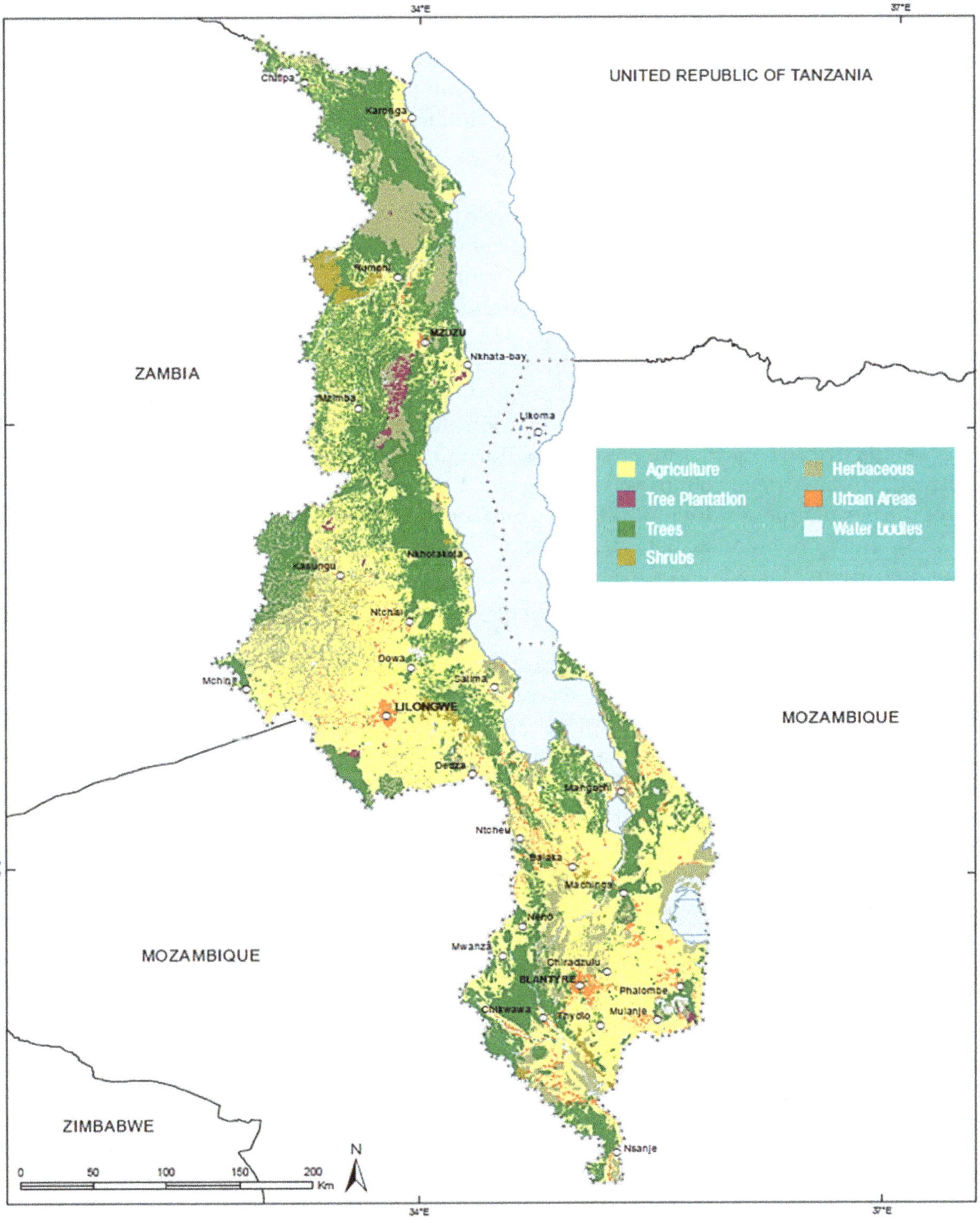

Figure 1. Vegetation map of Malawi. Source: Adapted from FAO (2013).

There are many conflicting estimates of Malawi's forest cover, land tenure regimes in forested areas, and overall deforestation rates. These discrepancies and their potential implications are presented in Appendix A. The deforestation rates above are much less grave than other widely cited estimates. Nonetheless, deforestation continues and the persistent drivers of deforestation provide reason to believe that its consequences will continue to threaten the livelihoods of Malawians who are heavily dependent on forest ecosystem services [7].

The available estimates for the land tenure of forested areas indicate a nearly equal distribution in national parks and game reserves, forest reserves, and on customary land [6,8]. National parks, game reserves, and forest reserves are public, government-controlled lands. Customary land is under Tradition Authority, an indigenous geo-political and socio-economic jurisdiction with customary sovereignty under authority of a Chief [9]. Essentially, all land that is not public or privately owned is customary, and customary land accounts for approximately 85% of the total land area in Malawi [10].

1.2. Drivers of Deforestation

The main drivers of deforestation in Malawi are a rapidly growing and extremely poor population converting forested land to support small-scale subsistence agriculture for food provisioning and income [11] and using wood as a primary energy source.

First, population. The total population of Malawi grew by 35% between 2008 (13,029,498) and 2018 (17,563,749). While this may seem startling, the gross increase in population equates to a current population density of 186 persons/km^2 in this comparatively small African country [12]. There are inconsistencies in various country rankings of population density, but 186 persons/km^2 currently places Malawi at approximately the 75th most densely populated country in the world, still far less dense than other small African countries (e.g., Rwanda, Burundi, The Gambia), but also less dense than some comparatively larger African countries (e.g., Nigeria, Uganda). A major difference between Malawi is that 84% of the population resides in rural areas, making Malawi one of the least urbanized countries in Africa [12,13]. The population is dispersed, placing greater pressures on the entirety of the landscape. Population density in Malawi has been shown to be a significant socio-economic explanatory variable of deforestation in Malawi [14], and population growth is a perceived underlying driver of land use and land cover change in Malawi [15].

Second, agriculture. Malawi's economy and the livelihood of Malawians are heavily dependent on agriculture, which "accounts for about 36% of the Gross Domestic Product (GDP), 87% of the total employment, and supplies more than 65% of the manufacturing sector's raw materials" [13] (p. 4). A 2009 report estimates an expansion of agriculture by 31% between 1975 and 1990, with the majority of expansion coming through the clearing of indigenous forest and woodlands. The report attributes these changes to the need to feed a rapidly growing population and desires to promote economic growth through expansion of agriculture production [7]. Indigenous forests on customary land have been the main source of additional agricultural land [16]. The overwhelmingly rural population derives its livelihood from small land holdings of 0.5–2 ha per household, but it is not agriculture on these small landholdings per se that is the threat to forests, but rather "poor husbandry techniques in the absence of alternative economic opportunities" [7] (Preface).

Third, poverty. Poverty is a critical and compounding factor which exacerbates the pressures of a growing population primarily dependent on subsistence agriculture. Malawi is the second poorest county in the world, as measured by annual per capita income [17]. A lack of income-generating opportunities leads many Malawians to extract natural resources directly from the immediate environment, both for daily household use and to sell for profit. Despite positive attitudes towards tree planting [18,19] and awareness of the negative consequences of a lack of social support for deforestation—making the intention to cut down forest trees generally low—extreme poverty and a lack of alternative income opportunities [20] continue to fuel deforestation.

Fourth, energy. The extraction of resources from forests is the main source of many essential products for the rural poor, most notably fuelwood. In fact, "over 90% of the energy demands of the country for domestic and industrial use are met from wood energy" [7] (p. 2). Trade in firewood and charcoal is the primary source of income for many rural poor and the only form of cooking fuel for 99% of the population [21].

Other extractive industries supported by forest resources also include poles and timber for home construction and hand tools, as well as non-timber forest products such as thatch, mushrooms, caterpillars, bushmeat, beeswax, medicinal plants, and materials for handicrafts [13,16]. Malawi also has a large pool (in terms of both diversity and quantity) of indigenous fruit tree resources [22–24]. These forest food "resources are an insurance against hunger and malnutrition. They provide much needed dietary diversity which avails both macro and micro nutrients necessary for good health" [13] (p. 30).

Although these goods are "free" to Malawians with access to forests, monetary value can be assigned to these products based on what a person without access will pay for it. Using this method, it was estimated that an annual supply of firewood equated to almost half a year's supply of maize (or 416 kg using 1996 prices) for a household, and an annual supply of poles equated to 120 days' supply of maize [25] (as cited in [13]). Using data from the National Statistical Office of Malawi, the Forestry Research Institute reports that "studies on micro-enterprises in Malawi have shown that most people who sell forest produce do so as individuals or as small family operations, start off with little, if any, capital outlay, produce small quantities of mainly unprocessed or crudely processed goods and make little profit" [13] (p. 33). As a result, little to no profit or savings are generated, as monies earned are used to meet immediate domestic needs rather than used for savings and/or investments.

On a national level, forests account for 12% of Malawi's natural capital [26]. A highly conservative underestimate is that Malawi's forests contribute 6.2% of GDP [27]. Another more comprehensive evaluation reveals that forests contribute 8% of GDP with a substantial total economic value and enormous economic contribution to livelihoods [26]. These underestimates do not take into account many environmental protection, goods, and ecosystem services provided by forests. Taking a small subset of the services provided by soil, forests, fisheries, and wildlife into account, it has been estimated that Malawi's GDP would be higher by 5.3% per year (2007 prices) were it not for unsustainable use of these resources. The discounted cost of damage over a 10-year period equated to 21.4% (in 2010) of GDP [28]. Over half (11%) of this 21.4% value was attributed to forests with only wood products, flood prevention, and air pollution services taken into account. The unsustainable use of forest resources and loss of economic value is negatively impacting Malawi's growth, where resources are limited and even small gains in natural and economic resources could have significant impacts on people's ability to meet basic needs.

1.3. Response to Deforestation

The national government recognizes the crucial regulating services (e.g., climate regulation, moderation of extreme weather, soil erosion, and pollination), habitat/supporting services (e.g., biodiversity), and cultural services (e.g., tourism) of forests [29]. Citizens are more aware of the provisioning services such as food, raw materials, and medicine that are acutely tied to their daily lives, but there is increasing awareness of the regulating services forests provide to maintain supplies of fresh water, prevent flooding, protect crops from wind damage, stabilize soil, and avoid excessive siltation of riverbeds downstream.

The largest threat to these services is occurring on customary lands, where ownership or usufruct rights (rights held by a member of the land-holding community in customary freehold [9]) "rest with individual villagers or group of individuals who the customary authority have appointed or delegated temporary ownership to, otherwise all authority rests with the customary leaders" [13] (p. 25). These lands are controlled by people residing in villages and the Traditional Authorities who oversee land tenure and land disputes on customary lands. Little research has been conducted at the local level on forests located on customary land [30]. In line with Malawi's move towards decentralization and community-based forestry, local level interventions are key to understanding country-level trends and policies that address deforestation [31].

1.4. The Misuku Hills Study

The Misuku Hills of Malawi have been recognized as one of the world's Key Biodiversity Areas [32]. There are three small forest reserves in the Misuku Hills, but it is the largest contiguous area of remaining forest in Malawi on customary land outside of public parks and forest reserves. This is most notably due to its comparatively low population. To maintain the integrity of this slowly dwindling forest, the policies that govern forest conservation must be understood for their applicability in this region and their pertinence to local control of forest resources and management of ecosystem services. There is a great need for local ecosystem service assessments in Africa that capture local resources and needs [33].

There is a long history of forestry conservation in Malawi, and the most recent forestry policies recognize deforestation as a cross-cutting issue with various drivers and impacts on a myriad of development goals. In line with the purpose of this special issue of *Forests*, we explore the policies aimed at forest protection in Malawi and discuss the challenges and opportunities faced in their implementation nationally and locally in the Misuku Hills bioregion. The research activities of this inter-disciplinary study are therefore threefold. First, an assessment of national forestry policies was conducted to determine how Malawi's forestry policies have evolved and how current policies aim to guide forestry conservation activities. Second, an inventory of the forestry conservation activities in the Misuku Hills region was conducted to determine how these policies are being applied to protect one of Malawi's last remaining indigenous forests on customary land. Third, an ecosystem services assessment was conducted using a Photovoice methodology in a remote village on customary land in the Misuku Hills.

The results reveal that Malawi has enacted a plethora of forestry policies that have been developed to meet international standards. While these policies should be lauded, the case of the Misuku Hills reveals that the full potential of these policies is underwhelming due to weak implementation and shortcomings in empowering communities to practice local control. The ecosystem services inventory conducted in Chikutu village reveals that community members have extensive knowledge of forest resources, and these resources are still readily available to the indigenous communities. These findings underscore the important role that community-based and indigenous land tenure regimes play in advancing national forest policies. The paper begins with a description of the materials and methods of analysis, followed by the results on national and local forest conservation policies and practices. The paper concludes with an examination of implications for forest management in the Misuku Hills forested region and proposals for future research.

2. Materials and Methods

2.1. Malawi Forestry Policy Evolution and Analysis

The search and retrieval of Malawi government forestry policy and guideline documents was conducted online and through personal communications with government officials, representatives of non-governmental organizations (NGO), and faculty at Mzuzu University. Policy document retrieval was further guided by a previous outline of forestry policies and by contemporary policies and guidelines that referenced previous policies. The search was focused on policies and guidelines directly related to forestry, although policies that included forestry conservation (e.g., environmental policy) were also reviewed. When the search for documents began to reveal no new information, and the body of policies could be considered comprehensive and complete, the policies were then reviewed for changes that have occurred over time. Some of the changes to previous policies were noted in the subsequent policies themselves. The evaluation of policy changes took into account changes in the form of government in Malawi and advancements and best practices in international forestry policies.

2.2. *Conservation Activity Inventory in the Misuku Hills Forested Region*

2.2.1. The Misuku Hills Forested Region Study Area

The Misuku Hills are part of the southern end and western branch of the East African Rift System within the Southern Rift Montane bioregion [34] and the Miombo woodlands ecoregion. The Misuku Hills forest, like other remaining forests in Malawi, is in upland and hilly areas and along the rift valley scarps [7].

The Miombo woodlands of the Misuku Hills are the most floristically diverse in the country [35], encompassing the vegetative/biotic communities of open canopy woodland of hills and scarps (*Brachystegia spp.*), open canopy woodland of plateau (*Brachystegia/Julbernardia/Isoberlinia*), closed canopy woodland of wetter uplands (tall *Brachystegia spp.*), and montane evergreen forest [36]. Within the Misuku Hills, there are three public forest reserves (FR): Matipa FR (1055 ha), Mughese FR (771 ha), and Wilindi FR (937 ha), all established in 1948 [6]. These forest reserves are largely composed of montane evergreen forest. The reserve land area comprises only a small fraction of the larger contiguous forest referred to in this study as the Misuku Hills forested region (MHFR). The MHFR spans a number of political and administrative boundaries, and data from these administrative units are presented below to provide context to the MHFR study area (Figure 2).

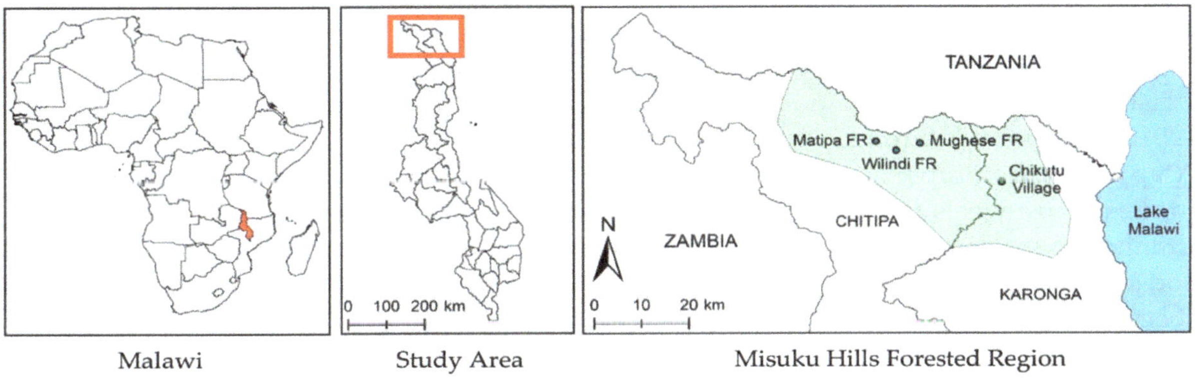

Figure 2. Location of the Misuku Hills forested region (MHFR) study area.

The area commonly identified as the Misuku Hills is within the administrative territory of Traditional Authority (TA) Mwenemisuku, but, from a biotic perspective, the contiguous forest of the MHFR spans two TAs and the country borders of Malawi and Tanzania (where they are referred to as the Umalila mountains). The focus of this study is confined to MHFR in Malawi.

Malawi is divided into three administrative regions (north, central, south), and the MHFR is one the most remote areas in the northern region of Malawi, where the most dramatic and highest concentration of hills and scarps are found. Since the country's independence in 1964, there has been a gradual migration of people to the northern region [37], but the effects of this migration on development in the region have been minimal. Only 13% (2,286,960) of the total population of Malawi lives in the northern region [12]. Not surprisingly, the northern region remains by far the most forested, with 48.7% tree cover as compared to the central (26.3%) and south (27.3%) regions [5]. Compared to Malawi's two other regions, the northern region has over double the rate of those reporting use of forest resources from their own land (9.9%), the highest proportion of forest resources use from communal land (18.2%), and the lowest proportion of those buying forest products from someone else (44.7%) [38]. Table 1 depicts the proportion of use and source of forest products in these two districts [38].

Table 1. Proportion and source of forest resources in study area districts.

District	Proportion of Enterprises Selling Forest Products	Own Land	Forest/Wild Park Reserve	Communal Land	Purchased from Someone
Chitipa	11.5	12.2	37.8	8.9	41.2
Karonga	12.3	15.0	27.7	41.6	15.7

Malawi's three regions are divided into a number of districts. The MHFR spans the borders of two of the seven districts within the northern region, Chitipa and Karonga. The population density of Chitipa (54/km^2) and Karonga (107/km^2) are both well below the country average of 186/km^2 [12]. This population density estimate for Karonga exaggerates the population density in its forested areas. Karonga is topographically unique, as it is split north to south between coastal lake plains to the east and hills and scarps to the west. The overwhelming majority of the population resides in the lake plains. The population density in the hills and scarps of Karonga is similar to that of Chitipa district.

It is difficult to accurately estimate the population of the MHFR study area, as it overlaps census units, but most of the MHFR is within the administrative unit of Traditional Authority (TA) Mwenemisuku. Field observations in the area of the MHFR outside TA Mwenemisuku confirm that it is very lightly inhabited. In 2018, the population of TA Mwenemisuku was estimated at 25,816 [12]. Strangely, this is approximately 2000 fewer people than the 2008 census estimate. The land area of TAs was not available, but TA Mwenemisuku is similar in size to other TAs in Chitipa. Since its population is about average for TAs in the district, its population density is likely very close to the very low population density for the district (54/km^2). A hot spot of change analysis reveals that while some districts in the northern region experienced significant forest and natural vegetation loss from 1990–2010, Chitipa and Karonga experienced no dramatic change [5].

Mapping these proportions allows us to view the MHFR as not conforming to administrative boundaries, and reveals that it has the largest contiguous area of community forest and woodlot opportunities in the country, high forest management of natural forest opportunities, and high priority for food security and biodiversity intervention [39]. This leads us to examine the interventions that have been conducted in the MHFR to pursue these opportunities and protect its forests.

2.2.2. Conservation Activities Inventory

The conservation activities inventory aims to remedy the lack of accurate accounting of the past and current conservation activities in the MHFR. The search for forestry conservation activities in support of forestry policies was conducted through an online search and through personal communications with government officials, representatives of various conservation-oriented NGOs, and faculty at academic institutions in Malawi and the United States (US). The inventory of conservation activities in the Misuku Hills was further guided by a historical summary of community-based forest management activities nationwide [40]. The most insightful sources of information came from NGO project reports and personal communications with representatives of NGOs and sponsoring agencies who have operated in the Misuku Hills, as well as a current US Peace Corps Volunteer who lives and works in the Misuku Hills performing environmental conservation activities.

Since the vast majority of the land area of MHFR is on customary land and not in the montane evergreen public forest reserves, an ecosystem services assessment was conducted in the representative village on customary land to determine the extent and variety of services reaped by residents in these typically overlooked areas within the MHFR.

2.3. *Chikutu Village Ecosystem Services Inventory*

Chikutu village is located in the MHFR to the east of public forest reserves and just outside the eastern boundary of TA Mwenemisiku in TA Kilupula in Karonga district. TA Kilupula, like the aforementioned Karonga district in general, is split north to south between coastal lake plains and hills and scarps. Chikutu is located deep in the lightly populated hills and scarps to the west of the lake plains. It is accessible by an unimproved, and at times impassible, dirt road. In ideal conditions it takes a 4 × 4 capable vehicle or motorbike approximately two hours to scale the hills to reach Chikutu from the main M1 highway paralleling the lake shore to the east.

This area around Chikutu is less populated than the area around the forest reserves to the west and represents one of the last hideouts in Malawi, where residents could directly reap ecosystem services from largely intact natural forests. If there is anywhere in Malawi where Malawians can still reap an array of benefits from a relatively intact natural forest on customary land, this is the place. It provides a lesson for the larger MHFR, the vast majority of which is on customary land.

The unique topographical and population dichotomy of northern Karonga makes Chikutu an interesting case. Chikutu, and other villages in the area, could be facing pressure to deliver forest goods to the much more highly populated areas in the coastal plains that are almost exclusively agricultural lands with greatly diminished forest resources.

Google Maps® designates this area as the Matipa Complex Forest. It is not clear how this designation was made, as there is no discernable local knowledge of this designation or reference to it in any forestry policy or conservation plans.

2.3.1. Ecosystem Services Assessment Photovoice Exercise

Photovoice is a praxis-based qualitative tool that enables participants to record, reflect, and produce knowledge on their community needs, experiences, strengths, and concerns through specified photographic techniques [41–43]. Participants use photography to represent their perspectives and lived experiences on a given topic and collectively discuss and analyze photos to inform community projects and advocate for their interests. Regarded as a tool to give agency to disempowered and marginalized groups in transforming their realities, Photovoice has been used in several disciplines, such as urban planning, education, public health, and sociology [44–49]. Because of the wide appeal of participatory photographic methods, it has been used with children, youth, and adults in various settings, ranging from youth programs, women's groups and organizations, to public health organizations [50–53].

In the context of this research project, Photovoice was used to provide deeper insight into the context of rural resource dependent communities and the ways in which groups derive value from their relationships with forest ecosystems. Few studies on ecosystem services assessments have utilized Photovoice as a research method [54,55], which makes this work unique and fills a significant gap in the literature.

Prior to conducting the Photovoice ecosystem services assessment, a number of necessary permissions were obtained. The Principal Investigator (PI) applied for and was granted the necessary permission to conduct research in Malawi from the National Committee on Research in the Social Sciences and Humanities under the Malawi National Commission for Science and Technology (Ref No: NCST/RTT/2/6). Once on site, the PI and co-PI met with TA Kilupula to inform him of our activities and seek his permission, which was granted. TA Kilupula oversees 16 Group Village Headmen (GVH), and hundreds of Village Headmen/women in northern Karonga, including the GVH in Chikutu. The PI and co-PI also informed the GVH in Chikutu of our arrival date and research prior to the data collection visit.

The co-PI is a Malawian who has lived in northern Karonga his entire life. He is an extremely well-respected member of the community, is fluent in the handful of languages spoken in the region, and is an employee of the Ministry of Health of Malawi. The PI has known the co-PI for nearly 20 years, two of which (1998–2000) were spent working together at a rural hospital in northern Karonga. Prior to fieldwork, the PI and co-PI met many times in Malawi to discuss the purpose of the study and to refine the Photovoice methods and procedures.

Photovoice involved community members photographing the lands and natural elements around their villages that they perceived as essential to their health and livelihood. The exercise replicated the Misuku Hills biodiversity and livelihood transect walk conducted in 2015 with Village Natural Resource Management Committee (VNRMC) members in the area around the forest reserves [56], but added photographs of ecosystem services identified by residents. The assessment was conducted throughout the mid–late summer months when many resources are in season and available, but data on resources not available during these months were also collected.

The Photovoice methodology was applied to the ecosystem services assessment to overcome the limitations in articulation with text-based research and the asymmetrical power balance inherent in other research techniques. While allowing research participants to drive the process might frustrate the answering of a narrow set of questions and evidence is often not generalizable, it is abundantly useful for "building participant-driven practical theory about how environments impact everyday people" [49] (p. 400). The Photovoice variation used in this study combining a walking tour with picture taking replicates the method used in an informal settlement in Lusaka, Zambia [49]. This method gives a significant level of control over to participants and roots the data in lived experiences.

Upon arriving in Chikutu, we began by meeting with the Village Headman and the study participants to explain the purpose of the visit and to receive informed consent. The Photovoice participants included four men and three women of ages ranging from adolescent to adult. The field research in Chikutu proved to be an opportunity to educate residents on informed consent standards and procedures. Before the Photovoice exercise, the participants, PI, and co-PI reviewed the informed consent form that had been translated into the local language (Chitumbuka). One representative from the participants read aloud each section to ensure that any illiterate participants understood what was being consented to, and, following each section being read aloud, the co-PI reiterated the information and asked if any participants had questions. Although Chikutu is very remote, residents have had occasional contact with government officials collecting data, but none of the residents had ever completed an informed consent. After the form was signed, the PI and co-PI encouraged residents to demand informed consent from future researchers so that they would be fully aware of the risks and benefits of participating in research.

Once the Photovoice ecosystem services inventory walk commenced, the participants pointed out a resource, a photograph was taken of the resource, and its local name and common uses were recorded. The participants were given slight prompts to provide a little more explanation when necessary, but at no point did the researchers independently point out a potential resource and ask "what is this?" The participants were given full control to point out the resources that were important to them for their health and livelihood and to end the exercise when they felt a full accounting had been collected.

3. Results

3.1. Malawi Forestry Policy

The data from this section are aimed at addressing the first of the three research questions: What are the forestry policy challenges and opportunities for forest conservation in Malawi?

Timeline of Malawi forestry policies and guidelines and selected other policies directly affecting forestry policies:

- 1942 *Forest Act*: Pre-independence command and control forest management
- 1994 *National Environmental Action Plan*
- 1996 *National Forest Policy*: First post-independence forest policy aimed at supporting the 1996 National Environmental Policy and the Environment Management Act
- 1997 *Forestry Act*: Enabling and enforcement legislation to support 1996 forest policy
- 2001 *National Forestry Program*: Strategic framework for linking policy and practice or to translate good intentions into real results
- 2001 *Forestry Rules*
- 2001 *Forestry (Community Participation) Rules*
- 2003 *National Forest Policy*
- 2003 *Forestry Amendment Rules*

- 2003 *Community Based Forest Management*: Supplement to the 2003 National Forestry Policy expanding on "aspects related to community based forest management, including access to resources, benefit sharing, the role of traditional leaders, and decentralization" [57] (p. 10)
- 2004 *National Environmental Policy*: Includes principles and strategies for sustainably managing forests
- 2013 *Standards & Guidelines for Participatory Forestry in Malawi*: Developed to guide the practice of participatory forestry management and establish standards for forestry extension service delivery and improved forest management
- 2016 *National Forest Policy*: Coordinates all natural resource management, including forest resources and environmental policy instruments in Malawi
- 2017 *National Forest Landscape Restoration Strategy*
- 2017 *Environment Management Act*
- 2018 *A Framework for Monitoring Progress on Malawi's National Forest Landscape Restoration*

There has been a clear evolution in forest policies under Malawi's three distinct forms of government that span its colonial demarcation as a country to its current multi-party democracy system. During the colonial era's command and control system of management (1890s to 1964), "forest guards were posted in every Traditional Authority ... to police forestry use and collect revenue for government from forestry products" [7] (p. 2). With independence from colonial rule in 1964, and under a new authoritarian one-party state, the focus turned to plantation timber production for local and international trade. Concurrently, forested land under the control of traditional leaders experienced accelerated deforestation and degradation as communities pursued extractive practices as a demonstration of political independence from a colonial system of forest management [7].

During Malawi's transition to a multi-party democracy in 1994, demand for forest goods and services far exceeded supply, putting further pressure on forest systems [58]. To address recognized environmental and forest issues in Malawi, the Forestry Policy was revised in 1996. The 1996 National Forest Policy was a departure from the traditional forest management approach, most notably with its marked move towards devolution of centralized powers to promote participatory management [57,59,60]. This new strategy emphasized "multi-stakeholder participation including local communities" [7] (p. 2). Unfortunately, democracy was also equated with deregulation and the deforestation of forest reserves for agriculture and fuelwood [61].

Many of the devolution objectives acknowledged forestry financing and enforcement challenges. To remedy these challenges, a market approach was adopted to provide economic incentives that promoted the sustainable utilization of forest resources by emphasizing local ownership and management of forests and small- and medium-scale forest-based industries. Local management of forest resources was designed to be achieved through community-based forest management practices embedded in traditional institutions and giving communities shared or exclusive decision-making rights [7]. Another notable change in the 1996 policy was the explicit recognition that forest conservation policy objectives were supporting quality of life measures for rural populations recognized as the most disadvantaged group in Malawian society [58]. The 1996 forest policy was also used to support the larger framework of the 2004 National Environmental Policy, which itself is aligned with Section 13(d) of the 1995 Constitution of Malawi outlining many environmental goals.

Subsequent enabling legislation, forestry policy updates, and participatory management guidelines all support the current 2016 National Forest Policy goal to provide "guidance to the management of forests, offer an enabling framework for all stakeholders to participate in the management of forests, and sustain the contribution of the national forest resources for the upliftment of the quality of life" [27] (Foreward). Although the recognized role of forests in supporting quality of life is still an objective of the 2016 National Forest Policy, its focus on the rural disadvantaged is no longer explicit. While the 1996 policy was emblematic of a new democracy facing a recognized threat, the 2016 policy is more outward looking and reflective of a young participatory democracy on a world stage. This is evident in its stated alignment with international agreements and conventions such as the

Rio Declaration, United Nations Framework Convention on Climate Change, the Montreal Protocol, United Nations Convention to Combat Desertification, United Nations Convention on Biological Diversity, United Nations Convention on International Trade in Endangered Species of wild fauna and flora, and the claim that the review of the policy was conducted by a wide range of stakeholders, including traditional authorities, district councils, the civil society, the private sector, statutory bodies, government departments, academia, and the general public [27].

The 2016 policy recognizes that forestry conservation is a cross-cutting issue which requires collaboration and broad participation to meet the goals of other focused policies, such as those addressing land, biodiversity, wildlife, water, energy, and population, but also the more comprehensive Malawi Growth and Development Strategy now in its third iteration. Among the policy outcomes aimed at protecting forests are financial benefits and other livelihood outcomes (e.g., food, biomass, shelter, health). Financial incentives to protect forests include eco-tourism and recreation, and also still include forest-based enterprises. The livelihood outcomes are realized in the goods residents reap on a daily basis and profit from to support their health and well-being.

The implementation and enforcement of these policies remains a significant challenge, but the Environment Management Act of 2017 aims to address these challenges. This act created an Environmental Protection Agency "with broad and substantial powers to strengthen environmental planning and risk management at national and decentralized levels" [62] (p. 2). If the same implementation challenges that have thwarted previous legislation can be overcome, this act "will be one of the most powerful legal instruments for environmental management introduced so far in Africa" [62] (p. 2). This act, like others, focuses heavily on local control of environmental resources.

3.1.1. Village Natural Resource Management Committees and Village Forest Areas

Locally developed and enforced customary laws have shown to have a greater impact on the protection of natural resources as compared to federally developed and enforced laws [63]. This is partially due to human and institutional resource constraints that continue to make government sponsored patrolling, enforcement, and prosecution a challenge. Customary control alleviates these resource constraints and aligns with the customs and rules that govern everyday life and natural resource management and sanctions in Malawi. "Locally developed and enforced resource-use rules which relate directly to the resource in question", are more easily monitored by other community members, and acknowledge a culture where subjective norms are well known within communities, weigh heavily on decision making, and traditional penalties are generally accepted [63] (p. 93).

Recognizing the necessity of local control and influence of customary law, key among the strategies to achieve the objectives of national forest policies is the establishment and support of Village Natural Resources Management Committees (VNRMC). These nationally registered committees receive technical advice from the Forestry Department officers on how to protect, control, and manage their forest resources [13]. Under Forest Rules 2001, the VNRMC has the authority to prohibit residing in protected areas, altering for agriculture, or damaging trees for any purpose (along with selected other powers). The local Forest Management Agreement created by VNRMC in consultation with Forestry Department officers governs the activities of demarcated Village Forest Areas [63].

Village Forest Areas (VFA) are areas on customary land that are actively managed by the VNRMC for forest resources or forest re-establishment. As with all customary land, and following customary law, the responsibility for allocating and overseeing the VFA lies with the traditional leadership of the Village Head, Group Village Head, or ultimately, the Traditional Authority.

During the colonial era, every village was required by law to have a VFA to oversee the conservation of wood products, water, biodiversity, and recreational facilities. "A total of 69,000 hectares of VFAs were set aside by 1940, under the control of local headmen and for the purpose of local use" [6] (p. 1). VFAs are no longer required by law, but with devolution to local level control, these VFAs remain critical to forest conservation and are often the only body overseeing forest conservation.

There were over 2000 "active" and another 1000 "trained" VNRMCs in Malawi in 2002 [6], and the scope of their activities have expanded in line with heightened knowledge of the critical role of forests in local and global ecosystems. Many VNRMCs are now tasked with reforestation, and tree planting activities often receive wide media coverage. Their influence has waxed and waned since the colonial era establishment of VFAs, but the move towards decentralization since 2008 provides guarded hope that their influence will return. Granted, it will likely take time to overcome decades of centralized control and the re-adoption of local responsibility. It is a promising sign that the large youth populations in Malawi are taking center stage as stakeholders in VFAs [64], although the national youth tree planting program has also experienced shortcomings in government administration [65].

3.2. Conservation Activity Inventory in the Misuku Hills Region

The data from this section are aimed at addressing the second of the three research questions: What conservation activities are taking place in the Misuku Hills in support of these policies?

It has been documented that forest management activities in Malawi often involve three parties: "a facilitator who catalyses the process (often coming from outside the community), the implementing agency (a local group or committee spearheading the change process) and the benefiting community" [7] (p. 6), and this holds true for the activities that have taken place in the Misuku Hills since 1999. The facilitators have often been Malawian NGOs backed by international aid organizations (Table 2). It is apparent in Table 2 that NGOs, and not government, have been the catalyst of forest conservation activities, with government officials acting in an advisory capacity. NGOs that operate in Malawi must be registered with the Registrar General in the Department of Justice and the NGO Board of Malawi. They may also voluntarily become a member of the Council of Non-Governmental Organizations in Malawi (CONGOMA).

Table 2. Inventory of sponsored forestry conservation activities in the Misuku Hills.

Date	Project	Implementing Organizations	Funders
1999–2004	COMPASS I	Development Alternatives Inc.	USAID [1]
2006–2009	COMPASS II	Development Alternatives Inc.	USAID [1]
2006–2010	IFMSLP I	LTS [2]	European Commission
	IFMSLP II	GOPA [3]	
2011–2014	Improved Livelihood and Biodiversity Conservation Project	Subcontract to CEPF [4], AfES [5]	European Commission
	Misuku Hills Indigenous Forest Project	Subcontract to CEPF [4], MBA [6], SDI [7]	
2013	Misuku Beekeeping Value Addition Project	MBA [6], SDI [7]	British High Commission
2013–2015	Promotion of Indigenous Forests in the Misuku Hills Area	MBA [6]	UNDP GEF [8]
2017	Misuku Hills Art Challenge	AfES [5], MBA [6], SDI [7]	CEPF [4]
2018	Small Producers Development Project	MBA [6], SDI [7]	IM-Swedish Development Partners

[1] United States Agency for International Development; [2] LTS historically stood for Land and Timber Services, but it is now just LTS as a stand-alone title; [3] Gesellschaft für Organisation, Planung und Ausbildung; [4] Critical Ecosystem Partnership Fund; [5] Action for Environmental Sustainability; [6] Misuku Hills Beekeepers Association; [7] During the time of the project, the organization was called Sustainable Rural Growth and Development Initiative (SRGDI), but the organization is now called Sustainable Development Initiative (SDI); [8] United Nations Development Program Global Environment Facility. COMPASS, Community Partnerships for Sustainable Resources Management; IFMSLP, Improved Forestry Management for Sustainable Livelihoods Program.

The remoteness of the Misuku Hills contributes to it being one of the last contiguously forested regions of Malawi not in a national park, but this feature has also caused it to be overlooked by conservation organizations and funding agencies operating in other regions of Malawi. Despite this, there have been a handful of forest conservation activities that have engaged the VNRMCs in the Misuku Hills charged with managing both VFAs and the three public forest reserves.

The two phases of the Community Partnerships for Sustainable Resources Management (COMPASS) project were national in scope with Misuku Hills as one among many intervention sites. The very limited documentation available on the activities and results of COMPASS I reveal that it was meant to improve natural resource management by emphasizing income generation, which is consistent with and supporting USAID/Malawi's Strategic Objective framework of sustainable increases in rural incomes [40]. Unlike COMPASS I, there is abundant documentation of COMPASS II [66]. COMPASS II "supported decentralized environmental management and capacity building in enterprise development in order to mainstream CBNRM [community based natural resource management] as a viable rural development strategy (COMPASS II Project, 2007)" [40] (p. 24). COMPASS activities in the Misuku Hills included the support of the Mzuzu Coffee Planters Cooperative Union (MCPCU) to engage in honey production. Honey production was promoted not only for the sale of honey and beeswax products, but also for pollination of coffee crops. The accomplishment of COMPASS targets were initially monitored across 15 districts in Malawi, including Chitipa, but then reduced funding in 2007 led to only seven districts being monitored, excluding Chitipa [67]. Therefore, the effectiveness of these programs cannot be evaluated, and this lack of evaluation proves to be a theme throughout subsequent forest conservation activities.

The Improved Forestry Management for Sustainable Livelihoods Program (IFMSLP) was a Government of Malawi, two-phase, national capacity-building exercise aimed at improving "the livelihoods of forest dependent communities through improved sustainable collaborative management of forests both in forest reserves and customary land" [68] (LTS webpage). IFMSLP I and II were aimed at the implementation of the National Forestry Policy and Program through community mobilization, institution building, and local forest management planning [69,70]. One of the IFMSLP intervention sites included the area within and around the three public forest reserves in the Misuku Hills [6]. IFMSLP II faced funding and implementation delays, but it was eventually pushed forward with a reduced set of strategies [70]. It was discovered through personal communications that the delays were due to suspected government corruption which caused the funding agency to halt the project, but it did eventually resume with "competitive grants for non-state actors to enhance their role in, and to accelerate, project implementation" [70] (p. 1924).

There is a conspicuous dearth of documentation available on the activities and outcomes of the two phases of the IFMSLP, but one available document was an evaluation of IFMSLP I completed in 2011 [71]. The evaluation was critical of the lack of monitoring and evaluation, which may help explain the lack of available documentation. Despite the program setbacks and the lack of data on program activities and outcomes, there were still some identifiable results.

On a national level, the IFMSLP led to the development of the national 2013 Standards and Guidelines for Participatory Forestry in Malawi [70], as referenced in Section 3.1.1. The IFMSLP activities and outcomes specific to the Misuku Hills intervention site were uncovered through communications with NGOs who received the "competitive grants for non-state actors" to pick up the pieces from the mixed successes of IFMSLP I and a halted IFMSLP II. One tangible outcome was the creation of the Matipa Forest Management Plan (in draft form and still under review by the Forestry Department) created under the Improved Livelihood and Biodiversity Conservation Project. This plan was designed to govern the activities of the VNRMCs in this area and create a greater sense of ownership by the surrounding communities to relieve unsanctioned pressures on the forest reserve resources. It was based on and serves the larger Strategic Forest Area Plan (SFAP) for the Matipa, Mughese, and Wilindi forest reserves, with a focus on three of the five priority objectives in the SFAP. These are to (1) increase in tree planting and natural regeneration, (2) conserve the forest for water catchment protection, unique biodiversity, and cultural heritage, and (3) regularize extraction of forest resource and products from forest reserves to uplift communities' livelihoods. The creation of forest management plans for all three forest reserves was originally the responsibility of the Department of Forestry in IFMSLP I, but only one of these plans has been prepared. The Matipa Forest Management Plan includes estimates of the products and income that could be generated from the sustainable use of forest resources and a monitoring plan for patrols by local residents [72]. Responsibility for the creation of plans for the two remaining forest reserves (Mughese and Wilindi) remained with the Department of Forestry in IFMSLP II, but these plans have yet to be prepared.

The IFMSLP II also supported the creation of a number of forest management agreements and beekeeping activities under the Misuku Hills Indigenous Forest Project [73]. Seven VNRMCs composed of members from 71 villages now have co-management agreements and licenses to oversee the entire 2762 hectares of the three forest reserves in the Misuku Hills. These agreements include tree planting and regeneration on 485.92 ha in the forest reserves. These VNRMCs simultaneously manage VFAs on customary land, and there were seven VFAs established under the program (Alther, Chipala, Chiwi, Kapiyira, Lupalang'ombe, Mwenga, Nangalamu) with four of these VFAs (Alther, Chiwi, Mwenga, Nangalamu) currently with completed VFA management plans. Although there is no readily available national accounting of VFAs or their level of current activity, a forestry officer from Chitipa reported that there are 68 VFAs around the Misuku Hills, and 21 of these VFAs have management plans.

The Misuku Hills Indigenous Forest Project also included activities in and around forest reserves focused on beekeeping, candle making, and selling non-wood products. The Misuku Beekeepers Association (MBA) was the lead organization in this effort and has proven to be a prominent force for forest conservation in the Misuku Hills. MBA is a registered company comprised of more than 1500 beekeeper members with a 2:1 ratio of men to women [74]. MBA has actively contributed to the creation of forest plans, and their forest-based enterprises were further enhanced through this project. Support from this project led to MBA being chosen to represent Chitipa district at the National Agricultural Trade Fair, where they established business connections and received beekeeping product orders. These activities were estimated to have increased the income of 350 households in the area by 80%. MBA members also increased capacity in project management skills and conservation science by participating in two workshops in Mbeya, Tanzania and Nairobi, Kenya. Their participation at these workshops led the MBA being nominated by the Tropical Biology Association for a site visit and learning exchange with Save Tanzania Forests.

The Misuku Beekeeping Value Addition Project (MBVAP) and Promotion of Indigenous Forests in the Misuku Hills Area (PIFMH) project occurred simultaneously with the Indigenous Forest Project discussed above. The MBVAP was aimed at continuing to build capacity in the forest-based enterprise of beekeeping among the estimated 2500 beekeepers in the Misuku Hills who produce honey at a subsistence level. Honey was being sold unprocessed and uncertified in unreliable markets in their area, and the beeswax was just thrown away. The project supported the training of 250 members of 50 beekeeping clubs in beekeeping techniques and honey and beeswax candle making; the provision of equipment and materials; certification of honey and candles; and linking farmers to markets [74]. The reported results of the project were "overwhelming." The 50 clubs that were trained were registered with MBA. MBA acquired certification from the Malawi Bureau of Standards and used the processing equipment to create professional and standardized packaging for their Misuku Hills Honey (Figure 3). The combination of professional packaging and certification was attributed with over an 80% increase in the value of their honey. MBA honey can now be found on grocery store shelves across the country, and it is being used as an ingredient in cough syrup by a pharmaceutical company. This has reportedly led to 300 households increasing their income by 50%. Although the wax processing and candle making plant was established and most of the clubs 50 clubs were trained, inadequate funding was given as a reason why the candle making venture has yet to be realized.

The PIFMH project also promoted beekeeping, but it was focused solely on the Mughese forest reserve and included the promotion of nutritional and medicinal products and "reconnecting cultural values with existing nature" [75] (webpage). Although no evaluation report on outcomes of the PIFMH was available, its status is listed as "satisfactorily completed" [75].

The relationships built between stakeholders and organizations in all of these projects led to further collaboration in the Misuku Hills Art Challenge (Figure 4) aimed at raising "awareness of the beauty and ecological, cultural, aesthetical and economic value of Misuku Hills Forest Reserve both locally and internationally" [76] (webpage).

Figure 3. Misku Hills Honey. Source: Sustainable Development Initiative.

(a) (b) (c)

Figure 4. Misuku Hills Art Challenge: (**a**) Beekeeping training; (**b**) painting submission; (**c**) prize winning sculpture. Source: Sustainable Development Initiative.

The Misuku Hills Art Challenge (MiHAC) was a national competition that brought together 12 Malawian artists, photographers, and film makers to showcase the remote and often overlooked Misuku Hills and bring greater national and international attention to this Key Biodiversity Area (KBA) threatened by logging, charcoal production, and agricultural expansion. A smaller art competition was also conducted for school children in 10 schools surrounding the Misuku Hills. MiHAC was widely advertised through the national media and on multiple social media platforms, and an exhibition of the 16 paintings, 3 sculptures, 68 photographs, 3 films, and 10 children's drawings was convened in the capital city of Lilongwe with cash prizes given to the top artists.

The impacts of the project not only raised awareness of the Misuku Hills among the public, but also engaged numerous government ministries to encourage the inclusion of the Misuku Hills into policies and plans as a KBA. The Misuku Hills is recognized internationally as a KBA [32], but it is not currently recognized by the Government of Malawi as such. Participation by the Environment Affairs Department (Biodiversity Focal Point) brought attention to the need to conduct a biodiversity assessment as the first step in it being recognized as a KBA in national policies and plans.

MiHAC also introduced tourism as a new approach to forest conservation in the Misuku Hills, and tourism brought in the Ministry of Tourism as a new player in addition to forestry. The only accounting of tourism in the area estimated that the Misuku Hills typically received three local and international tourists per month, but in the six-month period following MiHAC, this rose to eight tourists per month, nearly tripling the income of VNRMCs that charged small fees to tourists [77]. The MiHAC project also brought greater attention to the production of Misuku Hills Honey and Mzuzu Coffee as potential tourist attractions. In addition to the previously discussed success of marketing Misuku Hills honey, the Misuku Hills is also where 50% of the internationally-renowned Mzuzu Coffee is produced. A small number of tourists travel to the Misuku Hills to sample the coffee and learn about the community that grows the coffee.

A number of potential donor organizations were also invited to and attended MiHAC events, and the Small Producers Development Project resulted from the interactions with a donor. A pilot project was conducted to continue the work of supporting beekeepers with financing, certification, production, and marketing, but it was reported that the donor ultimately decided that the Misuku Hills were too remote to conduct monitoring and evaluation, so they ceased supporting beekeeping activities. Again, the double-edged sword of its remoteness being part of its attraction but also creating difficulties in accessibility.

Lastly, there appears to have been a small project funded by the Tilitonse Foundation aimed at strengthening VFA plans, but no documentation on this project was available.

Although not in direct administration of the conservation activities in the Misuku Hills listed above, the Wildlife and Environmental Society of Malawi deserves mention as a major player in conservation activities in the Misuku Hills and throughout Malawi. They were credited as a contributor to many of the activities and policies identified in this study.

3.3. Chikutu Village Ecosystem Services Inventory

The data from this section are aimed at addressing the final research question: What ecosystem services are residents using that are in need of protection?

The Photovoice ecosystem services inventory took approximately three hours to complete. The residents of Chikutu village identified 16 distinct forest products in the immediate vicinity around their village (Table 3). This included seven different types of fruits, roots used for medicine, fiber and timber used for construction, vegetation that prevented soil erosion, the soil itself, mushrooms, and grasses and leaves for domestic animal feed. Figure 5 displays selected photos of identified forest products.

A greater volume and variety of flora were identified in the Chikutu ecosystem service Photovoice exercise than were identified in a previous participatory assessment conducted as part of the Improved Livelihood and Biodiversity Conservation Project.

Forest fruits are an essential part of the diet for rural populations in Malawi as they are a source of critical dietary nutrients (vitamin A and C, calcium, fiber, minerals) and contribute to food security, especially as a supplement in times of famine. Mushrooms also provide nutritional benefits and are sometimes preserved for food security purposes [25]. Fruits and mushrooms can be sold to persons residing in surrounding areas where forests have been degraded and these resources are scarce [13].

Table 3. Inventory of forest products.

Local Name	Type of Forest Product	Local Uses	Notes
Ndilolo	Tree [1]	Food	Nut casing consumed and used for juice
Manga	Tree	Food	Mango fruit
Munyere	Shrub	Food	Wild avocado
Miyombo	Shrub	Construction	Stems used for rope
Mulungalunga	Tree	Medicine	Roots
Masuku	Herbaceous plant	Food	Fruit
Mushombe	Bamboo	Construction	
Mufiomi	Tree	Timber	Recognized for water conservation properties
Mwina	Tree	Timber	Noted as being very dense
Malina	Grass	Timber	Also animal feed
Dongo	Soil	Food	
Ntochi	Herbaceous plant	Food	Banana
Chighughu	Grass	Erosion control	Also animal feed
Popo	Tree	Food	Papaya
Chiwowa	Fungus	Food	Mushroom
Guaves	Tree	Food	Guava

[1] Almost all trees are also used for wood fuel.

(a) (b) (c)

Figure 5. Selected forest goods in Chikutu: (**a**) Ndilolo (cashew-like nut); (**b**) Chiwowa (mushroom); (**c**) Munyere (wild avocado).

The remoteness and inaccessibility of Chikutu village has contributed to the abundance of these forest products, but lack of connectivity to other settlements also creates great difficulty in accessing markets. This lack of access to markets makes these goods essential as a food source and, at the same time, creates great difficulty in selling these products for profit. When asked directly about the sale of forest products, it was confirmed that they were only consumed locally. One notable absence in the identified forest products was the harvesting of wild animals and insects.

A number of ecosystem service assessments were also conducted in three villages in the lake plains area of Karonga with much higher population densities and diminished forests. While residents in these villages still appeared to be knowledgeable about forest product ecosystem services, these services were less abundant and diverse. Essentially, the less forested area had fewer identified ecosystem services.

4. Discussion

Malawi is faced with many challenges to forestry conservation and the subsequent sustainability of ecosystem services. However, as the results above revealed, Malawi has formulated—in collaboration with an array of international organizations—a number of policies and plans to meet these challenges. There have been attempts to implement these policies, but there are significant challenges to realizing their full vision and potential. The first challenge is a lack of current, reliable, and comparable data on forest cover and change, but efforts to remedy this are currently underway (Appendix A). Discussed below are a number of other challenges, as well as potential remedies discovered through this research: The challenge of good governance, the potential remedies of increased location control (e.g., Village Natural Resource Management Committees), and the promotion of forest-based enterprises. As the results of the policy inventory revealed, there is no shortage of actionable forest policies, but shortcomings in good governance has limited their implementation. The forest-based activities that have taken place in the Misuku Hills demonstrate the potential for local initiatives and control when centralized government fails. The inventory of ecosystem services revealed a previously unrecorded wealth of goods that could be sustainably marketed through forest-based enterprises. The discussion concludes with suggested directions for future research.

4.1. *Good Governance*

Weak institutions have been identified as one of the many factors threatening the Miombo forests of Africa [78,79]. The Democracy Index categorizes the relatively young multi-party democracy of Malawi as a *hybrid regime* [80]. The hybrid regime-type of government is characterized by a number of conditions that affect the implementation of forest policy, including serious weaknesses in political culture, functioning of government, and political participation; widespread corruption; weak civil society and rule of law; and a non-independent judiciary. Many of these threats to good governance were confirmed in a separate, albeit ideologically-driven, analysis of economic freedom [81]. Combined with a lack of financial and human resources, these conditions certainly create challenges in the implementation and enforcement of forestry policy, and they are noted as a priority focus area in the most recent National Forestry Policy [27].

We see some of these challenges at the local level in the Misuku Hills. Among the reported lessons from one of the conservation projects in the Misuku Hills (intentionally not identified) was that corruption is a key challenge. It was reported, but not corroborated by the author, that government authorities accepted bribes to grant logging licenses to traders to cut down large trees in VNRMC co-managed forests without the consent of the VNRMC. These activities undercut the stewardship and disenfranchised the VNRMCs. Part of this can be explained by the lack of internalization of the co-management concept by government officials, despite the abundance of adopted national policies and guidelines promoting participatory forest management [82]. Throughout Malawi, residents and VNRMC members have demonstrated that they are willing to put in the effort, but they are bogged down by lack of support from the government officials with whom they co-manage forests. Even when drafts of local Forest Management Plans are created, there are significant delays in approving these plans, which demoralizes residents [82]. The Misuku Hills are a case in point. There has been a many years long delay by the Department of Forestry in approving the draft Matipa Forest Plan and finalizing the plans of the other two forest reserves, long after the local VNRMCs have done their part.

Despite these challenges, there are a great number of opportunities in good governance in Malawi. The national government has delivered on forestry policy, forest management acts and guidelines, and most recently a strategy for measuring progress on forest conservation. Furthermore, the total land area of public protected forests increased by 8% between 1998–2010 with more proposals for protected areas underway [29]. Malawi has also taken the bold step of temporarily deploying Malawi Defense Force soldiers to patrol the most threatened public forests [83]. Furthermore, it was reported that the National Tree Planting Season recently closed with 50 million trees planted on 25,000 hectares, just shy of the 60 million tree target [84].

Despite some negative and demoralizing interactions between VNRMCs and government officials, there have been positive interactions as well. The Misuku Beekeeping Value Addition Project reported "a lot of support" from the Ministry of Agriculture and Food Security, Community Development Officers, and Forestry Officers in the form of training and equipment and identifying markets [74]. The Forestry Department has served in advisory and administrative roles (see Indigenous Forest Project report), and their role will likely become more advisory with the continued movement towards central devolution to more local control.

4.2. Local Control

Malawi adopted a participatory forest management policy in 2001 (updated in 2003) and, as recently as 2013, adopted guidelines for participatory forest management to increase local control of forest conservation. There is ample guidance and authority for local control, but co-management challenges are still thwarted by a lack of policy implementation and capacity building at the local level [62,70]. This lack of implementation begins with government ministries and donors failing to fully engage and organize communities. Even in communities where VNRMCs have been organized, there have been many other challenges to their effectiveness in advancing local control. Two of these challenges are (1) a lack of empowerment to actively participate in decision-making and (2) a lack of downward accountability among leaders which has limited the devolution process [85].

Much of the community-based forestry management (CBFM) has been driven by donors, government, or NGOs and imposed on communities. While these activities may raise community awareness, they are not sustainable [70] and "undermine achievement of conservation and social goals" [86] (p. 687). They lack the community empowerment derived by involving community members in decision making and creating accountability mechanisms. COMPASS I and II are examples of imposed programs that do not properly empower communities or provide sufficient incentives for communities to continue the imposed project once the donors move on and funding ceases.

Empowerment in the decision-making process begins by understanding "the preexisting conditions and how communities understand and interpret the program" [87] (p. 338). One of these preexisting conditions includes the power relations in communities. In an evaluation of CBFM in southern Malawi, it was found that both the CNFM (community natural resource management) concept and implementation created new elites (forest committees) who largely operated as corrupt, unaccountable "village bureaucracies," alienating communities from CNRM. Widespread forest degradation and institutional breakdown ensued. Community management became committee management, and part of the problem. Rare "success" was associated with idiosyncratic leadership qualities of village heads, suggesting need for enhancing roles and leadership skills of traditional leaders in balancing the exercise of power among CNRM stakeholders, and for broad-based community empowerment so that members can demand accountability from local leaders [86] (p. 687).

Both the CBFM committee and community members must understand that the CBFM committee is working for, and accountable to, the community.

Despite these challenges, there are a great number of opportunities for local control and the local protection of forests in Malawi. In fact, with a dearth of government resources to manage and monitor forests effectively, local control is currently Malawi's only hope for forest conservation. It has been found that VFAs under participatory forest management (PFM) had higher tree species abundance and diversity than those without PFM [88]. This success was attributed to the regulation of access and the forest development work of communities who practice PFM in their designated VFAs.

As noted above, understanding the preexisting conditions is essential for the empowerment that would enhance local control. A fundamental preexisting condition that is ripe for success in its ability to engage communities and build understanding and trust is undertaking a participatory process in the inventory of local forest resources, such as the ecosystem services (ES) assessment conducted in this study in Chikutu. "Considerable indigenous knowledge and skills for managing forest goods and services are often available at village level" [16] (p. 6), and it is arguably remiss to protect forests without a locally-driven accounting of the resources that are available and in need of protection. These activities are in line with current policy. The Chikutu ES inventory supports the policy priority area five, strategy three of the National Forestry Policy noting indigenous knowledge acquisition and dissemination [27]. The Chikutu ES study found a number of services that were distinct from the more heavily studied area immediately surrounding the public forest reserves. It is only by engaging communities and respecting their knowledge to uncover these services that management priorities can be properly represented in local VFA plans.

4.3. Forest-Based Enterprises

Forest-Based Enterprises (FBEs)—interventions meant to empower communities to realize tangible benefits from forests—recognize that forest conservation and livelihood development are deeply intertwined [89,90]. FBEs tie people to local forests for their livelihoods, and this may be more realistic in the near-term as Malawi continues to defy the conventional development pathway and macro-level economic changes theorized as affecting forest conservation.

Usually, the transition from a low- to middle-income economy starts with an abundance of natural capital which is used to invest in infrastructure (produced capital) and education and health (human capital). At middle-income levels, produced capital roughly doubles its share and human capital grows rapidly to become the main asset. In Malawi, the opposite development occurred. Malawi is still highly dependent on its natural capital, which remained constant at 43% from 1995 to 2014, while human capital increased only slightly and produced capital shrank [62] (p. 5).

With human capital and produced capital near stagnation, Malawi has yet to follow the forest transition theory [91], which posits that economic development, industrialization, and urbanization causes an initial large decline in forest cover to fuel this development, followed by a slow increase in forest cover [92,93]. "In some places economic development has created enough non-farm jobs to pull farmers off of the land, thereby inducing the spontaneous regeneration of forests in old fields" [92]

(p. 23), but Malawi is still a largely agrarian-based economy, and FBEs may be asking people to be tied even more to the land instead of generating economic activity off of it, possibly perpetuating a "poverty trap" [92] when FBEs are not successful.

As Malawi continues to defy the conventional development pathway and the conditions necessary for a forest transition, FBEs may be a strategy only in areas where forests are still present. In the central and southern regions of Malawi, where the vast majority of the population lives, the near complete deforestation does not make FBEs a widespread solution to poverty alleviation and forest stewardship. The northern region (and the Misuku Hills, in particular), due to its lighter population and accompanying extant forest cover, is likely a better candidate for viable FBEs. Yet, FBEs may only be viable if the conditions that led to the deforestation of the central and southern regions are addressed; namely, a rapidly growing population of small landholders practicing largely subsistence agriculture on finite land. It is not until other developments that create technological separation, reduced population, urbanization, or all of these and other factors ensue before the people of the Misuku Hills can make FBEs more than a short-term solution. If larger external forces continue to diminish forests, there will be no resources remaining to sustain FBEs.

It is also critical to recognize that "households will not invest precious labour and time nurturing trees when there are more pressing needs for food security" [16] (p. 6). Sustainable agricultural practices must be implemented concurrently with the CBFM of "free" food stuffs from forests to ensure food security as a necessary precursor to the protection of forest resources.

There has been some successes of FBE in other regions of Malawi to protect the few remaining forests (e.g., Sustainable Management of Indigenous Forests program in Kam'mwamba, Neno District in southern Malawi, 1996–2006 (see [7])), and also in the Misuku Hills with honey production, but the remote location of the Misuku Hills is a proverbial double-edged sword as it attributes to the quality of its forest but also creates challenges in transporting goods to markets and attracting all but the most adventurous tourists. One of the goals of the Misuku Hills Art Challenge was to raise awareness of the beauty and potential of this region for tourism. Even a slight increase in tourism to this area could have a dramatic effect on increasing recognition and spurring greater infrastructure development.

There has also been some success in the changing attitudes towards the protection of goods used to support FBEs. With awareness raised as to the importance of forest conservation to FBEs, there have been community protests and bad press when trees are harvested. For example, the District Forestry Officer overseeing the Misuku Hills had to address outrage at tree harvesting in the press, even when trees were being harvested sustainably on customary land and the trees in question were planted by inhabitants [94]. This culture of protection does have its limits. For example, although charcoal production is illegal and has been widely recognized as a major contributor to deforestation in Malawi, there are many instances of persons selling charcoal in plain view of those responsible for enforcing this infringement. This burgeoning culture of conservation may be driven by not only the potential losses of cash value, but also by the highly understudied non-cash value of forests.

Non-Cash Value

Zulu 2013 offers *reciprocal altruism* as a meta-theory [95,96] to understand the non-cash incentives for forest conservation in Malawi. This theory involves the "the trading of altruistic acts in which benefit is larger than cost so that over a period of time both enjoy a net gain" [70] (p. 361). Since there are many conditions that do not make FBEs viable in most places in Malawi, non-cash benefits may be more appropriate to remedy the "generally poor results" stemming from the "heavy dependence" on cash incentives in contemporary co-management projects [70].

In his critique of the national IFMSLP (see Table 2), but also applicable to the many programs carried out in the Misuku Hills, we tend to agree with Zulu that "a narrow emphasis on cash incentives as the motivation for 'self-interested' users to participate in co-management overlooks locally significant non-cash motivations, inflates local expectations beyond ability to deliver, and often creates perverse incentives that undermine socio-ecological goals" [70] (p. 1919). The promise of cash incentives in

FBEs have been found to produce "modest early gains in institutions and capacity building and forest condition, but low and generally disappointing cash benefits." The disappointing results "burdened poor communities with conservation costs and created perverse incentives to overharvest, be dependent on the project/government, and to marginalize the local poor" [70] (p. 1919).

An additional assessment of CBFM in Malawi confirmed that many communities "also highly valued other non-cash based and environmental objectives and benefits, including the sustained access to firewood and NTFPs [non-timber forest products], social/religious services inside the forest, and continued water supply for consumption and irrigation" [82] (p. vii). Awareness of the non-cash value of forest ecosystem services appears to be high and might help explain a continued commitment to CBFM in forest reserves despite FBEs producing "disappointingly low financial benefits for poor communities burdened with conservation costs" [70] (p. 1936). In the ecosystem services assessment of Chikutu, no cash values were expressed, even when residents were explicitly asked about the potential to sell local forest products.

4.4. Future Research

This study is the first inventory and analysis of the policy potential, conservation activities, and ecosystem services in the Misuku Hills. This initial assessment aims to enable future studies to examine the impact of these policies and activities directed at protecting forests and the essential ecosystem services that forests provide to the indigenous populations in this region.

The primary suggestion on future research design is that the geographical scope of the Misuku Hills should be expanded to encompass the forested area on customary land. The larger Misuku Hills forested region on customary land has been completely ignored in research and largely ignored for inclusion in the conservation activities. Although a number of forest conservation studies were conducted in Malawi, very few have addressed the socio-ecological dynamics in remote areas such as the Misuku Hills, where forests are still abundant and small indigenous populations rely heavily on forest resources to meet daily needs.

Future research should also consider examining the dynamics between the Misuku Hills forested region and the highly populated area immediately to the east in northern Karonga. The lake plains to the east are almost completely deforested to support rice cultivation, but it is suspected that the communities in the lake plains rely heavily on the ecosystem services provided by the Misuku Hills. This dichotomy likely creates both opportunities and pressures on the Misuku Hills forested region that generate tensions that need further exploration.

Lastly, a recommendation regarding informed consent. As noted in Section 2.3.1. Ecosystem Services Assessment Photovoice Exercise, the residents of Chikutu were never previously exposed to the informed consent process. This study proved to be an opportunity to teach residents about their rights as subjects of human research and to demand informed consent from future researchers. Any future research should be prepared to confront a lack of awareness of informed consent protocols and empower residents through education about their rights.

5. Conclusions

Since independence, Malawi has enacted a plethora of forestry policies that have been developed to meet international standards, including policies that promote local control and community-based forest management. What appears to be lacking is the widespread implementation of these policies and the empowerment of communities to practice local control. Despite these challenges, the forest conservation activities in the Misuku Hills demonstrate the potential for rural communities to organize and assume stewardship of forests both on customary land and in public forest reserves. In fact, a focus on communities such as those in the Misuku Hills is necessary to protect the last remaining indigenous forests and the indigenous communities that depend on forests for their livelihood. The ecosystem services inventory conducted in Chikutu revealed that community members have extensive knowledge of forest resources, and these resources are still readily available in communities residing in Malawi's remaining forests.

Malawi continues to be an anomaly on the macro-level economic development pathway. Its poor and rapidly growing population and widespread practice of subsistence agriculture will continue to place pressure on Malawi's forests. Until there is a change in these social forces, an increase in the diversity of livelihood opportunities off the land, and improvements in state institutional capacity, local control and stewardship will be necessary to conserve and regenerate the forests of Malawi.

Author Contributions: Conceptualization, C.C.; Methodology, C.C.; Investigation, C.C.; Writing—original draft preparation, C.C.; Writing—review and editing, C.C., T.H., and A.J.

Acknowledgments: The authors would like to acknowledge the support of Hassan Milazi of the Ministry of Health of Malawi. This ecosystem services field research would not have been possible without his consultation on the methodological approach and assistance in logistical planning. The authors would also like to acknowledge the hospitality of the Milazi family and residents of Miyombo village that graciously hosted the author during the fieldwork. Information provided too by NGOs in Malawi, Action for Environmental Sustainability and Sustainable Development Initiative, was invaluable to form a complete accounting of conservation activities in the Misuku Hills. Lastly, the faculty of the Department of Built Environment at Mzuzu University were extremely gracious hosts. They provided invaluable guidance on this project and were extremely accommodating in allowing time away from campus to travel to the study site.

Appendix A

Forest Cover Data Discrepancies and Implications

There have been a number of noted discrepancies in estimates of Malawi's forest cover, rates of deforestation, and forest land tenure designations derived from outdated data sources [6,29]. The presentation of these discrepancies below is meant to instigate a critical evaluation of published forest cover estimates and to shed light on possible recent gains in forest cover (possibly due to enacted forest policies) and current efforts to remedy widely disparate forest cover estimates. It is certainly not meant to close the case on what led to these discrepancies, nor could it as much of the information on estimate sources and methodologies cannot be obtained.

Alarmist statements from as recent as 2019 that "over the last 40 years, more than half of Malawi's forests and woodlands have vanished" [62] (p. 1) are certainly dramatic, but they mask apparent recent gains. Admittedly, these gains are not easily discernable without a more deconstructed examination of trends in Malawi's forest loss over time that reveal a guarded optimism towards recent achievements in reforestation.

One cited estimate from 2016 is that Malawi is losing forest cover at a rate of 2.8% (250,000 ha) annually [27]. The 2.8% assessment appears to have been derived from a 1993 Biomass Assessment Report (assessment conducted in 1991), but this estimate masks significant changes over time and more recent declines in the rates of forest loss. This estimate appears to have been first cited in 2006 by a Forest Conservation Officer at the Food and Agriculture Organization (FAO) Regional Office for Africa. Despite its validity being called into question in 2010 due to its use of data demonstrating a loss of forest cover at a suspiciously unwavering 165,000 ha every 5-year period from 1990–2010 [29], it was still being perpetuated in 2016.

An oft-cited comparison of maps of Malawi forest cover from 1979–1999 reveals a dramatic loss of forest cover over this period, but if any validity can be given to these maps, it appears that this dramatic change happened prior to 1990. Malawi lost 41% of forest cover between 1972 and 1990 at a 2.3% annual rate [6], and this loss of forest cover occurred almost exclusively on customary and private land [97] (as cited in [6]). Post-1990 declines in deforestation rates are being masked when including pre-1990 forest cover loss estimates. Applying the FAO forest classification system and Malawi Forestry Department data (found in Table 8.3 from the cited report), the average annual deforestation rate from

1990–2010 was <1% per year [29]. Another report produced in the same year claims that Malawi lost 0.85% per year or a total of 16.9% (659,000 ha) of its forest cover over this period [98]. The report does not cite the source of this estimate. It may be from the report above, and it is still less than the 1990–2010 1.6% estimate of the same period cited in the introduction of this paper. It appears that pre-1990 estimates of rapid deforestation using a different classification as FAO are being incorporated into more recent estimates and masking 1990–2010 decreases in deforestation rates.

It could be that the rate of deforestation in Malawi after 1990 declined simply because Malawi forests had already been diminished, and there was little forest left to lose. This would be most apparent in the central and south regions, where there is less forest cover. In Dedza district (central region), 1991–2015, almost half of the forest was lost, but it only had 2.6% forest cover in 1991. This was almost a 2% annual loss over this period, but there was not much forest area to lose [15]. This is also pertinent to the districts in the northern region where the Misuku Hills are located. Since independence, there has been a gradual migration of people to the northern region [37], and "it is these areas that have also experienced the greatest amount of deforestation since independence" [14] (p. 274). An 1972–2009 estimate found that Karonga (−579 km^2; −28%) and Chitipa (−565 km^2; −20%) were among the districts that experienced some of the greatest declines in forest area [14]. Again, this estimate included pre-1990 conditions that could be masking more recent reductions in forest cover loss. A hot spot of change analysis revealed that, while some districts in the northern region experienced significant forest and natural vegetation loss between 1990–2010, Chitipa and Karonga experienced almost imperceptible changes [5]. These are the data cited in the introduction of this paper as being the most methodologically rigorous and forthcoming.

The 1972–2009 estimate reveals that nationally "there was a loss of 12,760 km^2 (36%) of original forested area but also 11,161 km^2 of new forest establishment, resulting in a relatively modest overall net loss of 1599 km^2 (5%)" [14] (p. 269). The districts of Chitipa (+54%, +622 km^2) and Karonga (+44%, +523 km^2) were among the districts that experienced the largest percentage and net gains in mosaic land cover (defined as a mixture of cropland, forest, woodland, grassland, scrubland, and other natural vegetation). In fact, in every district that had an overall loss in forest, there was an overall gain in mosaic land cover. Of course, this calls into question the ability of the mosaic land cover to replace the losses to biodiversity and ecosystem services brought about by the loss of indigenous forest cover.

Other estimates of national forest cover reveal apparent swings. The Biomass Assessment of 1991 showed that, in 1973, "Brachystegia forests occupied 45% of total land area of Malawi (36.5% if Lake Malawi is included) while in 1990/91 land under forest cover was estimated to be 25.3% (20.5% if Lake Malawi is included)" [29] (p. 143). The trusted estimate cited in the introduction of this paper of 26.8% is within the range of other estimates (18.2 to 28.7%) of total forest land cover [99], and greatly underestimates another as high as 34% (3.2 million hectares) [29]. There has undoubtedly been a loss of forest cover since 1973, but there has also been a gain (or at least a stagnation) since 1990. The difference between the 45% estimate in 1973 and the 26.8% estimate in 2010 equates to a 40% loss, but this loss occurred completely before 1990. As cited above, Malawi lost 41% of forest cover between 1972 and 1990 [6]. This is highly contradictory to the 5% net loss cited above that takes into account regeneration and new forest establishment [14]. One explanation that could attempt to reconcile these estimates is that previous estimates did not take into account regeneration, but this is far from clear.

Despite all of the data discrepancies and the opaqueness of how they were derived, the most recent estimates are cause for measured hope. Since the decline in forest cover and deforestation rates coincides with the beginning of the slew of forest policies in the 1990s, the larger unanswered question is the degree to which forest policy interventions led to the apparent halting of deforestation.

The academic exercise and arguable futility in trying to untangle the web of past forest cover and loss estimates are currently being addressed in future plans.

As part of the process for developing Malawi's National Monitoring Framework, the US Geological Survey (USGS) with support from USAID is developing national maps of land use and land cover, as well as maps documenting on-farm tree cover for baseline year 2017, the year the National FLR

Strategy was launched. These maps will provide data on the biophysical progress of FLR interventions in Malawi (e.g., percent of tree cover), which will serve to set a baseline for monitoring biophysical progress on the agricultural technologies, forest management, and community forest and woodlot restoration interventions [100] (p. 16)

These data are not just essential for forest monitoring, but also because they will inform other monitoring activities such as the Integrated Household Survey which considers community forests and woodlots in its assessment of local resources [100].

"In 2016, the Government of Malawi made a national pledge to the African Forest Landscape Restoration Initiative (AFR100) under the Bonn Challenge to restore 4.5 million hectares of degraded and deforested land by 2030" [100] (p. 3). Malawi has taken the bold step to not only make this pledge and have an increasing number of policies to support this goal, but Malawi now also has a 2018 framework to evaluate its progress towards reaching its pledge to "increase area of community forests and woodlots to 200,000 ha by 2020 and 600,000 ha by 2030" and "improve protection and management of 2 million ha of natural forest, restore 500,000 ha of degraded forest, and establish 100,000 ha of commercial plantations by 2030" [100] (p. 4).

References

1. Bare, M.; Kauffman, C.; Miller, D.C. Assessing the impact of international conservation aid on deforestation in sub-Saharan Africa. *Environ. Res. Lett.* **2015**, *10*, 125010. [CrossRef]
2. Ordway, E.M.; Asner, G.P.; Lambin, E.F. Deforestation risk due to commodity crop expansion in sub-Saharan Africa. *Environ. Res. Lett.* **2017**, *12*, 044015. [CrossRef]
3. Dimobe, K.; Ouédraogo, A.; Soma, S.; Goetze, D.; Porembski, S.; Thiombiano, A. Identification of driving factors of land degradation and deforestation in the Wildlife Reserve of Bontioli (Burkina Faso, West Africa). *Glob. Ecol. Conserv.* **2015**, *4*, 559–571. [CrossRef]
4. Rudel, T.K. The national determinants of deforestation in sub-Saharan Africa. *Philos. Trans. R. Soc. B Biol. Sci.* **2013**, *368*. [CrossRef] [PubMed]
5. Food and Agriculture Organization of the United Nations. *Atlas of Malawi Land Cover and Land Cover Change*; Food and Agriculture Organization of the United Nations: Rome, Italy, 2013.
6. Mauambeta, D.; Chitedze, D.; Mumba, R. *Status of Forests and Tree Management in MALAWI*; Coordination Union for Rehabilitation of the Environment: Blantyre, Malawi, 2010.
7. Food and Agriculture Organization of the United Nations. *An Integrated Approach to Improve the Management of Forests and Other Natural Resources: The Case of Malawi*; Food and Agriculture Organization of the United Nations: Rome, Italy, 2009.
8. Gowela, J.P.; Masamba, C.R. *State of Forest and Tree Genetic Resources in Malawi*; Food and Agriculture Organization of the United Nations: Rome, Italy, 2002.
9. Ministry of Lands and Housing. *Malawi National Land Policy*; Government of the Republic of Malawi: Lilongwe, Malawi, 2002.
10. The Program on Governance and Local Development. *The Local Governance Performance Index (LGPI) in Malawi: Selected Findings on Land*; The Program on Governance and Local Development: Gothenburg, Sweden, 2016.
11. Minde, I.J.; Kowero, G.; Ngugi, D.; Luhanga, J. Agricultural land expansion and deforestation in Malawi. *For. Trees Livelihoods* **2001**, *11*, 167–182. [CrossRef]
12. National Statistical Office. *2018 Malawi Population and Housing Census*; Government of the Republic of Malawi: Lilongwe, Malawi, 2018.
13. Forestry Research Institute of Malawi. *State of the World's Forest Genetic Resources. Country Report: Malawi*; Food and Agriculture Organization of the United Nations: Rome, Italy, 2013.
14. Bone, R.A.; Parks, K.E.; Hudson, M.D.; Tsirinzeni, M.; Willcock, S. Deforestation since independence: A quantitative assessment of four decades of land-cover change in Malawi. *South. For.* **2017**, *79*, 269–275. [CrossRef]
15. Munthali, M.G.; Davis, N.; Adeola, A.M.; Botai, J.O.; Kamwi, J.M.; Chisale, H.L.W.; Orimoogunje, O.O.I. Local perception of drivers of Land-Use and Land-Cover change dynamics across Dedza district, Central Malawi region. *Sustainability* **2019**, *11*, 832. [CrossRef]

16. Department of Forestry. *Malawi's National Forestry Programme: Priorities for Improving Forestry and Livelihoods*; Government of the Republic of Malawi: Lilongwe, Malawi, 2001.
17. World Bank GNI per Capita, Atlas Method (Current US$). Available online: https://data.worldbank.org/indicator/NY.GNP.PCAP.CD?locations=MW&view=chart&year_high_desc=false (accessed on 28 March 2019).
18. Meijer, S.S.; Catacutan, D.; Sileshi, G.W.; Nieuwenhuis, M. Tree planting by smallholder farmers in Malawi: Using the theory of planned behaviour to examine the relationship between attitudes and behaviour. *J. Environ. Psychol.* **2015**, *43*, 1–12. [CrossRef]
19. Meijer, S.S.; Sileshi, G.W.; Catacutan, D.; Nieuwenhuis, M. Agroforestry and deforestation in Malawi: Inter-linkages between attitudes, beliefs and behaviours. *Agrofor. Syst.* **2016**, *90*, 645–658. [CrossRef]
20. Meijer, S.S.; Sileshi, G.W.; Catacutan, D.; Nieuwenhuis, M. Farmers and forest conservation in Malawi: The disconnect between attitudes, intentions and behaviour. *For. Trees Livelihoods* **2016**, *25*, 59–77. [CrossRef]
21. Chavula, J. Smoke over Dzalanyama forest. *The Nation*, 20 April 2019; 8–9.
22. Kwesiga, F.; Mwanza, S. Underexploited wild genetic resources: The case of indigenous fruittrees in Eastern Zambia. In *Improvement of Indigenous Fruit Trees of the Miombo Woodlands of southern Africa*; Maghembe, J.A., Ntupanyama, Y., Chirwa, P.W., Eds.; PrimexPrinters, ICRAF: Nairobi, Kenya, 1995; pp. 100–112.
23. Maghembe, J.A.; Ntupanyama, Y.; Chirwa, P.W. *Improvement of Indigenous Fruit Trees of the Miombo Woodlands of Southern Africa*; PrimexPrinters, ICRAF: Nairobi, Kenya, 1995.
24. Campbell, B.M. The use of wild fruits in Zimbabwe. *Econ. Bot.* **1987**, *41*, 375–385. [CrossRef]
25. Lowore, J. *Miombo Woodlands and Rural Livelihoods in Malawi*; Center for International Forestry Research: Bogor, Indonesia, 2006.
26. Hecht, J.; Kasulo, V. *Development of Forest Valuation Systems Malawi*; Cardno Emerging Markets: London, UK, 2013.
27. Ministry of Natural Resources Energy and Mining. *National Forest Policy*; Government of the Republic of Malawi: Lilongwe, Malawi, 2016.
28. Yaron, G.; Mangani, R.; Mlava, J.; Kambewa, P.; Makungwa, S.; Mtethiwa, A.; Munthali, S.; Mgoola, W.; Kazembe, J. *Economic Valuation of Sustainable Natural Resource Use in Malawi*; Government of the Republic of Malawi: Lilongwe, Malawi, 2010.
29. Ministry of Natural Resources Energy and Environment. *Malawi State of Environment and Outlook Report*; Government of the Republic of Malawi: Lilongwe, Malawi, 2010.
30. Ngwira, S.; Watanabe, T. An Analysis of the Causes of Deforestation in Malawi: A Case of Mwazisi. *Land* **2019**, *8*, 48. [CrossRef]
31. Twongyirwe, R.; Bithell, M.; Richards, K.S. Revisiting the drivers of deforestation in the tropics: Insights from local and key informant perceptions in western Uganda. *J. Rural Stud.* **2018**, *63*, 105–119. [CrossRef]
32. KBA Partnership World Database of Key Biological Areas. Available online: http://www.keybiodiversityareas.org/site/factsheet/misuku-hills-forest-reserves-iba-malawi (accessed on 11 April 2019).
33. Wangai, P.W.; Burkhard, B.; Müller, F. A review of studies on ecosystem services in Africa. *Int. J. Sustain. Built Environ.* **2016**, *5*, 225–245. [CrossRef]
34. Droissart, V.; Dauby, G.; Hardy, O.J.; Deblauwe, V.; Harris, D.J.; Janssens, S.; Mackinder, B.A.; Blach-Overgaard, A.; Sonké, B.; Sosef, M.S.M.; et al. Beyond trees: Biogeographical regionalization of tropical Africa. *J. Biogeogr.* **2018**, *45*, 1153–1167. [CrossRef]
35. BirdLife International Misuku Hills Forest Reserves. Available online: http://datazone.birdlife.org/site/factsheet/misuku-hills-forest-reserves-iba-malawi (accessed on 2 April 2019).
36. Government of Malawi. *Malawi Vegetation/Biotic Communites*; Government of the Republic of Malawi: Lilongwe, Malawi, 1979.
37. Potts, D. Rural mobility as a response to land shortages: The case of Malawi. *Popul. Space Place* **2006**, *12*, 291–311. [CrossRef]
38. National Statistical Office. *Integrated Household Survey, 2016–2017*; Government of the Republic of Malawi: Lilongwe, Malawi, 2017.
39. Ministry of Natural Resources Energy and Mining. *National Landscape Restoration Strategy*; Government of the Republic of Malawi: Lilongwe, Malawi, 2017.
40. USAID. *Community Based Natural Resource Management: Stocktaking Assessment*; United States Agency for International Development: Washington, DC, USA, 2010.
41. Wang, C.; Burris, M.A. Empowerment through Photo Novella: Portraits of Participation. *Health Educ. Behav.* **1994**, *21*, 171–186. [CrossRef]

42. Wang, C.; Burris, M.A. Photovoice: Concept, methodology, and use for participatory needs assessment. *Health Educ. Behav.* **1997**, *24*, 369–387. [CrossRef]
43. Wang, C.C.; Yi, W.K.; Tao, Z.W.; Carovano, K. Photovoice as a participatory health promotion strategy. *Health Promot. Int.* **1998**, *13*, 75–86. [CrossRef]
44. Barndt, D. Making, naming, and connecting—Reclaiming lost arts: The pedagogical possibilities of photostory production. In *Participatory Practices in Adult Education*; Campbell, P., Burnaby, B., Eds.; Lawrence Erlbaum: Mahwah, NJ, USA, 2001; pp. 31–54.
45. Gallo, M. Immigrant workers' journeys through a new culture: Exploring the transformative learning possibilities of photography. *Stud. Educ. Adults* **2001**, *33*, 109–117. [CrossRef]
46. Luttrell, W. Making culture visible: Children's photography, identity and agency. In Proceedings of the Annual Meeting of the American Sociological Association, Montreal, QC, Canada, 11–14 August 2006.
47. McAllister, C.; Wilson, P.; Green, B.; Baldwin, J. Come and take a walk: Listening to Early Head Start parents on school-readiness as a matter of child, family, and community health. *Am. J. Public Health* **2005**, *95*, 617–625. [CrossRef] [PubMed]
48. Prins, E. Participatory photography: A tool for empowerment or surveillance? *Action Res.* **2010**, *8*, 426–443. [CrossRef]
49. Harris, J.C. Vulnerable Youth's Perspectives and Priorities for Informal Settlements: Photovoice Evidence from Lusaka, Zambia. *J. Plan. Educ. Res.* **2018**, *38*, 398–412. [CrossRef]
50. Wang, C.C. Photovoice: A participatory action research strategy applied to women's health. *J. Women's Health* **1999**, *8*, 185–192. [CrossRef] [PubMed]
51. Lykes, M. Creative arts and photography action research in Guatemala. In *Handbook of Action Research*; Reason, P., Bradbury, H., Eds.; SAGE Publications, Inc.: Thousand Oaks, CA, USA, 2001; pp. 363–371.
52. McIntyre, A. Through the eyes of women: Photovoice and participatory research as tools for reimagining place. *Gender Place Cult.* **2003**, *10*, 47–66. [CrossRef]
53. Chonady, J.; Ferman, B.; Amitrani-Welsh, J.; Martin, T. Violence through the eyes of youth: A Photovoice exploration. *J. Community Psychol.* **2013**, *41*, 84–101. [CrossRef]
54. Berbés-Blázquez, M. A participatory assessment of ecosystem services and human wellbeing in rural costa rica using photo-voice. *Environ. Manag.* **2011**, *49*, 862–875. [CrossRef]
55. Pereira, E.; Queiroz, C.; Pereira, H.M.; Vicente, L. Ecosystem services and human well-being: A participatory study in a mountain community in Portugal. *Ecol. Soc.* **2005**, *10*, 14. [CrossRef]
56. Action for Environmental Sustainability. *Misuku Hills Improved Livelihood and Biodiversity Conservation Project Biodiversity Survey Report*; Action for Environmental Sustainability: Blantyre, Malawi, 2015.
57. McConnell, R.; Sibale, B.; Utila, H. *Linking Forest Programmes and Poverty Reduction Strategies, Malawi*; Food and Agriculture Organization of the United Nations: Rome, Italy, 2007.
58. Ministry of Natural Resources. *National Forest Policy*; Government of the Republic of Malawi: Lilongwe, Malawi, 1996.
59. Kamoto, J.F.; Dorward, P.T.; Shepherd, D.D. *Decentralised Governance of Forest Resources: Analysing Devotion Policy Processes and Their Effects on Decision Making in Communal Forest Management in Malawi*; International Association for the Study of the Commons: Utrecht, The Netherlands, 2008.
60. Ministry of Local Government and Rural Development. *Revised Decentralized Environmental Management Guidelines*; Government of the Republic of Malawi: Lilongwe, Malawi, 2012.
61. Millington, S.; Kaferawanthu, M. *Analysis of Biodiversity Threats & Opportunities in Malawi*; United States Agency for International Development: Washington, DC, USA, 2005.
62. World Bank. *Malawi Country Environmental Analysis*; The World Bank: Washington, DC, USA, 2019.
63. Ministry of Mines Natural Resources and Environment. *Standards & Guidelines for Participatory Forestry in Malawi*; Government of the Republic of Malawi: Lilongwe, Malawi, 2013.
64. Reytar, K.; Ray, S.; Toh, L. Malawi is Putting its Money Where Its Forests Are. Available online: https://www.wri.org/blog/2018/12/malawi-putting-its-money-where-its-forests-are (accessed on 30 March 2019).
65. Mwale, J. Youth scheme in shambles. *The Nation*, 21 April 2019; 4–5.
66. USAID COMPASS ll: Community Partnerships for Sustainable Resources Management in Malawi. Available online: https://rmportal.net/library/content/tools/compass-ii (accessed on 3 April 2019).
67. USAID. *Final Project Report*; United States Agency for International Development: Washington, DC, USA, 2009.

68. LTS International Improved Forest Management for Sustainable Livelihoods Programme. Available online: https://www.ltsi.co.uk/project/malawi-improved-forest-management-for-sustainable-livelihoods-programme (accessed on 5 April 2019).
69. GOPA Worldwide Consultants. GOPA Project Database. Available online: https://www.gopa.de/en/projects/improved-forest-management-sustainable-livelihoods-programme-ifmslp-phase-ii (accessed on 6 April 2019).
70. Zulu, L. Bringing people back into protected forests in developing countries: Insights from co-management in Malawi. *Sustainability* **2013**, *5*, 1917–1943. [CrossRef]
71. TRANSTEC SA. *Final Evaluation of Improved Forest Management for Sustainable Livelihoods Programme*; TRANSTEC: Brussels, Belgium, 2012.
72. Action for Environmental Sustainability. *Matipa Forest Management Plan*; Action for Environmental Sustainability: Blantyre, Malawi, 2010.
73. Mazanga, G. *Misuku Hills Indigenous Forest Project Final Project Completion Report*; Sustainable Development Initiative: Blantyre, Malawi, 2016.
74. Misuku Hills Beekeepers Association. *Misuku Beekeeping Value Addition Project Completion Report*; Misuku Hills Beekeepers Association: Chitipa, Malawi, 2016.
75. UNDP Promotion of Indigenous Forests in the Misuku Hills Area. Available online: https://sgp.undp.org/spacial-itemid-projects-landing-page/spacial-itemid-project-search-results/spacial-itemid-project-detailpage.html?view=projectdetail&id=20503 (accessed on 11 April 2019).
76. Kumbani, P. Misuku Hills, hidden treasure. *The Nation*, 28 March 2017.
77. Sustainable Development Initiative. *CEPF Small Grant Final Project Completion Report*; Sustainable Development Initiative: Blantyre, Malawi, 2017.
78. Byers, B. *Conserving the Miombo Ecoregion*; WWF Southern Africa Regional Programme Office (SARPO): Harare, Zimbabwe, 2001; p. 32.
79. Campbell, B.M.; Angelsen, A.; Cunningham, A.B.; Katerere, Y.; Sitoe, A.A.; Wunder, S. *Miombo Woodlands—Opportunities and Barriers to Sustainable Forest Management*; CIFOR: Bogor, Indonesia, 2007; p. 41.
80. The Economist Intelligence Unit. *Democracy Index 2018: Me too?* The Economist Intelligence Unit: London, UK, 2019.
81. Miller, T.; Kim, A.B.; Roberts, J.M. *2019 Index of Economic Freedom*; The Heritage Foundation: Washington, DC, USA, 2019.
82. Remme, H.; Muyambi, F.; Kamoto, J.; Dengu, E. *A Technical Review of Community Based Forest Management on both Customary Land and Forest Reserves (Participatory Forest Management)*; TRANSTEC: Brussels, Belgium, 2015.
83. M'bwana, L. Malawi Government Will Continue Using Army Soldiers to Protect Forest Reserves from Further Depletion. Available online: http://www.maravipost.com/malawi-government-will-continue-using-army-soldiers-protect-forest-reserves-depletion/ (accessed on 14 April 2019).
84. Mkwapatira, M. Malawians plant over 50 million trees. *The Nation*, 22 April 2019; 3.
85. Chinangwa, L.; Sinclair, F.; Pullin, A.S.; Hockley, N. Can co-management of government forest reserves achieve devolution? Evidence from Malawi. *For. Trees Livelihoods* **2016**, *25*, 41–58. [CrossRef]
86. Zulu, L.C. Community Forest Management in Southern Malawi: Solution or Part of the Problem? *Soc. Nat. Resour.* **2008**, *21*, 687–703.
87. Chinangwa, L.L.; Pullin, A.S.; Hockley, N. Impact of forest co-management programs on forest conditions in Malawi. *J. Sustain. For.* **2017**, *36*, 338–357. [CrossRef]
88. Mtambo, C.; Missanjo, E. The Impact of Participatory Forest Management on Tree Species Abundance and Diversity in Selected Village Forest Areas in Kasungu, Malawi. *J. Ecol. Environ. Sci.* **2015**, *3*, 15–20.
89. Sunderlin, W.D.; Angelsen, A.; Belcher, B.; Burgers, P.; Nasi, R.; Santoso, L.; Wunder, S. Livelihoods, forests, and conservation in developing countries: An Overview. *World Dev.* **2005**, *33*, 1383–1402. [CrossRef]
90. Salafsky, N.; Wollenberg, E. Linking livelihoods and conservation: A conceptual framework and scale for assessing the integration of human needs and biodiversity. *World Dev.* **2000**, *28*, 1421–1438. [CrossRef]
91. Mather, A.; Needle, C. The Forest Transition: A Theoretical Basis. *Area* **1992**, *2*, 117–124. [CrossRef]
92. Rudel, T.K.; Coomes, O.T.; Moran, E.; Achard, F.; Angelsen, A.; Xu, J.; Lambin, E. Forest transitions: Towards a global understanding of land use change. *Glob. Environ. Chang.* **2005**, *15*, 23–31. [CrossRef]
93. Rudel, T.K. Is there a forest transition? Deforestation, reforestation, and development. *Rural Sociol.* **1998**, *63*, 533–552. [CrossRef]

94. Nyirongo, K. Human Activity Endanger Bird Species in Misuku. Available online: http://www.manaonline.gov.mw/index.php/national/environment/item/6955-human-activity-endanger-bird-species-in-misuku (accessed on 22 April 2019).
95. Plummer, R.; Fennell, D. Exploring co-management theory: Prospects for sociobiology and reciprocal altruism. *J. Environ. Manag.* **2007**, *85*, 944–955. [CrossRef]
96. Trivers, R.L. The evolution of reciprocal altruism. *Q. Rev. Biol.* **1971**, *46*, 35–57. [CrossRef]
97. Bunderson, W.T.; Hayes, I.M. Agricultural and environmental sustainability in Malawi. In Proceedings of the Conference on Sustainable Agriculture for Africa, Abidjan, Cote d'Ivoire, 17–21 July 1995.
98. Center for Environmental Policy and Advocacy. *National Forest Policy in Malawi*; Center for Environmental Policy and Advocacy: Blantyre, Malawi, 2010.
99. Haack, B.; Mahabir, R.; Kerkering, J. Remote sensing-derived national land cover land use maps: A comparison for Malawi. *Geocarto Int.* **2015**, *30*, 270–292. [CrossRef]
100. The Ministry of Natural Resources Energy and Mining. *A Framework for Monitoring Progress on Malawi's National Forest Landscape Restoration*; Government of the Republic of Malawi: Lilongwe, Malawi, 2018.

Permissions

All chapters in this book were first published in MDPI; hereby published with permission under the Creative Commons Attribution License or equivalent. Every chapter published in this book has been scrutinized by our experts. Their significance has been extensively debated. The topics covered herein carry significant findings which will fuel the growth of the discipline. They may even be implemented as practical applications or may be referred to as a beginning point for another development.

The contributors of this book come from diverse backgrounds, making this book a truly international effort. This book will bring forth new frontiers with its revolutionizing research information and detailed analysis of the nascent developments around the world.

We would like to thank all the contributing authors for lending their expertise to make the book truly unique. They have played a crucial role in the development of this book. Without their invaluable contributions this book wouldn't have been possible. They have made vital efforts to compile up to date information on the varied aspects of this subject to make this book a valuable addition to the collection of many professionals and students.

This book was conceptualized with the vision of imparting up-to-date information and advanced data in this field. To ensure the same, a matchless editorial board was set up. Every individual on the board went through rigorous rounds of assessment to prove their worth. After which they invested a large part of their time researching and compiling the most relevant data for our readers.

The editorial board has been involved in producing this book since its inception. They have spent rigorous hours researching and exploring the diverse topics which have resulted in the successful publishing of this book. They have passed on their knowledge of decades through this book. To expedite this challenging task, the publisher supported the team at every step. A small team of assistant editors was also appointed to further simplify the editing procedure and attain best results for the readers.

Apart from the editorial board, the designing team has also invested a significant amount of their time in understanding the subject and creating the most relevant covers. They scrutinized every image to scout for the most suitable representation of the subject and create an appropriate cover for the book.

The publishing team has been an ardent support to the editorial, designing and production team. Their endless efforts to recruit the best for this project, has resulted in the accomplishment of this book. They are a veteran in the field of academics and their pool of knowledge is as vast as their experience in printing. Their expertise and guidance has proved useful at every step. Their uncompromising quality standards have made this book an exceptional effort. Their encouragement from time to time has been an inspiration for everyone.

The publisher and the editorial board hope that this book will prove to be a valuable piece of knowledge for researchers, students, practitioners and scholars across the globe.

List of Contributors

Jéssica Chaves Cardoso
Natural Resources and Environmental Studies-Forestry, University of Northern British Columbia, Prince George, BC V2N 4Z9, Canada

Philip J. Burton
Ecosystem Science & Management Program, University of Northern British Columbia, Terrace, BC V5G 1K7, Canada

Che M. Elkin
Ecosystem Science & Management Program, University of Northern British Columbia, Prince George, BC V2N 4Z9, Canada

Gloria C. Blaise
Department of Natural Resources and the Environment, Cornell University, Ithaca, NY 14850, USA

Jeffrey A. Brown
Global Institute of Sustainability, Arizona State University, Tempe, AZ 85281, USA

Rebecca C. Jordan and Amanda E. Sorensen
Department of Community Sustainability, Michigan State University, East Lansing, MI 48824, USA

James McGinlay, Vassilis Gkoumas, Jens Holtvoeth, Andreas Kontoleon and Nikoleta Jones
Department of Land Economy, University of Cambridge Conservation Research Institute & Cambridge Centre for Environment, Energy and Natural Resource Governance, University of Cambridge, Cambridge CB2 3QZ, UK

Ruymán Federico Armas Fuertes
Grupo Tragsa, Avenida Quinto Centenario, Edifico San José, Local 4, 38800 San Sebastián de La Gomera, La Gomera, Canary Islands, Spain

Elena Bazhenova
Kullaberg Nature Reserve, County Administrative Board of Skåne, Italienska vägen 323, 26377 Mölle, Sweden

Alessandro Benzoni and Stefano Santi
Prealpi Giulie Nature Park, Resia, Piazza del Tiglio 3, 33010 Resia, Italy

Kerstin Botsch
Department of National Park Planning, Regional Development and Tourism, National Park Black Forest, Schwarzwaldhochstraße 2, 77889 Seebach, Germany

Carmen Cabrera Martel
Centro Administrativo del Parque Nacional y Parque Natural de Sierra Nevada, Carretera Antigua de Sierra Nevada km 7, 18191 Pinos Genil, Granada, Spain

Cati Carrillo Sánchez and Francisco Soriano Rico
Subdirección General de Patrimonio Natural y Cambio Climático, Dirección General del Medio Natural, Consejería de Agua, Agricultura, Ganadería, Pesca y Medio Ambiente, Región de Murcia, Plaza Juan XXIII, s/n. Desp. C1.10, 30008 Murcia, Spain

Isabel Cervera
Las Batuecas-Sierra de Francia Natural Park, Casa del Parque Natural Las Batuecas-Sierra de Francia, Ctra/ Las Batuecas, nº 22, 37624 La Alberca, Salamanca, Spain

Guillermo Chaminade
Söderåsen National Park, County Administrative Board of Skåne, Skäralid 747, 26453 Ljungbyhed, Sweden

Juliana Doerstel and Michael Lammertz
Wald und Holz NRW, Nationalparkforstamt Eifel, Fachgebiet Kommunikation und Naturerleben, Urftseestraße 34, 53937 Schleiden-Gemünd, Germany

Concepción J. Fagundo García
Garajonay National Park, Centro de Visitantes Juego de Bolas, La Palmita C.P., 38330 Agulo, La Gomera, Canary Islands, Spain

Angela Jones
Snowdonia National Park Authority, Penrhyndeudraeth, Gwynedd LL48 6LF, UK

Kaja Lotman
Environmental Board of Estonia, 15172 Tallinn, Estonia

Majda Odar and Mojca Smolej
Triglav National Park, Ljubljanska cesta 27, 4260 Bled, Slovenia

Teresa Pastor and Carol Ritchie
EUROPARC Federation, Waffnergasse 6, 93047 Regensburg, Germany

Holly Waterman
Peak District National Park Authority, Aldern House, Bakewell DE45 1AE, UK

Tomasz Zwijacz-Kozica
Tatra National Par, Ku´znice 1, 34-500 Zakopane, Poland

Panayiotis G. Dimitrakopoulos
Department of Environment, University of the Aegean, 81100 Mytilene, Greece

Giovanni Ávila-Flores, Juan Manuel López-Vivas and Oscar Arizpe-Covarrubias
Academic Department of Marine and Coastal Sciences, Autonomous University of Baja California Sur, Baja 23037 California Sur, Mexico

Judith Juárez-Mancilla and Plácido Cruz-Chávez
Academic Department of Economics, Autonomous University of Baja California Sur, 23037 Baja California Sur, Mexico

Gustavo Hinojosa-Arango
CIIDIR Oaxaca, National Polytechnic Institute, 95060 Santa Cruz, Mexico

Oreoluwa Ola and Emmanuel Benjamin
Department of Governance in International Agribusiness and Department of Agricultural Production and Resource Economics, Technical University of Munich (TUM), 85354 Freising, Germany

Manoela Sacchis Lopes and Dejanira Luderitz Saldanha
Programa de Pós-graduação em Sensoriamento Remoto, Centro Estadual de Pesquisas em Sensoriamento Remoto e Meteorologia, Universidade Federal do Rio Grande do Sul (UFRGS), Porto Alegre 91501-970, Brazil

Bijeesh Kozhikkodan Veettil
Department for Management of Science and Technology Development, Ton Duc Thang University, Ho Chi Minh City 758307, Vietnam
Faculty of Environment and Labour Safety, Ton Duc Thang University, Ho Chi Minh City, Vietnam

Javier López-Upton, J. René Valdez-Lazalde, Aracely Ventura-Ríos and J. Jesús Vargas-Hernández
Graduate School of Forest Sciences, Colegio de Postgraduados, Km. 36.5 Carr. México-Texcoco, Montecillo, Edo. de México 56230, Mexico

Vidal Guerra-de-la-Cruz
Experimental Site Tlaxcala, INIFAP, Km. 2.5 Carr. Tlaxcala-Santa Ana Chiautempan. Tlaxcala, Tlaxcala 90800, Mexico

Špela Pezdevšek Malovrh
Department of Forestry and Renewable Forest Resources, Biotechnical Faculty, University of Ljubljana, 1000 Ljubljana, Slovenia

Alessandro Paletto
Consiglio per la ricerca in agricoltura e l'analisi dell'economia agraria, Research Centre for Forestry and Wood (CREA), 38123 Trento, Italy

Stjepan Posavec
Department of Forest Inventory and Management, Faculty of Forestry, University of Zagreb, 10000 Zagreb, Croatia

Zuzana Dobšinská
Department of Forest Economics and Management, Technical University in Zvolen, Faculty of Forestry, 96001 Zvolen, Slovakia

Ilija Đorđević
Department of spatial planning, GIS and Forest Policy, Institute of Forestry, 11030 Belgrade, Serbia

Bruno Marić and Mersudin Avdibegović
Faculty of Forestry, University of Sarajevo, 71000 Sarajevo, Bosnia and Herzegovina

Emil Kitchoukov
Department of Management and Alternative Tourism, Faculty of business management, University of Forestry-Sofia, 1756 Sofia, Bulgaria

Aleksandar Stijović
Institute of Forestry of Montenegro, 81000 Podgorica, Montenegro

Pande Trajkov
Department of Forest Management, Faculty of Forestry, Ss. Cyril and Methodius University in Skopje, 1000 Skopje, North Macedonia

Tomislav Laktić
Ministry of the Environment and Spatial Planning, Water and Investments Directorate, Cohesion Policy Division, 1000 Ljubljana, Slovenia

Maurizio Marchi
CREA—Research Centre for Forestry and Wood, I-52100 Arezzo, Italy

Iztok Sinjur and Marjana Westergren
Slovenian Forestry Institute, Vecna pot 2, 1000 Ljubljana, Slovenia

Michele Bozzano
European Forest Institute, 53113 Bonn, Germany

Marija Maruna and Milica P. Milojević
Faculty of Architecture, University of Belgrade, 11000 Belgrade, Serbia

Tijana Crnčević
Institute of Architecture and Urban & Spatial Planning of Serbia, 11000 Belgrade, Serbia

Tomasz Gawęda
State Forests National Forest Holding, Bielsko Forest District, 43-382 Bielsko-Biała, Poland

Ewa Błońska
Department of Forest Soil Science, Faculty of Forestry, University of Agriculture, 31-425 Krakow, Poland

Stanisław Małek
Department of Forest Ecology and Reclamation, Faculty of Forestry, University of Agriculture, 31-425 Krakow, Poland

Erin K. McGrew, William E. Grant and Hsiao-Hsuan Wang
Department of Wildlife and Fisheries Sciences, Texas A&M University, College Station, TX 77843, USA

Tomasz E. Koralewski
Department of Ecosystem Science and Management, Texas A&M University, College Station, TX 77843, USA

Thomas D. Byram
Department of Ecosystem Science and Management, Texas A&M University, College Station, TX 77843, USA
Texas A&M Forest Service, College Station, TX 77843, USA

Tisha Holmes and April Jackson
Department of Urban and Regional Planning, Florida State University, Tallahassee, FL 32306, USA

Christopher Coutts
Department of Urban and Regional Planning, Florida State University, Tallahassee, FL 32306, USA
Department of Built Environment, Mzuzu University, Mzuzu 2, Malawi

Index

www.ingramcontent.com/pod-product-compliance
Lightning Source LLC
LaVergne TN
LVHW060531210926
848485LV00005B/1180